The Chemistry of Phosphate and Nitrogen Compounds in Sediments

by

Han L. Golterman
Association Leiden – Camargue,
Arles, France

KLUWER ACADEMIC PUBLISHERS
DORDRECHT / BOSTON / LONDON

A C.I.P. Catalogue record for this book is available from the Library of Congress.

ISBN 1-4020-1951-3

Published by Kluwer Academic Publishers,
P.O. Box 17, 3300 AA Dordrecht, The Netherlands.

Sold and distributed in North, Central and South America
by Kluwer Academic Publishers,
101 Philip Drive, Norwell, MA 02061, U.S.A.

In all other countries, sold and distributed
by Kluwer Academic Publishers,
P.O. Box 322, 3300 AH Dordrecht, The Netherlands.

QE
471
.2
.G65
2004

Cover image:
Bio-assay demonstrating the influence of nitrogen on the growth of Scirpus.
(Photo courtesy of H.L. Golterman)

Printed on acid-free paper

Contents

List of Figures

Acknowledgments for figures

Figure 1.1 Reprinted with kind permission of SIL (Society for International Limnology). See also http://www.schweizerbart.de

Figure 1.2 Reprinted with kind permission of UNESCO. Golterman, Sly & Thomas: 'Study of the relationship between water quality and sediment transport. A guide for the collection and interpretation of sediment quality data', Technical Paper in: *Hydrology*, **26**. © UNESCO, 1983.

Figure 1.4 Reprinted with kind permission of the publisher of: *Mem. Ist. Ital. Idrobiol.* **29**, 37–95.

Figure 1.5 Reprinted with kind permission of UNESCO. Golterman, Sly & Thomas: 'Study of the relationship between water quality and sediment transport. A guide for the collection and interpretation of sediment quality data', Technical Paper in: *Hydrology*, **26**. © UNESCO, 1983.

Figure 2.5 Reprinted with kind permission of SIL (Society for International Limnology). See also http://www.schweizerbart.de

Figure 3.3a Reprinted with kind permission of SIL (Society for International Limnology). See also http://www.schweizerbart.de

Figure 3.16 Reprinted with kind permission of the author.

Figure 4.1 Reprinted with kind permission of SIL (Society for International Limnology). See also http://www.schweizerbart.de

Figure 4.2 Reprinted with kind permission of SIL (Society for International Limnology). See also http://www.schweizerbart.de

Figure 4.4 Reprinted with kind permission of SIL (Society for International Limnology). See also http://www.schweizerbart.de

Figure 4.17 Reprinted with kind permission of the author and the American Society for Microbiology.

List of Tables

Preface

Phosphate and nitrogen compounds are the two plant nutrients involved in the process of eutrophication. In their cycles sediments are of paramount importance. In freshwater eutrophication remains the severest process of pollution, while coastal marine environments are now becoming seriously affected too (Orive, Elliot & De Jonge, 2002). Not only developed countries encounter these problems; developing countries have also begun to suffer. For a long time wetlands were almost not studied or even mentioned; this state of affairs has nowadays much changed. Not only were wetlands, in the past, considered to be unimportant, but, moreover, the close contact between sediment and water owing to the shallowness of the water-layer kept the P-concentration in the water low by adsorption, thus delaying the consequences of loading. Since wetlands have been recognized not only by the happy few, but also by water managers, they receive more attention.

Golterman & de Oude (1991) reviewed relevant literature on mechanisms and processes of eutrophication and paid some attention to the role of sediments, albeit only in their discussion concerning the influence of sediment on the phosphate in- and output balance. A few equations were discussed, but not the chemical mechanisms by which sediment influences the P-cycle. Nitrogen has always received much less attention, although the sediment layer in lakes and wetlands is the site of major changes and processes in the N-cycle. One of the most important processes is denitrification, by which nitrogen is lost from the system, for which reason it has been considered to be advantageous. It entails, however, a situation favourable for blue-green algal blooms.

Nowadays sediment cannot be left out of the quantification of the eutrophication process and even less out of studies on and measures against it.

Processes governing P- and N-cycles in waterbodies are of a chemical and bacteriological nature and can only be studied with a good comprehension of these processes and an appropriate methodology. The need for detailed chemical training and knowledge, which the researcher often lacks, is the raison d'être of this book. Several times a year I receive articles submitted for the sediment issues of *Hydrobiologia* in which a serious lack of knowledge caused a wrong set-up of the experiments and a wrong interpretation. For example, one cannot study the release of phosphate from sediments without an accurate knowledge of possible pH changes. Much work is being done with great zeal that would have been valuable if only such factors had been taken into account. Sediment studies are difficult; when I started my career in limnology I was told by one of my advisers, the well-known bacteriologist Kluyver, not to study the sediment layer for the first 10 years, "because this system is too difficult". This is still true to some extent, although our equipment and methods have much improved since.

The book is, therefore, written with the chemical processes as the 'golden thread'; around this thread literature has been woven to support it. It is of course impossible to discuss all the sediment literature – a literature search for phosphate and sediment reveals several thousands of articles and to a lesser extent the same is true for nitrogen and sediment. Preference is, therefore, given to articles around this golden thread and not to local studies, important as they may be for the management of the lake in question. The book is directed to beginning limnologists and environmental engineers appointed to include sediment in their studies and to many water managers, and must be seen as a complement to textbooks of limnology, which generally treat sediments only superficially.

All too often one encounters outdated concepts like: "Sediment is a sink or a source of phosphate", which is not true: sediment is an integrated part of lakes and a large part of the sediment phosphate remains bioavailable. This knowledge seems, however, to penetrate in limnology with difficulty. So, Reynolds and Davies (2001) wrote in the abstract of their paper about the British Perspective of Phosphate Availability: "Most other forms (*than o*-P), including phosphates of the alkaline earth metals, aluminium and iron are scarcely available at all" and later in the abstract: "The phosphorus transferred from arable land to drainage remains dominated by sorbed fractions which are scarcely available", as if there were not a mountain of literature on this subject showing the contrary.

In principle there are two main chapters, one on phosphate and one on nitrogen. For practical reasons the phosphate chapter has been cut into two parts, fractionation and availability on the one hand and release mechanisms on the other. As there is no good textbook on sediments as yet, two chapters have been added, the first on the nature and properties of sediments and the second on the chemistry of the overlying water layers – either deep or shallow. In order to prevent older knowledge to be forgotten, much old literature has been used, in the hope of preventing to some extent the rediscovery of the wheel. Much information is still used from the first symposium on the "Interaction between Water and Sediments", where the study of sediments started seriously (Golterman, 1977).

I hope this book will stimulate colleagues to read the quoted articles themselves. As editor of the sediment issues of *Hydrobiologia* I have too often found that authors quote articles without having read these very articles. In this way legend forming in science is becoming a serious problem. Some examples can be found in this book.

Baas Becking (1934), one of the early ecologists, who influenced the Amsterdam School of Botany strongly, wrote: "Everything is everywhere, but the environment selects", expressing earlier concepts by M. W. Beyerrinck, the Delft microbiologist. The statement is still true and the environment and its chemistry remain important. For Baas Becking the environment meant factors like: O_2, CO_2, pH and temperature, NaCl concentration, etc. Nowadays, the environment has been found to be far more complex and the chemistry becomes more and more difficult. Baas Becking furthermore wrote: "Field-ecology can never give us a complete description of the environment; this privilege belongs to the laboratory". May this book help to improve the understanding of this chemistry and the interaction between field and laboratory studies.

Acknowledgement

I want to express my sincere thanks in the first place to several colleagues who helped me during my studies. Specially named are Dr R. S. Clymo and Ir. N. T. de Oude, who were always ready to give advice. Dr L. Serrano-Martin read the manuscript critically and Dr Ph. Ford gave useful advice on the physical aspects and helped to improve the structure of the book.

In the second place I want to thank several of my former students, now colleagues, who helped with the practical work, often under difficult conditions; among them I mention Dr F. Minzoni, Dr C. Bonetto, the late Dr C. J. de Groot and the Doctorandi Mr S. Bakker, Mr P. Bruijn, Mr H. Schouffoer and Mrs E. Zwanenburg-Dumoulin, who also helped to trace the many mistakes in earlier drafts and to improve the structure.

I thank most warmly my wife, Mrs J. C. Golterman–Hardenberg, who has not only improved the English with great care and attention, but while doing so even found mistakes in the chemistry. This book, like all my publications during the last 20 years, has benefited from her constant help and encouragement.

I also thank Mr R. Hoksbergen whose skill as 'desk-editor' has given the book an attractive form.

May they all be aware that without their co-operation this book would not have been written.

Arles, 15/08/2003
H. L. Golterman
Association Leiden–Camargue

Ex pondere et numero veritas
(Kolthoff *et al.*, 1969)

Glossary

Autotrophic bacteria Bacteria having the ability to synthesize all of their carbon compounds from CO_2.

Bed load The amount of suspended matter moving just above the bottom of the river. (The distance is not properly defined.)

BOD Biological Oxygen Demand, i.e. the amount of O_2 taken up by a sample under influence of micro-organisms. Usually the BOD is measured after 5 days, but the process continues long after that. It is obsolete and its use should be abolished.

Cation Exchange Capacity The amount of cations (NH_4^+ or Ba^{2+}) a sediment can hold when a salt solution buffered at a standardized pH is leached through or equilibrated with the sediment.

Chelate(s) Binding at two or more points between a metal ion and an (organic) ligand forming a soluble complex.

Chemo-autotrophs Bacteria using inorganic compounds as energy source.

Chemotrophic bacteria Bacteria that derive their energy from organic compounds.

COD Chemical Oxygen Demand, i.e. the amount of O_2 used by a sample for a complete oxidation to CO_2. Usually as oxidant an excess of $K_2Cr_2O_7$ is used. $KMnO_4$ has been used in the past; it does not always give a complete oxidation, although at high pH, under boiling, it will not be far off. Auto-oxidation then becomes a problem. It is still in use for brackish waters, because of the oxidation of Cl^- by $K_2Cr_2O_7$.

Consolidation The process by which particulate matter becomes more solid or compact.

Debris The remains of anything (organic or inorganic) broken down or destroyed; fragmentary material accumulated from the breakdown of rocks.

Detritus (Detrital) Matter produced by detrition, i.e. material eroded or washed away. Debris of any kind, often dead organisms.

Diagenesis Physical and chemical changes undergone by a sediment after deposition; the recombination of two minerals to form a new mineral.

EDTA Ethylene diamino tetra acetic acid (see Figure 1.8).

Eutrophic Nutritional status of a freshwater rich in nutrients, most often N & P. In limnology usually: too rich in nutrients.

Flocculation The clogging together of smaller particles forming bigger ones under the influence of the chemistry of the carrying liquid.

Gel filtration Filtration over a column filled with a gel, e.g. Sephadex, having small pore size.

Gneiss A foliated usually coarse-grained metamorphic rock in which bands of granular minerals alternate with bands of flaky or prismatic ones; typically consisting of feldspar, quartz and mica.

Hard water Freshwater carrying a relatively high concentration of Ca^{2+}. Usually around 40 mg ℓ^{-1}; higher in non-equilibrium with air.

Heterotrophic bacteria Bacteria not having the ability to synthesize all of their carbon compounds from CO_2 and requiring an organic source of carbon.

Hysteresis The phenomenon whereby changes in some property of a system lag behind changes in the phenomenon causing it.

Intrusive rocks Intrusion is the influx of molten rock into fissures or between strata.

Lithotrophic bacteria Bacteria that derive their energy from inorganic compounds.

Load The amount of matter transported by a river (mass per volume; mg ℓ^{-1}).

Loading the amount of an element or a compound entering a lake per unit of surface per unit of time. (Mostly g m^{-2} y^{-1}.)

Mineral Inorganic natural substance with recognizable chemical composition of neither animal nor vegetal origin.

NTA Nitrilo triacetic acid.

Primary production The amount of vegetation produced by photosynthesis followed by conversion into biological material.

Reynolds numbers or ratio the ratio, found by Reynolds, between C, N and P (C/N/P = 106/16/1) in marine phytoplankton and later used to express phytoplankton composition as $C_{106}H_{263}O_{110}N_{16}P$. The numbers suggest a much too great precision and have no meaning for other organic matter, such as dead organic matter in sediment.

Schists A coarse-grained metamorphic rock with parallel layers of various minerals and splitting in irregular plates.

Shear stress The force exerted by two layers sliding on top of each other (dynes cm^{-2} or N m^{-2}).

SiO$_2$ Silicon dioxide, is found in nature as *quartz*, *tridymite* or *cristobalite*, with different crystalline forms. It is also found in a hydrated form (silica gel), *opal*, and in earthy form as *kieselguhr* or *diatomaceous earth*.

Slates A fine-grained metamorphosed sedimentary rock typically dark grey, blue or green in colour characterized by easily splitting.

Abbreviations used

dw.	= dry weight
ww.	= wet weight (i.e. the weight of a certain volume of sediment as it is taken from the water system)
BD	= the extraction solution of dithionite in Na-bicarbonate
CDB	= the same with citric acid
z	= lake depth
H_2O	= chemically pure water
water	= naturally occurring water
o-P	= dissolved inorganic phosphate
P_{aa}	= phosphate available for algae
$Fe(OOH){\approx}P$	= phosphate adsorbed onto iron hydroxide
$CaCO_3{\approx}P$	= phosphate bound onto $CaCO_3$, but for calculations involving the solubility product the formula for apatite, $Ca_5(PO_4)_3.OH$, is used
org-P	= organic phosphate
res-P	= residual P remaining in the sediment after extractions
P_{extr}	= extractable phosphate (the extractant, if specified, is indicated by $\longrightarrow$)
org-$P_{\rightarrow ac}$	= org-P soluble in acid (previously "ASOP")
org-$P_{\rightarrow alk}$	= org-P soluble in alkaline (mostly 1.0 M NaOH)
nr-$P_{\rightarrow alk}$	= P not reactive with the molybdate method, but extractable with (mostly 1.0 M) NaOH
P_{sed}, P_{sest}	= phosphate bound in sediment, seston, etc.
NH_4^+	= used as symbol for the sum of NH_3 and NH_4^+ the ratio of which depends on the pH

Units

The choice of which units to use was difficult. Obviously, the used of m(μ)mol per g dw. is the best, but this is, regrettably, uncommon in limnology. So I decided to follow the general habit and use m(μ)g per gram dw.. Only in cases where stoicheiometry of reactions is demonstrated, the mmol is used. Concentrations in sediments are expressed per unit of dw., except when authors quoted did not provide this unit.

Remark

Words marked with an * in the text can be found in this glossary.

Chapter 1

Nature and properties of sediment[1]

1.1 Introduction

Continental waters are the major agents in the weathering of rocks and the subsequent changes of the landscape. Lake sediments begin their life cycle at the first contact between rocks and weathering processes, in which the water has two clearly different functions, as it promotes chemical changes, by dissolving inorganic and organic matter, and provides physical transport for energy and mass.

The chemical composition of water, plus its sediment, results from three factors: 1) type of rock (mineralogy and grain size); 2) climatic factors, such as rain fall, flow rate, temperature; and 3) flow conditions, time of contact between rock and water. Erosion and weathering work like a blacksmith, whose hammer is here replaced by rapidly changing physical conditions, while his fire is replaced by water. The physical processes mainly cause a fragmentation of rock particles which then will become hydrated by the water, converting stony rock particles into mud. These processes are depicted in Figure 1.1.

The weathering process produces two kinds of matter, dissolved and particulate. All natural waters contain varying amounts of particles, called suspended sediment, which finally may become a consolidated* sediment. 'Sediment' is a general term which is used to describe both suspended and deposited or settled material. In aquatic systems, sediment consists of inorganic and organic compounds and includes all particulate matter that is washed or blown into lakes or rivers (allochthonous), or formed in the water-body itself (autochthonous). Diagenesis*, alterations which occur in the sediment, produces new compounds (authigenic). Diagenetic processes also cause changes in the chemical composition of pore fluid, the fluid between the settled particles.

The organic matter consists of micro-organisms (bacteria, phytoplankton, zooplankton),

---

[1]This introduction is meant as a general orientation for the non-geologist reader and not as a new review. It is partly based on Golterman *et al.*, 1983 and Golterman, 1984.

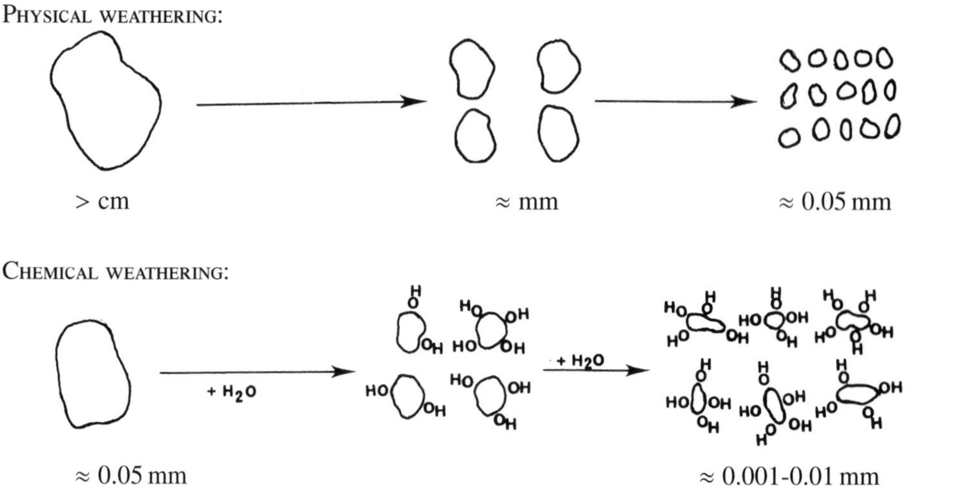

PHYSICAL WEATHERING:

> cm   ≈ mm   ≈ 0.05 mm

CHEMICAL WEATHERING:

≈ 0.05 mm   ≈ 0.001-0.01 mm

FIGURE 1.1: Simplified schematic representation of physical and chemical erosion (with indication of particle size).

the remains of macrophytes and other large sized organisms, together with the detritus derived from decaying material. The inorganic matter consists of erosion products from rocks in the watershed (rock particles and clays), together with compounds which may be brought into the water or formed within it from soluble products or compounds such as Fe_2O_3 (hematite), FeOOH (goethite, derived from $Fe(OH)_3$), SiO_2 (sand, quartz and other minerals*), and $CaCO_3$ (calcium carbonate or limestone).

Natural sediment is a mixture of different compositional types:

a) minerals and small pieces of rock derived from the fragmentation of source material, i.e., the eroded original rock material;
b) clay minerals;
c) precipitates and coatings;
d) organic matter.

Sediment composition is largely controlled by the composition of the rock from which it is derived by erosion and weathering; it is moreover influenced by climatic regime (weather and hydrologic conditions), land form, land use and time in transit. In addition, man-made debris* may result in the production of sediment having characteristics similar to those of natural origin.

1.2   Quantitative aspects of sediment transport

Sediment influences water use and availability, for both men and aquatic organisms, in several ways. Quantitatively, wide variations in the amount of material and the particle size, such as gravel, sand, and silt, may cause deposition in river/stream and lake systems or excessive erosion (channel instability). High sediment loads may also cause technical problems, such as turbine failures in hydroelectric power stations, and clogging of screens and intake filters in treatment plants. The presence of sediment may have a significant influence on biological communities such as plankton, benthic organisms and even fish. High concentrations of suspended matter may result mainly in low algal growth or primary production[*] because of restricted light penetration, causing a reduction in general productivity. The availability of O_2 may be limited by high sediment O_2 demand, either COD[*] or BOD[*]. The presence of high concentrations of suspended matter may adversely affect respiration and may result in the burial of existing substrates. Negative aspects of high suspended matter concentrations have been reported of fish. Robbe (1975) mentioned that at concentrations above 200 mg ℓ^{-1} a thickening of bronchial cells of freshwater fish may occur, decreasing life expectancy.

Some rough estimates about the total transport of particulate material can be given. The total amount of particulate material transported by all rivers of the world has been estimated to be about $15-20 \cdot 10^9$ t y^{-1} (Sunborg, 1973; Meybeck, 1976, 1977); with an estimated river flow of slightly more than 30,000 km^3 y^{-1}; this gives a world average of about 500 g m^{-3} (= 0.05%). A considerable portion of the total amount is transported by only a few rivers; thus, the Amazon is estimated to carry about $0.5 \cdot 10^9$ t y^{-1} or about 3% of the world total. There is some evidence (Table 1.1) that the proportion of suspended matter is about 1.5 to 2 times that of the dissolved load. Only in hard water rivers, like the Rhine, the dissolved components appear to be relatively more important (Table 1.1). The annual variation in quantities of suspended matter is, of course, considerable: thus, the Rhine, having an average of 57 g m^{-3}, spans a range of 50 to 500 g m^{-3}. Certain rivers may have very high values during severe flood; some examples are given in Table 1.1. The annual mean values in the Rhône are usually between 50 and 300 g m^{-3} at a flow rate of 1700 m^3 s^{-1}, but reached about 6 kg m^{-3} when the flow rate was 9500 m^3 s^{-1} (Moutin *et al.*, 1998). Man-made hydraulic works in rivers have changed, or will greatly change, sediment transport and its grain size: because of the many barrages in the Rhône the sediment transport has been much changed from its original values, with the major effect that finer particles now dominate the suspended matter and change the sedimentation in its coastal area. The year to year variation is also considerable; the Rhine sediment, for example, has ranged between $2 \cdot 10^5$ and $4 \cdot 10^5$ t y^{-1} between 1930 and 1975 (Golterman, 1975a).

Müller & Förstner (1968a,b) found that the amount of suspended matter transported by a river depends upon the flow and can be described by the equation:

$$C_s = a \cdot Q^b,$$

where C_s = concentration of suspended matter (mg l^{-1}), Q = flow rate (m^3 s^{-1}) and a and b are constants. Constant a can vary from one river to another between 0.004 and 80,000, the latter exceptional value being valid in Rio Puerco at Bermuda (Northern Mexico). It depends on climate, vegetation, rock erodability and the morphological stability of the valley system. Constant b can vary between 0.0 and 2.5 and is a measure of erosional forces. This equation can be used for predicting rate and quantities of long-term transport in a given river. The relation between C_s and Q may be different if the flow rate is de- or increasing. Hysteresis[*] effects will cause the decreasing concentrations to remain relatively high. Sometimes rather complicated patterns are found, such as e.g., by Walling (1977) in the Humber river (Ont.,

TABLE 1.1: Hydrological river and sediment data (Golterman *et al.*, 1983).

	Catchment area (10^3 km^{22})	Runoff (10^3 km^3 y^{-1})	Average conc. dissolved load (g m^3)	Flux, diss. load (10^9 t y^{-1})	Average conc. suspended load (g m^{-3})	Flux, susp. load (10^9 t y^{-1})	Ratio: suspended/ dissolved load
Colorado River	637	0.021	760		1,930		
Pacific slopes, California	303	0.072	167		970		
Western Gulf	829	0.049	770		1,88		
Missipi river	3240	0.555	248	0.137	510	0.3	2
South Atlantic and Eastern Gulf	735	0.291	171		131		
North Atlantic	383	0.188	128		156		
Columbia	679	0.250	85	0.021	56	0.014	0.67
US totals and weighted means	6807	1.500	153	0.32	295	0.442	1.9
North America total		4.56 – 6.43	142 – 189	0.646 0.572			
Rhine	220	0.070	200 – 250 (2) / 600 – 900 (3)	0.015 (2) 0.303 / 0.068 (3) 1.36	57 – 67	0.003 – 0.004	0.2
Rhone	100	0.053			50 – 300		
Europe total		2.5 – 3.0	101-182	0.455 0.303			
Asia		11.1 – 12.3	111–142	1.57 1.36			
Nile	3000	0.09					
Africa total		5.9 – 6.0	96 – 121	0.715 0.581			
Australia		0.32 – 0.61	59 – 176	0.019 0.107			
Amazon	6300 – 7000	5.5	52.9 (1)	0.292	100	0.53	1.5
Rio de la Plata	3200	0.73	114	0.07		0.13	1.85
South America total		8.0 – 8.1	69 – 71	0.0552 0.575			
World total		32.3 – 36.5	88 - 120	3.5 4.0	$\approx$ 500	15	

(1) Earlier data 60–70 g m^{-3}.
(2) Before 1900
(3) Present time

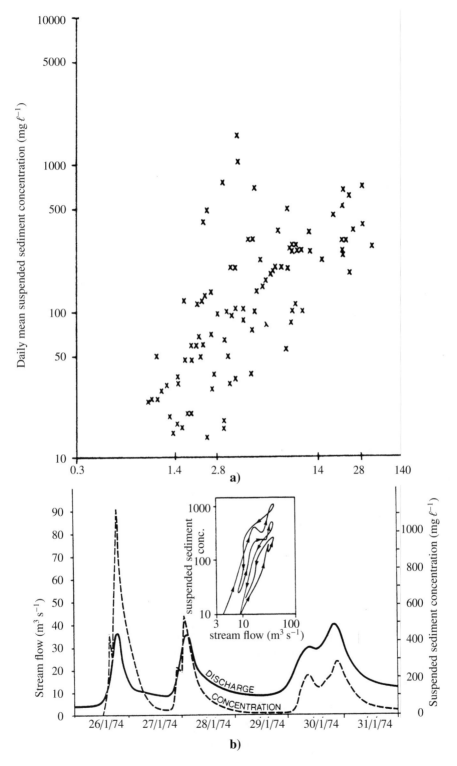

FIGURE 1.2: Discharge-sediment concentration relationship for a) the Humber River (Ont.); and b) the river Dart (UK). (Reprinted from Golterman *et al.*, 1983, modified after Walling, 1977.)

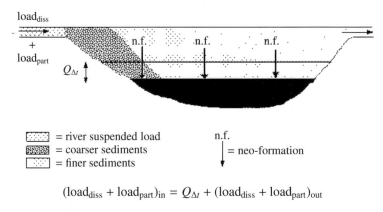

$$(load_{diss} + load_{part})_{in} = Q_{\Delta t} + (load_{diss} + load_{part})_{out}$$

FIGURE 1.3: Budget for dissolved and suspended loads.

USA) and the river Dart (UK). (See Figure 1.2.) The relationship cannot be extrapolated from one river to another and does not hold for the lower reaches of rivers, where the decrease in flow rate and the occurrence of flocculation* cause the suspended matter to settle.

The erosion or denudation of rocks and soil in a watershed depends in principle on four factors:

a) size of the basin, which has a damping effect on differences caused by local effects and events;
b) gradient or relief in the watershed, which influences both rainfall and stream velocity and, therefore, the quantity and size of material transported and eroded;
c) source material, which controls character and availability of material for erosion;
d) climate (quantities and distribution of rain and temperature) and vegetation, which strongly influences weathering.

Meybeck (1976, 1977) related total erosion in watersheds to their surface size, and arrived at the following relationships:

1) $500\,t\,km^{-2}\,y^{-1}$ for small alpine rivers;
2) $200\,t\,km^{-2}\,y^{-1}$ for middle-sized basins (Area $\approx 100,000\,km^2$)
3) $140\,t\,km^{-2}\,y^{-1}$ for large basins (Area $> 400,000\,km^2$).

His estimate of a world figure is $20 \cdot 10^9\,t\,y^{-1}$, or five times the dissolved load; this estimate is somewhat higher than those made previously. A global mean denudation of 75 to $100\,t\,km^{-2}\,y^{-1}$ can be calculated, for comparison, from Schuiling's (1977) data.

In aqueous systems the total quantity of eroded material in sediment budgets should be equal to the sum of the dissolved and particulate fractions of water-transported material (including bed* and suspended loads). (See Figure 1.3.) When studying specific watersheds it is, therefore, important to measure the ratio of dissolved to particulate material.

Some estimates of the total quantities being eroded can be obtained on a local scale. Lal & Banerji (1977) were able to estimate the amount of silt entering Indian reservoirs, as opposed to estimations from soil loss in a watershed. In their example, based on data from Indian reservoirs, they found that the silting can be described by the equation:

$$S = \frac{1}{100} \left(\frac{C}{I}\right)^{0.22} \left(\frac{I}{W}\right)^2,$$

where

S = siltation index ($m^3\,km^{-2}\,y^{-1}$);

C/I = capacity to inflow ratio (with C the original capacity and I the average annual inflow);

I/W = inflow per unit watershed area ($10^3\,m^3\,km^{-2}$).

The coefficients of C/I and I/W depend of course upon geology, climate and land use, and must be measured separately for different geological units, but the equation is useful for first estimates, especially in tropical areas.

1.3 Size of sediment particles

Sediment in mid-latitude rivers and streams or lakes are often typified by a bulk composition in which we find the following components in decreasing concentrations: mineral and rock fragments > clay > organic matter > precipitates/coatings, except for hard* waters such as in some Swiss lakes and marl lakes, where $CaCO_3$ usually forms the bulk material. In contrast, in regions where sediment transport is limited and evaporation is high (some low latitude areas), the typical composition may be: precipitations/coatings > clay > mineral and rock fragments > organic matter. Organic matter may be predominant in marshes or wetlands especially if macrophyte covering is dense.

The size of individual particles varies greatly; the coarsest material includes boulders, cobbles, gravel and sand; the finest material includes silt and clay (mud being an undefined mixture of silt and clay), and colloids (i.e., small particles that do not settle).

There is no satisfactory definition of the boundary between the size fractions of silt and clay. Some authors consider all particulate matter in rivers to be silt, while others define silt as being all particles with a grain size smaller than $16\,\mu m$ or even $50\,\mu m$. The use of sizing has developed as much in response to the availability of techniques as to the identification of precise property boundaries. In addition, there remains the problem of defining exactly what one is measuring at small particle size, after treatment with dispersants. Although these difficulties remain, size fractionation is essential, not only because of the need to differentiate physical characteristics influencing, e.g., sedimentation, but also because chemical properties, such as adsorption capacity, depend strongly on particle size.

Size fractionation is nowadays done by automatic particle counters, in which a sample is sucked through a small hole. During the passage of a particle the electric current caused by a set voltage is interrupted. These apparatuses can distinguish between different size classes, which can be preset. In older literature it was usually done by measuring the amount of matter sedimenting over a certain distance during a certain time. By means of existing tables the size class could then be read.

There are several different systems of classification. The United States Department of

Agriculture has been using the following criteria:

Coarse gravel	1000	to	2000 μm
Fine gravel	100	to	1000 μm
Medium sand	25	to	500 μm
Fine sand	50	to	100 μm
Silt	5	to	50 μm
Clay		<	15 μm

Geochemists have been using the following limits, based largely on the Wentworth classification:

Sand		>	63 μm
Silt	4	to	63 μm
Clay		<	4 μm

Most of the clay minerals have a maximum diameter of 4 μm, although some agriculturists distinguish between clays having diameters of < 16 μm and those having diameters < 4 μm. For all practical purposes (settling experiments), and for the purpose of contaminant studies, it is useful to consider portions of both the silt-size fraction and the clay-size fraction. This is because the larger particles have a limited adsorption capacity.

Metal hydroxides occur in particles having a near to colloidal size, at the lower limit of discrete resolution; as an example, freshly prepared diluted Fe^{3+} solutions will pass through a 0.45 μm filter, but will be retained on a 0.1 μm filter. For practical purposes, however, a pore size of 0.45 μm is commonly preferred and its use is common for sediment studies. For more precise studies of 'soluble' fractions, filtration over a filter with pore size = 0.45 μm must be followed by one over a filter with pore size < 0.1 μm. After some time a 0.45 μm filter will retain finer particles as well, because of clogging.

Particle size has a controlling influence on the water content of sediment. Physically, the smaller the particles, the less water. This is, however, counteracted by the (chemical) hydration capacity due to the chemistry of the smaller particles, especially clays. It is, therefore, not possible to predict water content from the particle size. Water content plays a dominant role in diffusion from and into sediment layers. (See Section 5.3 for a further discussion.)

1.4   Sedimentation and consolidation[*]

Assessing sedimentation is different for lakes or rivers, with different degrees of complexity. In lakes all suspended matter will eventually reach the bottom, the finer or even colloidal particles being entrapped in the flux of the coarser material. Sedimentation can be assessed by taking core samples. In rivers the situation is quite different; sedimentation is the result of factors such as specific density, grain size and water flow rate. Grain size is an important factor in sedimentation, in addition to its influence on the adsorption capacity of sediment. Terwindt (1977) has shown that the rate of deposition depends not only on factors such as depth of flow and initial concentration and type of suspended matter, but also upon the shear stress[*], exerted by the fluid flow, on the bed. Below a critical value of shear stress, particles in the moving fluid cannot remain in suspension, and sink gradually to the bed. Above a certain value, the rate of deposition is determined by the initial concentration in the fluid.

The bulk of river sediment will be deposited in the estuaries, where under suitable geological conditions sedimentation is enhanced by the decrease in water velocity and increase in salinity, causing a consolidation of the sediment. Regarding water velocity, Postma (1967) has shown that unconsolidated particles smaller than $10\,\mu m$, are deposited at flow velocities lower than about $10\,cm\,s^{-1}$ (measured just above the river-bed), while the velocity at which erosion occurs is near $80\,cm\,s^{-1}$ for consolidated clay and silt with a water content of less than 50%. Thus, when sediments are consolidated a higher water velocity is needed than the one at which deposition occurred. As size increases, water content becomes less critical as a factor controlling erodability; at a size of about $100\,\mu m$ there is no significant effect. The erosional velocities for unconsolidated sediment are about 1.5 to 2 times their depositional velocities.

Bottom erosion and transport of sediment in lakes as a function of wind velocity is different from that in rivers and different for mineral and organic matter. Huttula (1994) measured wind induced sediment resuspension and transport by currents in the shallow lake Säkylän Pyhärjävi (Finland). He presented equations for the calculation of this difference based on shallow water waves, wind induced currents and critical stress. The basic equation gives the change in concentration of suspended matter as a function of critical stress, either in the case of deposition or of resuspension, water depth and the coefficient of proportionality, K_e. The parameter K_e was obtained from sediment trap data and traditional equations for erosion flux. The article gives a good overview of the possibilities and the relevant literature. It was shown that the critical water depth for minerals was 3 m, while that for organic matter was 10 m. The consolidation process, i.e., the squeezing out of water due to the superimposed load, is often irreversible in the case of clay. In organic rich sediment consolidation may not take place at all; in shallow, eutrophic* lakes, wind erosion may have a strong influence on resuspension and become an important factor in possible P-release. (See Section 3.8.2.)

Deposition under the influence of chemical flocculation*, which is related to the increase in salinity, is most important for sedimentation in marine environments. Organic matter and Ca^{2+} also have an influence on flocculation in freshwater environments. Fine unconsolidated bed material with a high water content, which usually contains the highest pollutant concentrations, is re-eroded before material having a low water content, such as dried mud of a river bank or reservoir shore.

Sediment transported by some rivers finally arrives in lakes, where quantities are expressed as 'accumulation', the normal values of which fall between 0.1 and $\approx 5\,mm\,y^{-1}$. Kemp *et al.* (1976, 1977, 1978) gave good examples of estimates for Lake Erie, Ontario, and Superior, and specified the difficulties to obtain these data. Dominik *et al.* (1981) used radioisotope methods to measure sedimentation rate in Lake Constance (Germany/Switzerland) and assessed an alternative method of ^{210}Pb dating, using ^{137}Cs. In these radioisotope methods, a radioactive tracer is used to determine the age of a given sediment layer. The ^{210}Pb method of sediment dating is based on the decay of the nuclide enclosed, unsupported by its mother isotope, in a sediment column. A semi-logarithmic plot of nuclide excess activity versus time (t $\approx$ depth) permits the calculation of sedimentation rate. ^{137}Cs can be used as a time marker in sediment through its relation with its delivery pattern by nuclear explosions (1954,

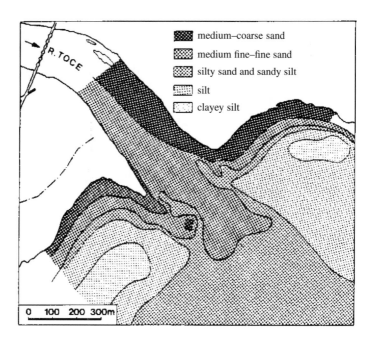

FIGURE 1.4: Distribution of recent sediment of the Toce Delta (Lago Maggiore, Italy). (Reprinted from Golterman, 1984.)

1963). Dating with ^{137}Cs was shown to be fairly accurate for high rates of deposition, but questionable for low rates. The rate of deposition in the main basin of L. Constance (Obersee) varied between $0.6 \, \text{kg m}^{-2} \, \text{y}^{-1}$ (central part) and $1.3 \, \text{kg m}^{-2} \, \text{y}^{-1}$ (eastern part). These deposition values are 5–10 times lower than the value used in Table 3.1 to calculate the P-balance of this lake, which is in agreement with the fact that $\approx 90\%$ of the sediment load of L. Constance remains in the delta area of the incoming river. The value of the total sediment load in combination with the P-concentration of the sediment is, therefore, a powerful means to calculate past and present P-loading.

Sediment arriving in lakes is not distributed homogeneously, as sorting according to particle size occurs. A typical example is given by Damiani (1972) for the river Toce delta in Lago Maggiore (Italy). Smaller particles are transported farther than larger particles, while rough relief material may be settling in areas of non-deposition. (See Figure 1.4.) Wieland *et al.* (2001) demonstrated the time- and space-dependent variations of settling particles in Lake Zürich (Switzerland). They showed an increasing vertical transport of biogenic and abiogenic material in the shoreward direction, an increased sedimentation rate at greater depth (causing 'funneling' or 'focusing'), patchy and episodic events in $CaCO_3$ precipitation and the formation of a patchy nepheloid layer in denser bottom waters consisting of fine-grained particles.

One or two samples are, therefore, not sufficient to characterize the entire lake bottom. In

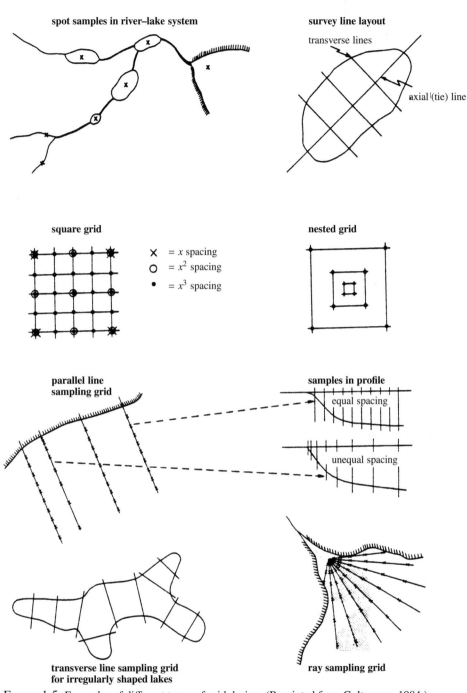

FIGURE 1.5: Examples of different types of grid design. (Reprinted from Golterman, 1984.)

TABLE 1.2: Chemical composition of some important rocks in order of decreasing solubility.

solubility	Minerals and rocks	Chemical composition
	Non-silicates	
XXXXXXXXXXXXXXX	Rock salt (halite)	NaCl
XXXXXXXXXXXX	Gypsum	$CaSO_4$
XXXXXXXXXXX	Calcite	$CaCO_3$
XXXXXXXXXX	Dolomite	$MgCO_3.CaCO_3$
XXXXXXXXX	Silicates	
XXXXXXXX	Feldspar	$KalSi_3O_3$, orthoclase
XXXXXXX	Soda-lime feldspar	A series from $NaAlSi_3O_8$ plagioclase, albite to $CaAl_2Si_2O_8$ anorthite
XXXXXX	basalt	$(Fe, Mg)_2SiO_4$, olivine $NaAl(SiO_3)_2$ alkali-pyroxene $R^{2+}SiO_3$ pyroxene (R = Ca, Mg or Fe) $NaAlSiO_3O_8$ plagioclase
XXXXX	Granite	$Kal_2Si_3O_{10}(OH, F)_2$, K-mica or K-muscovite and more complex structures
XXXX		$(OH)_2Ca_2Na(Mg, Al)_5[(Al, Si)_4O_{11}]_2$ or $(OH)_2Ca_2(Fe^{2+})_5[(Fe^{3+}, Si)_4O_{11}]_2$, hornblende or amphibole
XXX		$K(Fe, Mg)_3Si_3AlO_{10}(OH)_2$, biotite
XX		$Al_2Si_2O_5(OH)_4$, kaoline
X		$KaAlSi_3O_8$ or $NaAlSi_3O_8$, feldspar

Golterman *et al.* (1983), statistically correct procedures are given which may be tedious. In order to meet statistical criteria, samples should be taken from each typical area at a spacing depending on depth. (See Figure 1.5.) For shallow areas (10–20 m) the spacing should be 100–300 m, while for deeper waters (> 40 m) this may be increased to 100–3000 m because of the greater homogeneity. For a not even very large lake like Lake Geneva, this would mean an impossible number of samples; only common sense may help to reach a reasonable compromise. Much depends on the purpose of the study.

The major problem for evaluating the transport of N and P containing sediment is that there is no general model yet. Individual cases need their own quantitative studies. Some examples of the importance and the complexity of these processes in estuaries are given in Chapter 3 and Chapter 4.

1.5   Rocks as source material for sediment

Rocks are generally composed of a number of minerals, but some may consist of only one. The chemistry of some of the more important rock-forming minerals is outlined in Table 1.2, in order of decreasing solubility. Rock may be divided into the following major groups, depending upon their formative origin: igneous rocks, sedimentary rocks, metamorphic rocks and minerals formed under the influence of chemical processes. The formation of these groups is dominated by physical processes.

TABLE 1.3: The chemical composition of some igneous rocks with high and low SiO_2 concentration (as oxides), and crustal average. (Data from Dapples, 1959.)

	granite	olivine gabbro	Average of 16 km crust
SiO_2	70.18	46.49	59.08
TiO_2	0.39	1.17	1.03
Al_2O_3	14.47	17.73	15.23
Fe_2O_3	1.57	3.66	3.10
FeO	1.78	6.17	3.72
MgO	0.88	8.86	3.45
CaO	1.99	11.48	5.10
Na_2O	3.48	2.16	3.71
K_2O	4.11	0.78	3.11
H_2O	0.84	1.04	1.30

Large rock particles can be identified easily by visual examination, but the visual identification of smaller particles such as a fine sand and silt, even by petrographic microscopy, is much more difficult. Mineral species in the clay size range may be specified by the technique of differential thermal analysis or, more generally, by X-ray diffraction methods.

1.5.1 IGNEOUS ROCKS

Igneous rocks are formed as a product of the cooling and consolidation of a magma. Their major constituents are Al- and Mg-silicates or SiO_2. The composition of igneous rocks can be expressed either chemically or mineralogically. The chemical composition is normally expressed as percentage of the oxides of all elements present. (See Table 1.3.) Often the individual significance of each oxide may be interpreted in several ways; the same element may occur in different types of rock, e.g.:

1) SiO_2 present in quartz, chert, flint, silicate minerals and clay;
2) Al_2O_3 present in feldspar, mica, silicate and clay minerals;
3) Fe_2O_3 present in ferromagnesium silicates and sedimentary iron cores;
4) FeO present in various silicates and sedimentary iron cores;
5) MgO present in ferromagnesium silicates, in dolomitic limestone and as magnesite replacing dolomite and limestone;
6) Cl present in rock salt and chlorapatite.

Common minerals include quartz, feldspar, mica, olivine, pyroxene and amphibole groups (see Table 1.2 and Table 1.4). During the cooling of a magma reservoir, chemical differentiation takes place resulting in the successive formation of rocks having a different chemical composition, typically changing from basic, SiO_2 poor, to acidic, SiO_2 rich. Firstly, heavy dark minerals are formed; these include the olivines with a low SiO_2 content, and a ratio of 1 SiO_2 to 2(Mg, Fe)O. These sink to the bottom of the magma reservoir and, as they contain less SiO_2 than the average magma, the remaining magma will be relatively enriched in SiO_2 and some other elements. Progressively, rocks with a higher concentration of

SiO_2 solidify. These later-stage rocks still contain some dark minerals, such as pyroxenes ($SiO_2/(Ca, Mg, Fe)O = 1/1$, amphiboles, micas (glimmers) and, in smaller quantities, some olivines, but they contain increasing amounts of light minerals. The light minerals include feldspars (plagioclase, orthoclase and microcline) and finally quartz, which fills the interstitial spaces between the earlier-formed minerals. Variations, due to the presence of Fe or Al in the melt, occur. Lakes in this kind of watershed will have an extremely low conductivity and a relatively low pH. (See Section 2.2.1.)

1.5.2   METAMORPHIC ROCKS

Metamorphic rocks may be formed by the alteration of either igneous rocks or sedimentary rocks which have been subjected to high temperatures and/or high pressure. They may be of different metamorphic grade, depending on formative pressure or temperature (such as the sequence: slates*, schists*, gneiss*). Thus, changes occur in the texture, appearance and mineralogical composition of the original rock type. The new rocks may contain such minerals as amphiboles, chlorites and micas in addition to feldspar and/or quartz. High grade rocks may be formed, containing new minerals such as garnet and kyanite. These rocks often have a layered structure and are called schists.

Under conditions of severe folding and thrusting, part of the earth's crust may be subjected to intense temperature and pressure, in which sediments depressed into the deeper parts of the folds become permeated by the underlying magma and form gneisses. Gneiss differs from schist by having a coarser and larger-scale banding of light and dark minerals. Limestone rocks which undergo metamorphism become marble; quartz, sand or sand-stones are altered into quartzites. Minerals produced by intense metamorphism are generally resistant to weathering. Waterbodies in such areas usually contain little suspended matter; they have, therefore, a small capacity for chemical reactions such as adsorption.

1.5.3   SEDIMENTARY ROCKS

Sedimentary rocks are formed after weathering, in which some of the components of igneous and metamorphic (and previously formed sedimentary) rocks are set free. The largely unaltered rock fragments and mineral particles are known as clastics; after erosion and re-deposition, the clastics may form a wide range of new sedimentary rocks such as breccias and conglomerates, pebble beds, sand-stones and mud or clay stones.

In temperate regions chemical weathering may produce various forms of hydrous Al-silicates (clay, see Table 1.4) and under tropical conditions Al and Fe may remain in the form of hydroxides $Al(OH)_3$ and $Fe(OH)_3$, or rather $FeOOH$, during so called lateritic weathering.

Released components include the alkaline ions Na^+ and K^+ which are easily dissolved and ultimately appear in the hydrosphere, although some K^+ will be adsorbed onto clay. Ca^{2+} will also go into solution and may later form carbonate rocks with the CO_2 of the atmosphere through the $CO_2/HCO_3^-/CO_3^{2-}$ system. Despite chemical weathering, clastic carbonates are a frequent form of sedimentary deposit.

Englund *et al.* (1977) described the weathering of late Precambrian and Cambro-Silurian

TABLE 1.4: Minerals reported from freshwater lake sediments. (Modified from Jones & Browser, 1978.)

	Type of source		
	allochthonous	autochthonous	diagenetic
1.4.1 Non-clay silicates			
Quartz	X		
Potash feldspar	X		
Plagioclase	X		
Mica	X		
Amphibole	X		
Pyroxene	X		
Opaline silica (diatoms)		X	
1.4.2 Clays			
Illite: $K_{0.8}Mg_{0.35}Al_{2.26}Si_{3.43}O_{10}(OH)_2$	X		
Smectite: $X_{0.3}Mg_{0.2}Al_{1.9}Si_{3.9}O_{10}(OH)_2$	X		
Clorite: $Mg_5Al_2Si_3O_{10}(OH)_8$	X		
Kaolinite: $Al_2Si_2O_5(OH)_4$	X		
Mixed layers clays, 'vermiculite'	X		?
Palygorskite: $(Ca, Mg, Al)_{2.5}(Si_4O_{10}(OH)_4H_2O$	X		?
Nontronite: $X_{0.5}Fe_2Al_{0.5}Si_{3.5}O_{10}(OH)_2$			X
4.3 Carbonates			
Calcite, $CaCO3$	X	X	x
Dolomite, $Ca, Mg(CO_3)_2$	X		?
Aragonite, $CaCO_3$	x	X	
Mg-Calcite intermediate		x	X
Rhodochrosite, $MnCO_3.H_2O$			X
Monohydrocalcite, $CaCO_3.H_2O$		X	?
Siderite, $FeCO_3$	?		?
4.4 Fe-Mn oxides			
Goethite, $Fe(OOH)$	X	x	X
Magnetite, Fe_3O_4	X		
Hematite, Fe_2O_3	X		?
Birnessite. $(Na, Ca)Mn_7O_{14}.3H_2O$	?		X
Ilmenite, $FeTiO_3$	X		
4.5 Phosphates			
Apatite, $Ca_5(PO_4)_4.3(OH, F)$	X		x
Vivianite, $Fe_3(PO_4)_2.8H_2O$			X
Ludlamite, $(Fe, Mn, Mg)_3(PO_4)_2.4H_2O$			X
(?) lipscombite, $Fe_3(PO_4).2(OH)_2$			X
(?) phosphoferrite, $(Mn, Fe)_3(PO_4)_2.3H_2O$			X
(?) anapaite, $Ca_3Fe(PO_4)_2.4H_2O$			X
4.6 Sulphides			
Mackinawite, $FeS_{0.9}$		x	X
FeS			X
Pyrite, FeS_2	X		x
Greigite, Fe_3S_4			X
Sphalerite, ZnS		x	
4.7 Fluoride			
Fluorite, CaF_2			X

sedimentary rocks in the watershed of Lake Mjøsa (Norway) which is dominated by sandstones, shales and limestone. The authors found a clear correlation between the regional distribution of different rock types and the chemical composition of the water. The Cambro-Silurian calcareous rocks produced the highest concentrations of minerals in the water, while the lowest values were obtained from rivers draining sandstone areas and Precambrian gneisses and granites. Table 1.4 provides a summary of the mineralogical composition of sediments which have been deposited in typical freshwater environments and distinguishes between those derived from allochthonous or autochthonous sources and those formed by diagenetic processes.

1.5.4   ROCKS AND MINERALS FORMED UNDER THE DOMINANCE OF CHEMICAL PROCESSES

Sedimentary rocks formed mainly by chemical processes include non-clastic carbonates, evaporites and hydrolysates. After burial under heavy layers of sediment a certain degree of recrystallization may occur, which further consolidates the deposit.

1.5.4.1   Carbonates

These rocks are mainly $CaCO_3$ (limestone) and $CaMg(CO_3)_2$ (dolomite) and are typically formed by reaction between atmospheric CO_2 and Ca^{2+} and Mg^{2+}, while biogenic activity is particularly important in their formation. Algal and coral limestones form a substantial part of this rock type. Precipitation of dolomite in natural freshwater is unimportant for the control of carbonate equilibria, but dolomite underlying extensive areas may cause a water composition different from that of water in equilibrium with $CaCO_3$ (Stumm & Morgan, 1981).

1.5.4.2   Evaporites

Occasionally, salt deposits are formed by evaporation in closed sea or lake basins. The most important deposits include rocksalt, halite ($NaCl$), gypsum ($CaSO_4$) and calcite ($CaCO_3$). Ideally, the order of crystallization of salts in a lake or sea is as follows: calcite, gypsum, $NaCl$, Na_2SO_4, $Na_2SO_4.10H_2O$ followed by the double sulphates of Na, K and Mg. The last salts to be formed are generally those of KCl and $MgCl_2$, provided that the evaporite sequence is completed.

In carbonate and evaporite areas karstic phenomena occur, i.e., the dissolution of rock (carbonate or gypsum) by rain forming sinkholes, caves, springs and enhanced underground flow. About 15% of land surface is karst, among which carbonate rocks are dominant. Karstic water in these areas contains high concentrations of Ca^{2+}: 1 mmol ℓ^{-1} if the water is in equilibrium with air, and more if free CO_2 renders the pH lower. This system is further discussed in Section 2.2.2. Dissolution and industrial disposal of gypsum causes relatively high concentrations of Ca^{2+} and SO_4^{2-} (e.g. 1 mol ℓ^{-1} in the Rhône), changing the ratio Ca^{2+}/HCO_3^-. (See Section 2.2.2.)

1.5.4.3 *Precipitates and coatings*

Hydrous metal (Fe, Mn) oxides can be formed from a variety of sources, but mainly by the weathering of various minerals. These elements enter waterbodies from both ground and surface waters and occur in both dissolved and particulate state, but in the former case only in reduced form. Upon contact with water containing O_2 and in the alkaline or slightly acid pH range, Fe^{2+} is rapidly oxidized to Fe^{3+}, which will then hydrolyse to the insoluble FeOOH precipitate. A relatively higher pH for equivalent rates of oxidation is needed in the case of Mn^{2+}. Because of adsorption of Mn^{2+} onto the formed hydroxides, MnO_2 is seldom found; it probably occurs only under high pH or redox values.

In anoxic sediment, both Fe^{2+} and Mn^{2+} migrate through the interstitial water until they come into contact with O_2, where precipitation of FeOOH and $Mn(OH)_{3-4}$ then occurs. In lake water, considerable concentrations of Fe^{2+} and Mn^{2+} are often found in anoxic hypolimnia, up to several mg ℓ^{-1}. Because of upward migration of the metals and downward diffusion of O_2 a contact zone will be established and precipitation of the oxides will occur. These precipitates may, under the correct conditions, nucleate and form freshwater ferromanganese concretions (Damiani *et al.*, 1977). In the same way particles of other substances, e.g., $CaCO_3$, may be coated by thin layers of these hydroxides. (For further reading, see Stumm & Morgan, 1981.)

1.5.4.4 *Clay particles*

The term 'clay' refers either to a series of minerals with a sheet structure as lattice, or to a group of particles whose size is $< 4\,\mu m$. When referring to their chemical composition, it is usual to specify their names, otherwise the term 'clay' is assumed to refer to particles of a certain size in which clay minerals constitute an important part of the material. Quartz and feldspar may dominate in the particles in the $1-4\,\mu m$ range and clay dominates in the $< 1\,\mu m$ range. Clay minerals are essentially Al-silicates, but with various specific mineral forms. Clay minerals occur in almost all types of sediment and sedimentary rocks, in which they are often the most abundant minerals, sometimes as much as 40%. Fifty percent or more of the clay in the earth's crust is illite, followed in order of relative abundance by montmorillonite and mixed layers of illite-montmorillonite, chlorite and mixed layers of chlorite-montmorillonite, and kaolinite.

The weathering process is schematized in Figure 1.6. In this figure, arrows do not symbolize distinct chemical reactions but represent mixtures of reactions, the nature of which is not fully understood. It is not yet possible to predict what kind of clay will prevail in a certain river or lake, even if the nature of the surrounding rock material is known. In the sediment of Lake Mjøsa, Englund *et al.* (1977) found a high concentration of illite and chlorite, in agreement with their presence in the surrounding rocks.

Gradual hydration converts rock material into a spongy mass. Part of this process takes place during transport: rivers not only transport sediment, they alter the sediment and add properties to it. In addition lake bottoms receive autochthonous materials, mostly consisting of $CaCO_3$, FeOOH, SiO_2 and organic matter, all of which are powerful adsorbents owing to their H_2O, H^+, or OH^- groups.

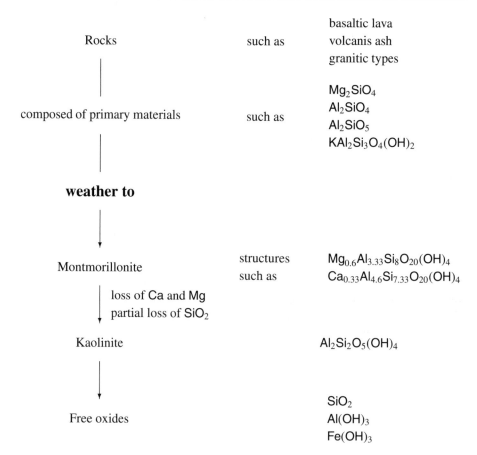

FIGURE 1.6: Schematic representation of physical and chemical erosion, leading through clay to free oxides.

Weathering is essentially hydrolysis of primary and secondary minerals. The extent of hydrolysis depends on the type of rock, climate and relief. The most important stage of the weathering process in diagenesis of clay is the conversion of a primary into a secondary mineral (a disproportionate or incongruent dissolution of Al-silicate). An atmospheric influence is necessary for this dissolution to occur, since it provides H_2O (rainfall is an important factor in the weathering process) in the presence of such gases as CO_2, H_2S, SO_2 and even HCl from volcanic action. In some areas kaolinite may be formed directly in igneous rocks by late-stage geothermal effects during magmatic cooling (kaolinization).

The breakdown process is accompanied by release of both cations and SiO_2. The water becomes alkaline by the reaction with H_2CO_3 as proton donor. The solid residue becomes, therefore, more acidic than the Al-silicate. The $-O_{10}(OH)_2$ unit is typical of structures formed during this process. In this way feldspars become kaolinites, montmorillonites and micas, all of which will be transported by the rivers as solids, in addition to the soluble products. Weathering can be simplified as follows:

Primary mineral	Secondary mineral
$6KAlSi_3O_8 + 4CO_2 + 28H_2O \longrightarrow$ (orthoclase)	$4K^+ + 4HCO_3^- + 2KH_2Al_3Si_3O_{12} + 12Si(OH)_4$ (muscovite)
$6KAlSi_3O_8 + 6CO_2 + 33H_2O \longrightarrow$ (orthoclase)	$6K^+ + 6HCO_3^- + 3Al_2Si_2O_5(OH)_4 + 12Si(OH)_4$ (kaolinite)
$6KAlSi_3O_8 + 6CO_2 + 48H_2O \longrightarrow$ (orthoclase)	$6K^+ + 6HCO_3^- + 6Al(OH)_3 + 18Si(OH)_4$ (gibbsite)

$Si(OH)_4$ will be removed largely in soluble form. Clays are formed in the solid phase in temperate zones. $Al(OH)_3$, $Mn(OH)_{2-4}$, $Fe(OH)_3$, $FeOOH$ and quartz (SiO_2) are formed by lateritic weathering in tropical zones. The hydrolysis does not follow one single reaction. Faust & Aly (1981) listed 10 reactions for the hydrolysis of $Na_2O - Al_2O_3 - SiO_2 - H_2O$ and 14 for the $K_2O - Al_2O_3 - SiO_2 - H_2O$ systems with different energy yields. Some reactions are not important quantitatively.

Clays have significant adsorption capacities through which they can bind and transport several compounds including many pollutants. The basic structural units are layered silicates (SiO_2) combined with brucite ($Mg(OH)_2$) or gibbsite ($Al(OH)_3$) sheets. All clays are built up of SiO_4^{2-} tetrahedra, connected at three corners in the same plane, which may be condensed into hexagonal networks (see Figure 1.7). The brucite or gibbsite sheet consists of two planes of hydroxyl ions, between which lies a plane of so-called intergrown Mg^{2+} or Al^{3+} ions. This unit is known as the octahedral sheet. The sheets are combined so that the oxygens at the tips of the tetrahedra project into a plane of hydroxyl ions in the octahedral sheet and replace two-thirds of the hydroxyl ions. H-bridges are found between the two layers.

The substitution of a cation with a lower charge for a cation with a higher one, in both the octahedral and tetrahedral sheets, gives the clay layer a net negative charge which is satisfied by interlayer cations (K^+ being the most common one, but Na^+ and Ca^{2+} are found as well). Expanded, or expandable, clay minerals have loosely bound cations and layers of water between the sheets. Na^+, Ca^{2+}, H^+, Mg^{2+}, Fe^{3+} and Al^{3+} are the most naturally occurring interlayer cations in these cases. Much of the interlayer water can be driven out only at temperatures between $120\,°C$ and $200\,°C$. Therefore, the dry weight and loss on ignition cannot be determined easily if these clays are present. (See Section 4.1.)

The layer electric charge (i.e., the electric charge like the positive on Na^+ or the negative on Cl^-) ranges from 0.3 to 1.0 per $-O_{10}(OH)_2$ unit of structure. Montmorillonite falls in the low-charged (0.3 to 0.6) group of expanded minerals. Kaolinite minerals are all pure hydrous Al-silicates of the basic structure $2Al(OH)_3.2Si(OH)_4$ which, on losing water, turns into $Al_2Si_2O_5(OH)_4$. All clay minerals have a so-called Cation Exchange Capacity (C.E.C.), but Fe- and Al-oxides, which are present on the surface of clays (so-called coatings) may influence the measured C.E.C. of clay because their $-OH^-$ groups may adsorb cations, or be exchanged against anions. In montmorillonite, the C.E.C. is closely related to the proportion of the expanded layers. There is an excellent linear relation between the C.E.C. and surface area (one monovalent cation per $0.67\,nm^2$, against 0.75 to $1.0\,nm^{-2}$ for montmorillonite). Part of the C.E.C. is also due to broken bonds at the edges of flakes. The charge is reversible, being negative in a basic environment and positive in an acidic one. C.E.C. is important in the N-cycle, as it is responsible for the NH_4^+ adsorption onto sediment, delaying the diffusion into

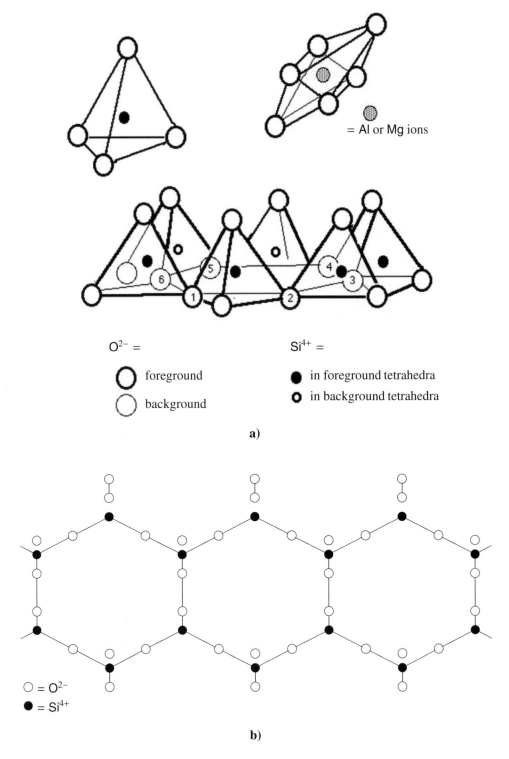

FIGURE 1.7: a) Tetrahedral silicate structures. b) Chain of hexagonally oriented SiO_2 tetrahedra, projected on ground plane.

the overlying water. The quantity of NH_4^+ adsorbed is determined by suspending the sediment in a solution of Ba^{2+} or K^+ ions and measuring the quantity of NH_4^+ released. Negative ions such as phosphate anions may be adsorbed, but the mechanism is more complicated, as will be discussed further in Section 3.2.1. (For further reading, see Weaver & Pollard, 1975.) See Section 1.7.2 and Section 3.2.1.

1.6 Organic matter

Sediment and suspended matter may contain up to a few percent of organic compounds, viz., allochthonous, derived from land by water or wind, and autochthonous, produced in the water by all organisms (see Section 4.1). These compounds include, besides natural products, variable loads of pollutants and their degradation products. Often, both products are adsorbed onto inorganic material such as $FeOOH$, clays or $CaCO_3$. Compounds which are slowly degraded are called refractory; temperature and adsorption onto sediment may greatly influence the extent and rate of degradation.

The organic compounds are mainly humic substances, but all biochemical compounds (lipids, proteins, carbohydrates, nucleic acids) have been found, though in small quantities – sometimes the total recovered amounts to only a few per cent of all organic matter present. Humic compounds, either in dissolved or particulate form, will form chelates* with most cations. These include bivalent or higher charged ions, such as Ca^{2+}, Mg^{2+} and Fe^{2+}, but also pollutants like Hg^{2+} and Cd^{2+}. Cu^{2+}, in low concentrations not a pollutant but a nutrient, is also adsorbed and, in high concentrations, made less toxic by this chelation.

It is difficult to distinguish between natural and man-made compounds. Eglington (1975) demonstrated that a fluorocarbon or a chlorocarbon may have a natural origin, although it is mainly a man-made pollutant. Some aromatic hydrocarbons are believed to be formed naturally.

Cranwell (1975, 1976a, 1976b, 1977) was one of the first to distinguish between autochthonous and allochthonous organic material – in both water and sediment –, using an organic solvent extraction and comparing it with the corresponding extracts from soils and aquatic detritus*. Separation of the extract gave several classes, differing as to functional groups. The abundance of each class, and the composition in terms of percentage, may be related to the source of the organic matter. Sediments from two lakes, differing in trophic state, illustrated the influence of the source material on the composition of organic matter.

During the downward transport of organic matter with sediment considerable release or mineralization may occur. Stabel & Münster (1977) found considerable concentrations of humic matter in the interstitial water, at different Ca concentrations and pH in interstitial waters, whence they could diffuse into the overlying water. The determination of the molecular weight fraction of these compounds by gel-filtration* can be used to follow these processes.

The organic constituent of particles can be distinguished into easily biodegradable (especially in polluted rivers) and refractory, although the limit between the two groups is vague because the breakdown of organic matter is closely related to the rate of microbial activity. In cold and fast running water little degradation takes place. In warmer and less turbulent water such as in lakes, degradation is more rapid; much of this takes place during the sedimentation of particles to the bottom, especially in lakes deeper than 10–25 m.

In shallow lakes sediment may, therefore, provide an important substrate and food source for microbial populations, as well as a source of heavy metals which can be released as toxic substances following biochemical conversions, e.g., mercury into methyl mercury. Phosphate can be released from organic compounds (see Section 3.8.2 and Section 3.8.4.7) because of the high concentrations in interstitial waters, but it may be rebound onto Ca- or Fe-compounds before a complete transfer can take place into the overlying water.

In deeper lakes refractory material will reach the bottom. Organic matter in Lakes Erie and Ontario has a concentration of $\approx$ 30% fulvic and humic acids and 60% humins. These three together are called humic substances and form complex organic materials with hetero-geneous mixed polymer systems. Fulvic acids (comprising hydrophilic and polyhydroxide aliphatic and aromatic acids) are the most soluble material and provide an important mecha-nism for the transport of metal ions in solution; if the metal is Fe, it may even play a role in the phosphate cycle. Humic acids are mostly insoluble and immobilize heavy metals and perhaps even phosphate in the sediment; they are partly soluble in 0.1–1 M NaOH. Eutrophic lakes which may have a high pH ($\approx$ 9) during daytime due to photosynthesis, may contain high concentrations of humic matter and, when they occur in peaty areas, may contain $CaCO_3$ as well. This may lead to distribution of phosphate over numerous compounds. Humins have the highest molecular weight, are highly polymerized and insoluble. The presence of humic substances plays an important role in metal processes in lakes, while their role in the P-cycle remains dubious (see Section 3.4.2). Whether or not they play a role in the N-cycle is stil under discussion and depends on the definition of humic compounds. During the symposium *Humic substances, their structure and function in the biosphere* (Povoledo & Golterman, 1975) the organic chemists studying humic substances held the opinion that they do not con-tain N-compounds, while ecologists considered that the real humic matter contains (adsorbed) amino acids, without which it would lose much of its ecological properties. This situation has not changed very much. (For further reading, see: *Limnology of humic waters*, Keskitalo & Eloranta, 1999.)

1.7   Chemical processes

1.7.1   THE BONDING PROCESS

Sediment may provide a means of transport, and of removal from the dissolved state, for both nutrients and toxic substances by means of adsorption, ion-exchange, co-precipitation or complexation and chelation. Not all reactions are possible in all sediments. Several of these reactions are pH sensitive and, when pH changes, the natural sediment may release the con-taminants, such as phosphate. Some compounds may be released under reducing conditions (Hg, Fe, Mn).

Theoretically, there are two extreme forms of bonding, defined as physical or chemical adsorption. While in the physical adsorption a small amount of energy is produced by the adsorption, in chemical adsorption this is more, although less than by a chemical ionic reaction. They may be described as follows:

$$\text{Clay}^- + \text{X}^+ \longrightarrow \text{clay.X}$$
$$\text{Clay-H}^+ + \text{X}^+ \longrightarrow \text{clay.X}^+ + \text{H}^+.$$

Normally they are not identified separately and it does not seem very likely that physical adsorption plays a role in sediment/water equilibria, except perhaps in the bonding of humic material onto $FeOOH$ or $CaCO_3$ particles, but certainly not in phosphate bonding.

1.7.2 PHYSICAL ADSORPTION

Clay minerals and their intergrown metal-hydroxides have reversible adsorption properties determined by their structure and surface and by the chemical composition of the solution in which they are suspended. The residual charge of a clay mineral particle is usually negative; therefore it attracts cations, in competition with the charge of the hydrating ions. The C.E.C. is measured at pH = 7. Montmorillonite, smectite and vermiculite have large C.E.C.'s (100–150 mmol per 100 g of clay, further called units). Kaolinite has 3 to 18 units. Intermediate values (10 to 40 units) are found for illites and chlorites. The C.E.C. is, however, strongly influenced by coatings. These coatings, often of $Fe(OH)_3$, are probably responsible for the adsorption of the negative anions of phosphate, more than the $-O_{10}(OH)_2$ unit.

The mutual exchangeability of cations depends on their charge density and their relative contraction in solution. Thus the larger ion K^+ will be less easily exchanged than the smaller Na^+ because the charge density of Na^+ is greater (i.e., the charge per unit of volume). A high charge density causes a stronger binding of a cation onto its hydration-shell complex in solution. In order for an ion to be adsorbed it must lose part of its hydration-shell, which demands some energy. In calcareous clays Ca^{2+} is the most exchangeable cation; in lime-free clays the H^+ ion replaces it. The adsorption of cations will form an outer layer of these cations. Through weak bondings (perhaps H-bridges) humic material may be loosely bound. This humic material is probably the humic material extracted by chelating* extractants such as Ca- or Na_2-EDTA*.

1.7.3 CHEMICAL ADSORPTION

In this process the cations are exchanged for the H^+ ions of $Si(OH)$, $Al(OH)$, $Al(OH)_2$ and $FeO(OH)$ groups of clay minerals and metal-hydroxides, and the $-COOH$ and $-OH$ groups of humic substances. The chelating process represents a special case of H^+ exchange in which the metal exchanged for the H^+ also forms a bond with the N-atoms of the complex. EDTA and NTA are strong chelating compounds while glycine is a weak one (see Figure 1.8). Naturally-occurring humic substances are moderate chelators, but it must be realized that different values are obtained for the different metals. These humic/Fe complexes are stronger than the chelates mentioned above and are probably the complexes dissolved in the hot $NaOH$ extraction step of the phosphate fractionation procedure (see Section 3.4.2). Whether the humic/Fe/P complex really exists is not yet proven.

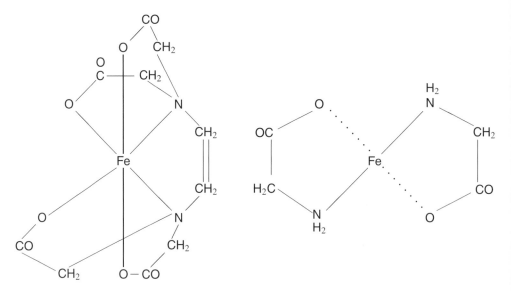

FIGURE 1.8: Molecular structure of the Fe complexes of EDTA and glycine.

P-adsorption onto FeOOH (Fe(OOH)≈P)is stronger and is the major source of iron bound phosphate in sediment. More energy is released than by physical adsorption. The mechanism will be discussed in Section 3.2.1. The adsorption onto clay is more complex. In the first place, charged sites caused by substitution of Al or Mg for Si may fix phosphate anions (Van Wazer, 1961, pp. 1663 and 1676). Another form of phosphate binding may occur according to the equilibria:

$$R\text{-}OH_2^+Cl^- + H_2PO_4^- \Longleftrightarrow R\text{-}OH_2^+.H_2PO_4^- + Cl^-$$

Or:

$$2(R\text{-}OH_2^+).OH^- + H_2PO_4^- \Longleftrightarrow (R\text{-}OH_2^+)_2.HPO_4^{2-} + 2OH^-.$$

These equilibria are originally proposed for soils. It has not yet been shown that they play a role in sediment, where FeOOH is a stronger adsorbent and is present in higher concentrations; in soils it will often dry up to Fe_2O_3, losing its adsorption capacity. The FeOOH particles will also adsorb humic compounds, now by exchange adsorption. These are probably the complexes which in the P-fractionation appear in the hot NaOH extract, together with fulvic acids and phytate. Phytate is adsorbed onto FeOOH even more strongly than phosphate. See Section 3.3. The separation between chemical and physical adsorption is not sharp and is more interesting to the theoretical chemist than to the practical limno-chemist.

Chapter 2

Chemical composition of freshwater

The chemical composition of a given freshwater body depends strongly on the chemical compounds entering with the water itself, both in dissolved and suspended state, and thus on the geology of the watershed, but the lake sediment may influence to a certain degree the chemical composition of the overlying water and vice versa. Before dealing with the chemistry it is necessary to describe the physical properties, such as melting and boiling points, heat of vaporization and fusion, dielectric constant and viscosity.

2.1 Physical properties

In a water molecule the two hydrogen-oxygen bonds form an angle (about $105°$). Therefore, the water molecule forms a dipole and larger molecular aggregates are formed. For this reason many of the physical properties of H_2O, such as melting and boiling point, are much higher than one would expect from a consideration of the H compounds of the elements near to oxygen in the periodic table of elements (Table 2.1).

In the context of limnology, even more important than the abnormal melting and boiling points is the heat of vaporization, which is much greater than that of related compounds and

TABLE 2.1: Melting and boiling points, and their differences, of H-compounds of elements near to oxygen in the periodic table of elements.

	melting	boiling	difference		melting	boiling	difference
H_2O	0	100	100	HF	−83	19.5	103
H_2S	−83	−60	23	H_2O	0	100	100
H_2Se	−64	−42	22	H_3N	−78	−33	45
H_2Te	−48	−2	46	H_4C	−183	−161	22

which plays an important role in the heat budget of a lake. In a lake at $50°$ latitude, $60\,cm$ of water (and thus $6\ell\,dm^{-2}$) may evaporate during the period March–October (200 days). This is equivalent to an average heat loss of $6 \times 2500/200\,kJ\,dm^{-2}\,d^{-1} = 75\,kJ\,dm^{-2}\,d^{-1}$. Thus, $\approx 40\%$ of the solar energy input during this period (170–$200\,kJ\,dm^{-2}\,d^{-1}$) is dissipated in the evaporation of water. If, in the first 100 days of this period, a $2\,m$ deep lake warms up $20°C$, the heat necessary for this temperature increase is only $17\,kJ\,dm^{-2}\,d^{-1}$. It is proportionally greater for deeper lakes. Under tropical conditions nearly all the solar energy absorbed is lost by evaporation. Heat budgets may be important for the calculation of eddy diffusion. (See discussion of Benoit & Hemond (1996) in Section 5.3.)

The formation and melting of ice also has a great influence on the temperature of lake water. In spring, when the surface ice melts, it absorbs heat (the latent heat of fusion of ice) from the lake water. This has a moderating influence on the rate at which the water warms up after winter, and the lake organisms are thus protected against sudden temperature increase. The same heat-exchange mechanism operates in reverse during winter ice formation, when latent heat is released into the underlying water layers. This compensates for sudden falls in temperature which might otherwise damage the lake organisms. The melting of $1\,m^3$ of ice diminishes the warming up of $10\,m^3$ of water with $8°C$.

Other physical consequences of the molecular structure of water are the abnormal values of its density, viscosity, and dielectric constant. H_2O aggregates dissociate with increasing temperature, so these properties change markedly with changing temperature. The relation between density and temperature is well known (see Figure 2.1). The fact that water is densest at $4°C$ is of utmost importance for shallow aquatic ecosystems, as water at this temperature sinks and does not freeze during cold periods even though there may be ice at the surface, as ice, with a density of ca. 0.92, floats. Wintering of organisms can thus take place at the bottom of these systems, even though the upper layers are frozen. The fact that water layers at $4°C$ are denser is also of importance in the stratification of lakes. When surface water cools to $4°C$ it sinks. The relatively low density of ice is of an importance that hardly need be elaborated – it protects the rest of the lake against further cooling by mixing and against severe winter conditions.

Viscosity also changes greatly with temperature:

Temperature	Viscosity (cp.)		Temperature	Viscosity (cp.)
0	1.8		15	1.1
4	1.6		20	1.0
8	1.4		30	0.8

(See also Figure 2.1.) The viscosity of water affects the speed with which suspended matter (dead algae, mineral particles, etc.) sinks. Upward movement of cells occurs when the density of the cells is less than that of the water or when currents of water carry particles upwards. In both cases the viscosity is important. The sinking rate of particles may approximate Stoke's law, which holds for spherical particles. For the evaluation of the influence of temperature on sinking rate, Stoke's law (for spheres) gives a first approximation:

$$V = \frac{kr^2 \cdot \Delta r}{\eta},$$

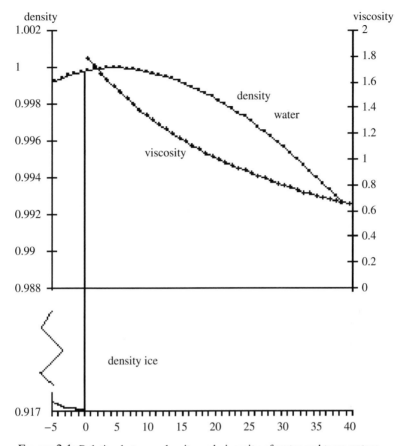

FIGURE 2.1: Relation between density and viscosity of water and temperature.

where V = sinking rate ($L\,t^{-1}$), r = radius of sphere (L); Δr = difference in density of sphere and medium ($M\,L^{-3}$), and η = viscosity. If a particle sinks from a warm water layer into a colder one, its velocity will be determined (within the range 0–30 °C) more by changes in water viscosity than by those in water density, firstly because the change in viscosity is larger than that in density and secondly because the sinking rate depends on the difference in density between particle and water, Δr. For a given temperature change within the range of freshwater, this difference changes less than the difference in the densities themselves. As the sinking rate is lower at lower temperatures, sedimenting particles tend to accumulate in the thermocline (or metalimnion), the water layer between the warmer upper layer of a lake, the epilimnion and the cooler, therefore heavier, lower layer of the lake, the hypolimnion. (See Figure 2.2.)

Grossart & Simon (1993, 1998) showed that aggregates of $\approx$ 3–5 mm (with a dry weight of 3–100 μg agg^{-1}) accumulated at two thermoclines at 6 and 15 m depth in Lake Constance from July till September. Formation of this 'lake snow' depended mostly on wind-induced turbulence and reflected the plankton populations in the lake to a large extent. It was estimated that lake snow could represent 20–40% of detrital[*] org-C_{part}. The particles showed a high concentration of amino acids and a high bacterial protein production; bacterial density

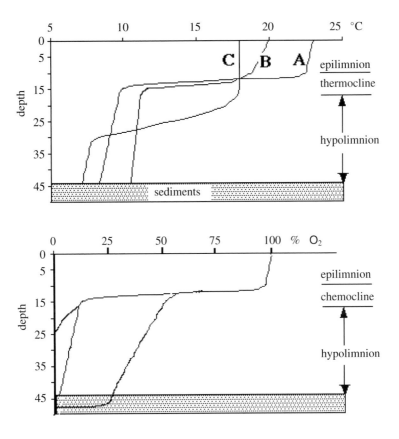

FIGURE 2.2: Physical and chemical stratification (schematized). The curves **A**, **B** and **C** describe the cooling off in autumn, and the sinking of the thermocline.

varied between 1.10^6 and 22.10^6 agg^{-1}. Concentration of o-P in the matrix water (= interstitial water) of these particles was enriched to 1500–2500 times that of the surrounding lake water. There is no information on the FeOOH and Ca^{2+} concentrations in the matrix water, knowledge of which might clarify their chemical action or function in P-transport and recycling. Kleeberg & Schubert (2000) found in the eutrophic softwater Lake Scharmützel (N-E Germany) that such aggregates of smaller particles may accumulate at high concentrations in steady-state in both thermocline and hypolimnion under anoxic conditions. This also influences the P-sinking rate, which is treated in Section 3.8.1.2. In these situations larger particles may contribute more to the sinking flux than smaller ones; these particles seem to be an important downward mechanism in stratifying lakes, but quantitative data are still scarce.

The difference between Lake Constance and Lake Scharmützel is probably due to the mineral content, and therefore to the density, of the particles; Lake Scharmützel, being a softwater lake, will have no CaCO$_3$ particles, while in the Lake of Constance CaCO$_3$ certainly forms the major part of the sinking material. The accumulation of particles in and just below the thermocline is important for the positioning of sediment traps (see Section 3.8.1), while the sinking rate influences the time a particle remains at a certain depth and thus its degree of mineralization.

Turbulence caused by temperature differences, wind, or fish movements cannot be dealt with quantitatively, but it is of great importance in relation to the sinking of particles. The high viscosity of water ensures that particles are carried with the water movement and are more or less prevented from sinking rapidly. Sinking live phytoplankton cells have a greater possibility of incorporating nutrients, preventing depletion of nutrients in the layers around the cells.

Lastly, the dielectric constant of liquid water is also much higher than that of other comparable liquids, making water an excellent solvent for several salts, because they will be dissociated in ions as a function of the energy effect. The dissociation of ions, which needs energy, is not an automatic process. The energy needed is provided by the energy produced by the hydration of the formed ions.

2.2 Dissolved compounds

2.2.1 GENERAL COMPOSITION (QUANTITIES AND ORIGIN)

The elements and compounds that occur in natural waters in dissolved state, can be allocated for convenience to the categories: major, minor, trace elements, gases, and organic compounds. Furthermore, several compounds occur in water in particulate form. As we have seen, rivers carry large quantities of silt into lakes where, due to decreased water movement, sedimentation will take place. The terms 'major' and 'minor' refer to quantities found and not to their biological significance. A summary of this quantitative classification is given in Table 2.2. Dissolved gases such as O_2, CO_2, N_2, H_2, CH_4 will be discussed in Chapter 3 and Chapter 4 as far as their role in the N- and P-cycles in sediments is concerned.

In Table 2.2 the percentages of the major elements are calculated for a 'mean' global freshwater composition (Rodhe, 1949). It is clear that globally $Ca(HCO_3)_2$ is the most abundantly occurring dissolved salt.

The sum of the cations in Table 2.2, except those placed between brackets, should be equal to the sum of the anions. Fe^{2+} and NH_4^+ normally occur only in anaerobic water, and are, in alkaline water, not present in the ionic state. H^+ is of course always present, but it contributes only significantly to the ionic balance in soft water when pH = 4 or less. OH^- may occur in strongly alkaline waters but only together with much larger concentrations of HCO_3^- and CO_3^{2-}. It will therefore not contribute significantly to the ionic balance.

The relative quantities are calculated as though a mean freshwater composition exists. Unlike the major constituent composition of seawater, there is no standard freshwater composition, because the concentrations are always changing, in both time and place. As the above calculations were made before the water composition of the large African lakes was known, the given percentages are different from those which would be calculated now. Lake Tanganyika, containing roughly 20% of the world's freshwater, has 8 mmol ℓ^{-1} cations plus anions, of which Mg^{2+} = 48%, Na^+ = 33%, K^+ = 12%, Ca^{2+} = 7%, while HCO_3^- = 89% and

TABLE 2.2: Major, minor, trace elements, organic compounds, and gases normally found in natural waters (the percentages refer to relative quantities calculated from 'mean' freshwater composition).

Major elements		Minor elements		Trace elements and organic compounds
Ca^{2+} (64%)	HCO_3^- (73%)	N	(as NO_3^- or NH_4^+)	Fe, Cu, Co, Mo, Mn, Zn, B, V
Mg^{2+} (17%)	SO_4^{2-} (16%)	P	(as $HPO_4^{2-} \Leftrightarrow H_2PO_4^-$)	
Na^+ (16%)	Cl^- (10%)	Si	(as $SiO_2 \Leftrightarrow HSiO_3^-$)	
K^+ (3%)				
H^+				Organic compounds such as humic compounds, excretion products, vitamins, metabolites
(Fe^{2+})				
(NH_4^+)				
Concentrations usually found				
0.1–10 meq ℓ^{-1}		(< 1 mg ℓ^{-1})		$\mu g \, \ell^{-1}$
(mean 2.4)		(Si sometimes higher)		Org: $\mu g \, \ell^{-1} \longrightarrow$ mg ℓ^{-1}

Cl^- = 10% (Talling & Talling, 1965). Other areas studied more recently are some Australian lakes and other inland waters, especially in Queensland (Bayly & Williams, 1972; Williams & Hang Fong, 1972). In many of these waters, both dilute and saline, Na^+ dominates the cations and Cl^- the anions. In other regions, however, divalent cations and HCO_3^- dominate, whilst in an intermediate type the Cl^- concentration is roughly equal to the HCO_3^- concentration. Including data from Lake Tanganyika, Golterman & Kouwe (1980) calculated that HCO_3^- is still the dominating anion in the new 'standard' composition; its percentage changed from 74% to 82%. Ca^{2+} and Mg^{2+} are present in roughly the same absolute amounts, but have also changed, to 34% and 33% respectively, instead of about 64% and 17%. Na^+ and K^+ have much increased.

The importance of standard freshwater is twofold. Just as there is a standard seawater to be used to check the quality of laboratory analyses, the availability of standard freshwater might largely improve the quality of freshwater analysis. Secondly, for the culturing of many algal species it is useful, if their preferred medium is not known, to start using this standard freshwater and then to modify it as required.

The presence of Ca^{2+} and HCO_3^- is important in determining and buffering the pH of water.

The electric conductivity of any water is roughly proportional to the concentration of its dissolved major elements, because of the formation of their ions. It varies enormously, ranging in value between < 10 and > 100,000 $\mu S \, cm^{-1}$. An approximate estimate of the quantity of dissolved ionic matter in mg ℓ^{-1} in a water sample may be made by multiplying the specific conductivity by a factor varying from 0.55 to 0.9, depending on the nature of the dissolved salts. A similar estimate in mmol ℓ^{-1} may be made by multiplying the conductance in $\mu S \, cm^{-1}$ by 0.01.

Talling & Talling (1965) gave examples of many African waters with exceptionally low or high mineral concentrations. They divided lakes in Africa into three classes according to their conductivity. Class I had a conductivity $< 600\,\mu S\,cm^{-1}$, Class II $600-6000\,\mu S\,cm^{-1}$, and Class III $> 6000\,\mu S\,cm^{-1}$. Some of the lowest conductivities were found in the Congo Basin and in lakes whose inflow drains through swampy regions. These lakes are frequently dark coloured and contain a high concentration of organic compounds, like in South America the black waters of Amazonia (Sioli, 1967, 1968). Class II included great lakes such as Lake Rudolf, one of the most saline in this group, which occupies a closed basin. The salinity in these lakes is largely due to Na^+, Cl^-, and HCO_3^-. Class III includes lakes in closed basins without visible outflow; under these conditions salts accumulate; the highest conductivity which Talling and Talling measured was $160,000\,\mu S\,cm^{-1}$, with alkalinities up to $1,500-2,000\,meq\,\ell^{-1}$ being recorded. Lake Tchad is another example of such accumulation (Carmouze, 1969).

The high salt concentrations influence both viscosity and density in the same direction. The influence on the sedimentation rate is, therefore, negligible. They do, however, have a strong influence on adsorption constants (see Section 3.2.2) and solubility products. This influence can only be estimated theoretically, which so far has not been done.

Cations and anions in lake waters originate mostly from mineral sources such as eroding rock and soils. Goldschmidt (1937) estimated that $160\,kg$ of rock has been eroded per cm^2 of soil surface or $600\,g$ per kg of seawater (equivalent to a layer of $400-600\,m$ depth over the whole earth). The oceans are now in a state of equilibrium, but for freshwater this is certainly not the case. Long-term changes in freshwater are, however, too small to be measurable over the period during which analyses are available. Some recorded changes are probably due rather to a change in methodology than to real changes. In a few cases human input has changed freshwater composition drastically. From the study of the chemical and physical characteristics of sediment layers from lake bottoms, however, wide differences in erosion patterns may be inferred. (See Mackereth, 1966.)

Concentrations of HCO_3^- normally exceed those of the other anions present, and in most cases the major cation is Ca^{2+}, although in African waters Na- and Mg-bicarbonate may reach high concentrations. If $NaHCO_3$ is the main product of erosion, it will gradually precipitate any Ca^{2+} from the solution owing to the low solubility of $CaCO_3$. High concentrations of soluble phosphate are then possible (Lake Rudolf and Lake Albert, see Talling & Talling, 1965).

Cl^- may come from natural erosion or, near the coast, from the sea. In the latter case the Cl^- is carried in by wind; thus, in the English Lake District Gorham (1958) found that the Cl^- concentrations in different tarns are a function of the distance to the coast. The following relation was found:

$$c = 0.71d - 0.33,$$

where c = concentration (meq ℓ^{-1}) and d = distance (km).

Gorham showed that atmospheric precipitation (by both wind and rain) is of the greatest importance as a source of many ions especially for bogs, upland tarns on insoluble rock, and more humus-rich waters. He showed that the relative proportion of ions which are derived from rain varied with the total salt concentration. While SO_4^{2-}, K^+, Mg^{2+} and Ca^{2+} decrease

TABLE 2.3: Concentrations of rain water in Amazonian and marine/coastal regions and ratio's between low and high values in both regions.

ion	continental		amazonian	marine/coastal		ratio	
	low	high		low	high	low/low	high/high
Ca^{2+}	0.2	4	0.03	0.2	1.5	1	0.38
Mg^{2+}	0.05	0.5	0.013	0.4	1.5	8	3.00
Na^+	0.2	1	0.048	1	5	5	5.00
K^+	0.1	0.5	0.023	0.2	0.6	2	1.20
NH_4^+	0.1	0.5	0.068	0.01	0.05	0.1	0.10
SO_4^{2-}	1	3	0.187	1	3	1	1.00
Cl^-	0.2	2	0.128	1	10	5	5.00
NO_3^-	0.4	1.3	0.192	0.1	0.5	0.25	0.38
pH	4	6	4.9	5	6	1.25	1.00

with increasing Cl^- concentration, Na^+ did not change in relation to Cl^- as, in all waters involved, it is only of marine origin. On the other hand, K^+ also comes from rain in the more dilute waters, but there is a steadily decreasing proportion as total salt concentration rises. Cl^- may enter inland waters near the coast through brackish river estuaries, while in low-lying areas it may even arrive by leakage or seepage from below.

Berner & Berner (1987) give data for the chemical composition of continental and marine/coastal rain water. The major ions tend to fall in the range of 0.1–$5 \, mg \, \ell^{-1}$, with Ca^{2+} equal or smaller in marine coastal rain, but with Mg^{2+}, Na^+ and K^+ between 2–5 times larger. SO_4^{2-} and NH_4^+ concentrations are so strongly influenced by local pollution that they may reach tens of $mg \, \ell^{-1}$. A summary of his data are presented in Table 2.3. In the column 'ratio' it can be seen that the ratio between the low values of the two different kinds of rain differs from the ratio of the highest values and varies largely with the kind of component.

Lastly, large amounts of dissolved salts may come from human, mainly industrial, activities. A classical example has been the river Rhine, which received huge quantities of $NaCl$ from the potassium mines. Concentrations well over $300 \, mg \, \ell^{-1}$ have been found in the past. Golterman & Meyer (1985) have shown that in the Rhine the disposal of industrial sulphate increased the sulphate concentration to $50 \, mg \, \ell^{-1}$, whereas in the river Rhône (France) such a concentration is natural.

F^- is also a naturally occurring element. Wright & Mills (1967) found $10 \, mg \, \ell^{-1}$ in the Madison river in the Rocky Mountains; in some parts of Wales several $mg \, \ell^{-1}$ are commonly found. F^- is eroded mainly from acidic silicate-rich rocks and from alkaline intruded* rocks (Fleischer & Robinson, 1963). It also occurs in gases and waters of volcanic origin; in volcanic gases in the Valley of Ten Thousand Smokes (Alaska, USA) up to 0.03% F^- is found, so that $200,000 \, t \, y^{-1}$ of HF enter the geochemical cycle there. Its solubility is determined by the solubility product of CaF_2, which is $4 \cdot 10^{-11}$. In waters with $1 \, mmol \, \ell^{-1}$ of Ca, only $4 \, mg \, \ell^{-1}$ of F^- can dissolve. Phosphate further decreases its solubility, since fluor apatite is even less soluble than CaF_2. The high concentrations of phosphate present in the Rhine in the past limited the solubility to $2 \cdot 10^{-19} \, g \, \ell^{-1}$. Soluble F^- is found in many tributaries of the

Rhine, but it is no longer detectable in a dissolved state in the Rhine itself.

Many relations exist between the different ions. The product of the concentrations of Ca^{2+} and CO_3^{2-} can theoretically not exceed the solubility product (see Section 2.2.2), while the same is true for Ca^{2+}, PO_4^{3-}, OH^- or F^- (see Section 3.2.3). Comparable interrelations exist between phosphate and iron (see Section 3.2.2). Therefore the effect of one or two elements should never be studied separately without reference to the whole chemical balance. This can be done by measuring the sums of cations and anions, and these totals should be equal ('ionic balance'). An estimate of the accuracy of the analytically determined ionic balance may be made by comparing it with conductance measurements; errors or omissions in the analysis or calculations may sometimes be traced if the results do not agree closely. In the Rhine and Rhône rivers, Golterman & Meyer (1985) found that few ionic balances had errors < 5% in the Rhine and < 7% in the Rhône; in several cases errors of up to 30% were noticed. The same is true of several other published ionic balances. In principle the error in an ionic balance must be < 2%. The presence of organic acids is regularly proposed as the cause of these errors, but this is wrong, as they are automatically included in the alkalinity titration.

2.2.2 THE CALCIUM BICARBONATE/CARBONATE SYSTEM

Bicarbonate ions, HCO_3^- have two important functions in lake-water. In the first place they provide the main buffer system for regulating the H^+ concentration in water, while secondly they provide the CO_2 for photosynthesis. For this reason photosynthesis has an influence on the pH of water, causing the pH to rise and CO_3^{2-} ions to be formed in the light, the process being reversed in the dark, or during mineralization of dead plankton.

A clear understanding of the bicarbonate/carbonate system is therefore essential and is best obtained by first considering the solubility of CO_2 in water. The solubility follows Henry's law and shows a linear relationship with the partial pressure of CO_2 in the gas phase (0.03% for CO_2 = 0.23 mm Hg) and with temperature. Under normal conditions the solubility of CO_2 is $0.4\,mg\,\ell^{-1}$ at 30 °C, $0.5\,mg\,\ell^{-1}$ at 20 °C and $1.0\,mg\,\ell^{-1}$ at 0 °C. The solubility is independent of the pH. A small amount of CO_2 will react with water to form HCO_3^- ions:

$$CO_2 + H_2O \Longleftrightarrow H_2CO_3 \Longleftrightarrow H^+ + HCO_3^-. \tag{2.1}$$

Larger quantities of HCO_3^- occur in mineral waters, since CO_2 is often present in excess. This excess CO_2 will increase the H^+ concentration sufficiently to dissolve carbonates:

$$CaCO_3 + H_2O + CO_2 \Longleftrightarrow Ca^{2+} + 2HCO_3^-. \tag{2.2}$$

The HCO_3^-/CO_3^{2-} system can be described by the following reactions, which each present an equilibrium between the ions and molecules that together regulate the system as a whole, i.e., H_2O, H^+, OH^-, CO_2, HCO_3^-, CO_3^{2-} and Ca^{2+}. The latter will be considered, for the sake of simplicity, to be the only cation present:

$$H_2O \Longleftrightarrow H^+ + OH^-, \tag{2.3}$$

$$CO_2 + H_2O \Longleftrightarrow H_2CO_3^-, \tag{2.4a}$$

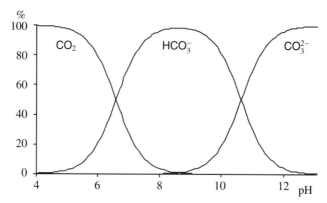

FIGURE 2.3: Relation between pH and percentage of total 'CO_2' as free CO_2, HCO_3^- and CO_3^{2-}.

$$H_2CO_3 \Longleftrightarrow H^+ + HCO_3^-, \tag{2.4b}$$

$$HCO_3^- \Longleftrightarrow H^+ + CO_3^{2-}, \tag{2.5}$$

$$Ca^{2+} + CO_3^{2-} \Longleftrightarrow (CaCO_3)_{solid}. \tag{2.6}$$

An equation may be written for each equilibrium. Because Σcations $= \Sigma$anions:

$$[H^+] = [HCO_3^-] + [OH^-] + 2[CO_3^{2-}] \tag{2.7}$$

if no mineral cations are present. There are therefore six simultaneous equations and these must all be satisfied at any one time. Furthermore one may define:

$$(CO_2)_{total} = (free\ CO_2 + H_2CO_3) + HCO_3^- + CO_3^{2-}. \tag{2.8}$$

The equilibrium constant for reactions (2.4a)–(2.4b) combined is called K_1, the apparent dissociation constant for the first ionization step for H_2CO_3. The dissociation constant for (2.5) is K_2. Both equilibria depend on temperature (Table 2.4) and also on the ionic strength of the solutions. Although the pK differences look small, the K values show great differences; therefore the temperature has an influence on this system. During summer lakes release CO_2, while during winter they adsorb it.

If some CO_2 is removed from the solution, for example by photosynthesizing algae, or by escape to the air, reaction (2.4a), and therefore (2.4b), will shift to the left, causing H^+ to decrease. Therefore, the reactions (2.3) and (2.6) will shift to the right, causing increasing OH^- and CO_3^{2-} concentrations. Because $K_1 \gg K_2$, more HCO_3^- will react following reaction (2.4a), (2.4b) than (2.5). The resultant increase in CO_3^{2-} is therefore less than the amount of CO_2 removed from the system. The amount of CO_3^{2-} formed can be estimated by measuring the change in pH and then using Figure 2.3.

As reaction (2.5) is of relatively little quantitative importance in most natural waters, it is useful to combine the others, (2.3) + (2.4a) + (2.4b), as:

$$HCO_3^- \Longleftrightarrow OH^- + CO_2. \tag{2.9}$$

It can be seen from (2.9) that, when $NaHCO_3$ is dissolved in H_2O, the solution has a slightly alkaline reaction (pH $= 8.3$ for 10^{-3} M $NaHCO_3$). In this case the amount of CO_2 formed

TABLE 2.4: Temperature dependence of the first (K_1) and second (K_2) ionization constant of H_2CO_3 and of the ionization product K_w of water ($pK = -\log K_w$).

Temp.	0	5	10	15	20	25
pK_1	6.58	6.52	6.46	6.42	6.38	6.35
pK_2	10.63	10.56	10.49	10.43	10.38	10.33
pK_w	14.94	14.73	14.54	14.35	14.17	14.00

TABLE 2.5: Measured pH values and calculated CO_2 concentrations in $NaHCO_3$ solutions.

	After 0 h		After 24 h	
	pH	CO_2 (mg ℓ^{-1})	pH	CO_2 (mg ℓ^{-1})
0.001 M in unboiled H_2O	7.2	6	7.94	1.1
0.001 M in boiled H_2O	8.1	0.8	8.1	0.8
0.003 M in boiled H_2O	8.35	1.1	8.6	0.8
0.01 M in boiled H_2O	8.44	3.7	8.95	1.1

according to (2.4a), (2.4b) is more than can dissolve in H_2O in equilibrium with CO_2 in the air, which is 0.5 mg ℓ^{-1} of CO_2 at 20 °C under atmospheric CO_2 partial pressure. Until the system is in equilibrium with air the excess CO_2 will escape. The disappearance of the CO_2 causes a decrease in H^+ (= increase in pH). As K_2 is much smaller than K_1, the solution becomes strongly alkaline when CO_2 is used during photosynthesis.

The overall reaction, the combination of reactions (2.4a) + (2.4b) + (2.6), may be written as:

$$2HCO_3^- \Longleftrightarrow CO_3^{2-} + H_2O + CO_2. \qquad (2.10)$$

This reaction will take place until the equilibrium between HCO_3^-, CO_3^{2-}, the 'equilibrium' CO_2, and the pH has re-established itself. 'Equilibrium' CO_2 refers to CO_2 in equilibrium with HCO_3^- and CO_3^{2-}. It is not, in general, in equilibrium with air. The equilibrium position depends on the K_1 and K_2 and on the pH of the solution. Figure 2.3 gives the percentages of HCO_3^-, CO_3^{2-} and CO_2 present in a solution in relation to the pH. In Table 2.5 are shown the pH values and the changes in pH with time of some $NaHCO_3$ solutions, and the amounts of free CO_2, calculated from the equation:

$$pH = pK + \log \frac{[HCO]}{[CO_2]}. \qquad (2.11)$$

The pH of fresh solutions varies, depending partly on the amount of CO_2 already present in the distilled H_2O, which is not negligible in weak solutions such as occur in nature; this variability is more marked in the very dilute solutions of $NaHCO_3$. For precise work, when making solutions to examine the system, CO_2 must first be removed. In most natural waters where Ca^{2+} is the predominant cation, the solubility product of $CaCO_3$ determines the amounts of Ca^{2+} and CO_3^{2-} (and thus the HCO_3^- concentration) which can co-exist in solution.

(The solubility product is a constant at a given temperature and is obtained as $[Ca^{2+}] \times [CO_3^{2-}]$ in a saturated solution. At 15 °C the value is $0.99 \cdot 10^{-8}$, at 25 °C it is $0.87 \cdot 10^{-8}$.)

Therefore there is an upper limit to the possible concentration of a $Ca(HCO_3)_2$ solution at a given temperature and partial CO_2 pressure. The $NaHCO_3$ system has one more degree of freedom as Na_2CO_3 is very soluble. In Na^+-dominated systems the limiting factor determining actual concentrations is the CO_2 in equilibrium with air, whereas in the Ca system it is the solubility of $CaCO_3$. In the absence of CO_2 the solubility of $CaCO_3$ is about 15 mg ℓ^{-1}. In water in equilibrium with air the maximum concentration is about $0.6 \cdot 10^{-3}$ M or 50–60 mg ℓ^{-1}. (This amount is calculated as though $CaCO_3$ is in solution; Ca^{2+} is of course balanced by HCO_3^- ions.) For further detailed discussion, see Weber & Stumm (1963) and Stumm & Morgan (1981).

In many lakes, however, metastable conditions may persist for a long time, with Ca^{2+} concentration greatly exceeding theoretical values. Golterman (1973a) found that the degree of supersaturation (i.e., the product $[Ca^{2+}] \times [CO_3^{2-}]$ being larger than the solubility product) is a function of pH. The higher the pH, the higher the supersaturation. He suggested that an electric double layer of CO_3^{2-} or OH^- ions stabilizes the colloidal particles. Wetzel (1972) has suggested that organic matter in lake-water may also have a contributory effect. The pH dependence of the supersaturation which can occur is then difficult to understand; especially so since the data for different lakes like Tjeukemeer (a humic-rich, eutrophic lake) and the humic-poor, oligotrophic Lake Vechten (the Netherlands) showed the same tendency. Golterman & Kouwe (1980) showed that several of the hardwater lakes studied during the "International Biological Programme" fell on the same curve as that for Tjeukemeer and Vechten. Golterman & Meyer (1985) showed that even in rivers like the Rhine and the Rhône the supersaturation, i.e., the ionic product (IP) of $[Ca^{2+}]$ and $[CO_3^{2-}]$ divided by the solubility product, K_s, can be described by:

$$\frac{IP}{K_s} = 8.73 \cdot 10^{-15} \cdot pH^{15.667}, \tag{2.12}$$

which gives a supersaturation with about a factor 7.5 for a pH = 9.

The supersaturation can be demonstrated by dissolving $CaCO_3$ in H_2O with CO_2 gas. After aeration to remove excess CO_2, supersaturation is always found and may last a long time.

The solubility of $CaCO_3$ can also be increased by the presence of an extra amount of free CO_2, which is then not in equilibrium with air. Tillmans & Heublein (1912) have given an approximate relationship between dissolved $Ca(HCO_3)_2$ and the amount of free CO_2 necessary to keep the $Ca(HCO_3)_2$ in solution (Table 2.6). This applies only to waters where the pH is lower than 8.3 because the concentration of free CO_2 which can occur at that pH or a higher one is negligible. These solutions are obviously not in equilibrium with the air.

The amount of free CO_2 present in excess is sometimes called 'aggressive' CO_2 because it is able to react with alkaline carbonates or metals. It is the compound important in the chemical erosion of silicates (see Section 1.5.4.4).

Table 2.6 indicates how much $CaCO_3$ will precipitate if a river supersaturated with CO_2 enters a lake and achieves equilibrium with the air. Assuming an alkalinity in a river of

TABLE 2.6: Relation between concentrations of $CaCO_3$ and free CO_2.

$CaCO_3$ in H_2O	Equilibrium CO_2	$CaCO_3$ in H_2O	Equilibrium CO_2
1 (mmol ℓ^{-1})	0.6 (mg ℓ^{-1})	3 (mmol ℓ^{-1})	6.5 (mg ℓ^{-1})
2 (mmol ℓ^{-1})	2.5 (mg ℓ^{-1})	4 (mmol ℓ^{-1})	15.9 (mg ℓ^{-1})

4 mmol ℓ^{-1}, it is clear that 3 mmol ℓ^{-1} will precipitate, or 3 M m^{-3} (or 150 g m^{-3}), because after equilibrium only 1 mmol ℓ^{-1} will remain in solution. If this process takes place in a water column of 10 m depth, 15 kg m^{-2} will precipitate. Entz (1959) estimated that in Lake Balaton 84,000 t of $CaCO_3$ per year are deposited by this mechanism. The amount of biogenic $CaCO_3$ formed may be even greater because photosynthesis may take up larger quantities of CO_2 than would escape by chemical processes alone (Wetzel & Otzuki, 1974). The occurrence of this effect can be observed on submerged water plants such as *Potamogeton* and *Elodea*, which are often covered with a thick layer of $CaCO_3$. *Stratiotes aloides* is reputed to accumulate so much $CaCO_3$ on its leaves during summer that it sinks to the bottom, although in spring it is a free floating plant. This kind of precipitation is the cause of the fact that the sediment of alpine lakes, such as Lake of Geneva, may contain up to 80% of $CaCO_3$.

The addition of other Ca^{2+} or (bi)carbonate containing solutions, e.g., $CaSO_4$, will change the above described system. This Ca^{2} may be derived from natural sources (Rhône) or from disposal by industry (Rhine). Golterman & Meyer (1985) have shown, with a data bank of 916 samples, that the relation between $CaSO_4$ added and the quotient Ca^{2+}/HCO_3^-, which theoretically must be 0.5, approached this value in most sampling stations of the Rhine and Rhône (although some exceptions were found).

For the Rhine the relation appeared to be:

$$Ca^{2+}/HCO_3^- = 0.70 + 0.50 \cdot (SO_4^{2-}) \tag{2.13}$$

and for the Rhône:

$$Ca^{2+}/HCO_3^- = 0.85 + 0.43 \cdot (SO_4^{2-}), \tag{2.14}$$

which is in agreement with the reaction:

$$\left\{ Ca^{2+} + HCO_3^- (\Leftrightarrow CO_3^{2-}) \right\} + Ca^{2+} + SO_4^{2-} \longrightarrow CaCO_3. \tag{2.15}$$

This reaction will make the calculation of the amount of $CaCO_3$ to precipitate under certain conditions more difficult than for the system in H_2O alone.

2.2.3 PHOSPHATE AND NITROGEN CHEMISTRY

Part of the P- and N-compounds enter waterbodies from soil erosion. Previously it was possible to give quantitative data; nowadays the quantities vary much under influence of over-fertilization. Some older data are given by Gächter and Furrer (1972) and Furrer and Gächter (1972) (see Table 2.7). Rekolainen *et al.* (1997) mention an output of Tot-P between 200

TABLE 2.7: N and P loss from soils (kg km^{-2} y^{-1}).

	Forest regions		Agricultural regions	
	P	N	P	N
Lower alps of Switzerland	0–4	82	69	1634
Swiss lowland	0–1	959	35	2102

and 2700 kg km^{-2} y^{-1} for Nordic European countries. The subject is treated in 16 national reports edited by Van De Kraats (1999), which publication gives, however, no conclusion or synthesis. The reports are not comparable because of differences in approach and statistics and it is not always clear in what kind of units the data are presented (e.g. P, P_2O_5 or PO_4). The following estimates can be extracted from one or two of these reports: P run-off appears to vary between 20 and 50 kg km^{-2} y^{-1} and N run-off between 1000 and 6000 kg km^{-2} y^{-1}, the variability depending on the types of soil and crop. Increase in P and N concentrations is evident in most of the participating countries.

Phosphor is the eleventh element in abundance in igneous rock at 0.07–0.13% of P, and occurs in about 190 minerals, of which only the apatites are quantitatively important. These encompass hydroxy-, fluor-, or chlor-apatites, i.e., $3Ca_3(PO_4)_2.Ca(OH, F, Cl)_2$. Volcanic rock may have higher concentrations. Van Wazer (1961) estimated that in the solid sphere of the earth about $1 \cdot 10^{19}$ t are present. In the sea and sea sediment about 10^{11} t are present, while Van Wazer considered only $3 \cdot 10^{10}$ t to be economically mineable. P-sedimentation in oceans has deposited about 10^{15} t, so that nowadays we can find sedimentary rocks rich in P. Mechanical and chemical weathering may dissolve part of this P again. In several lakes in Africa, Talling & Talling (1965) showed a striking coincidence of high concentrations of SiO_2 and o-P, indicating the weathering of phosphatized rocks which has produced these simple inorganic compounds in high concentrations even in nearly pristine systems. Golterman (1973b) went so far as to suggest that the ratio of SiO_2-Si/PO_4-P of natural erosion is about 110 (range 64–178), a value that might be used to estimate the concentration of P originating in natural erosion in systems charged in addition with human input. Processes involving phosphate and sediment will be discussed in Chapter 3.

Nitrogen enters the aquatic systems from different sources. The N concentration of igneous rocks is only 46 g t^{-1} (46 p. p. m.), although that of some sedimentary rocks may be ten times higher. Larger quantities come from erosion of both natural and artificially fertilized soils. In rivers heavily influenced by agriculture the NO_3^- concentration may be as high as 10 mg ℓ^{-1} of N, and it is often the main source of N for lakes. In contrast with P, N occurs in the aquatic system in several oxidation states, ranging from +V to −III, and there are several processes by which the different compounds are converted into each other. Many of these processes, denitrification ($NO_3^- \longrightarrow N_2$), nitrification ($NH_3 \longrightarrow NO_3^-$) and NH_3 production and release, occur in the mud. Processes involving N and sediment will be discussed in Chapter 4.

2.2.4   THE PHOSPHATE AND NITROGEN CYCLES

When nutrients enter a not too much enriched lake, they are likely to be taken up by phytoplankton during the growing season, as they are needed to sustain primary production in ecosystems. Either nitrogen or phosphate is generally the nutrient limiting this production. In a normal, i.e., non-polluted, situation the available concentrations of N and P are low compared to the demand and, as a result, by far the larger part of the nutrients is in the vegetation, i.e., mainly in the phytoplankton. The algae will incorporate these nutrients for a while, but after lysis and death, large parts are released as o-P and NH_3. Only small quantities of both are not mineralized and are finally incorporated into the sediment as org-P. This recycling is depicted in Figure 2.4. The dissolved nutrients re-enter the phytoplankton, and again a small part enters the sediment. But as this happens over and over again, the part entering the sediment eventually becomes considerable. Vicente & Miracle (1992) estimated that in the shallow, eutrophic Albufera of Valencia (Spain) about 30% of the biomass yield entered the sediment. The rest will leave the lake, or will increase the nutrient concentrations during the winter. Recycling after the death of macrophytes and phytoplankton is needed to maintain primary production. This recycling is not limited to phytoplankton, it occurs with nutrients taken up by macrophytes as well. Best *et al.* (1990) showed that in the first days of decomposition a small part of the phosphate from decomposing *Ceratophyllum demersum* was liberated as o-P, both under laboratory conditions and in litter bags in the field, while the larger part was liberated as small particles containing phosphate, which then entered the sediment. At the end of the experiment (16 days) most P had been taken up by the sediment. At lower temperatures, more P was liberated. Nitrogen, on the other hand, increased in the litter bags, probably as the decomposing organisms have a higher C/N ratio than *Ceratophyllum*. The same was described by Mann (1972).

Two major recycling processes can be distinguished: Nutrients from dying or dead phytoplankton will be remineralized in a few days – either by zooplankton or by decomposers – while the phytoplankton is still suspended in the water. This occurs on a nearly daily basis. Most of the P and N will re-dissolve in the water, while only a small part, the refractory material, will sediment. On the contrary, most macrophytes die late in autumn and sink to the sediment with which they are then mixed. The recycling processes passing through the sediment may take months, and for some fractions of organic nutrients years or decades.

In open water lakes 95–99% of primary production is phytoplankton production. Only in shallow lakes or marshes macrophyte production becomes quantitatively significant, but even in extreme cases it will never exceed 50%.

The main processes of the N- and P-cycles are represented in Figure 2.4. In the first instance macrophytes are not included – the main nutrient flux is through the phytoplankton.

In Chapter 3 I will discuss examples of P-studies from the literature, with some emphasis on the more generalized articles, and summarize some of my results obtained in the wetlands of the Camargue (Delta of the Rhône, Southern France); the N-cycle will be discussed in Chapter 4.

Processes that will be considered are:

for the P cycle: a) adsorption onto and incorporation into sediment;
           b) chemical composition, fractionation and bioavailability of P_{sed};
           c) release mechanisms;
for the N cycle: d) ammonification and mineralization, nitrification, denitrification, dis-

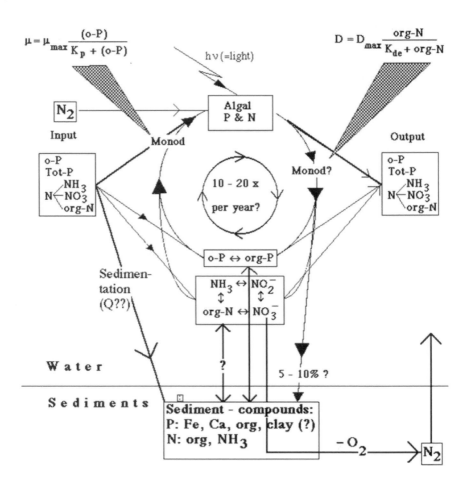

FIGURE 2.4: The nutrient labyrinth. Main pathways of N and P in freshwater systems (Golterman, 1995c).

similatory nitrate reduction.

The result will demonstrate the long way we still have to go before a quantification or a mathematical description of the N- and P-cycles can be developed.

2.3   Chemistry of the anoxic hypolimnion

2.3.1   INTRODUCTION

When physical stratification in a deep lake is well established, chemical changes in the hypolimnion will follow. Because of the sinking of organic matter, mostly derived from dead

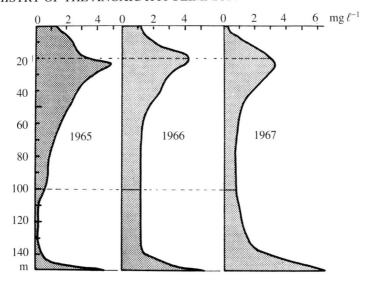

FIGURE 2.5: O_2 consumption between the beginning and the end of stratification in the hypolimnion and just above the sediment in Lake of Lucerne Station II (Golterman, 1975, modified from Ambühl, 1969).

phytoplankton, and the presence of heterotrophic[1] bacteria, O_2 will be consumed and CO_2 will be produced. The pH and the 'apparent redox potential' will decrease. (For apparent redox potential, see Section 5.2.4.) It is not yet precisely known how much of the dead phytoplankton is mineralized in the hypolimnion. Early estimates were 'a major part', but later it was shown that most of the phytoplankton, after its death, will already be mineralized in the epilimnion. Golterman (1976) estimated that $80 \pm 10\%$ of the algal cell material is easily biodegradable and is mineralized in the epilimnion; only the less degradable material enters the hypolimnion and subsequently the sediment. In the hypolimnion some further mineralization takes place, till O_2 and NO_3^- are depleted. Bacteriological and chemical processes will there continue to use O_2 and produce CO_2; the O_2 concentration will decrease and the apparent redox potential too, not because the O_2 itself continues to influence this potential, but because after O_2 depletion organic matter (normally broken down in the presence of O_2) becomes a H-donor – needs an electron acceptor.

Ambühl (1969) measured the O_2 consumption in the hypolimnion of the Lake of Lucerne between March/April (beginning of stratification) and October (end of stratification). The O_2-loss curves were differently shaped in space and time in several parts of the lake, but the most frequently occurring pattern showed maximal O_2 consumption at about 20 m depth (just below or even in the thermocline) and just above the sediment. (See Figure 2.5.) Total, cumulative, O_2 consumption was of course higher in the hypolimnion because of the larger water layer involved. The degree of O_2 depletion in a hypolimnion depends on the total amount of O_2 present, which increases with lake depth, but then decreases depending on the

---

[1] The classification of bacteria in heterotrophs, autotrophs, lithotrops and chemotrops is explained in the glossary.

amount of sinking dead organic material, which in its turn depends on the primary production.

The production of CO_2 in hypolimnion and sediment leading to a decrease in pH gives rise to a sequence of events coupled with the pH change (Section 2.3.2). The decrease in redox potential gives rise to a series of oxido-reduction events (see Section 2.3.3.)

2.3.2   THE pH SEQUENCE

As a result of the increase in the CO_2 concentration in the hypolimnion the pH will decrease and HCO_3^- will be formed, in soft water largely balanced by Na^+ ions, in hard water by Ca^{2+} ions.

Mortimer (1941, 1942), studying exchange processes between sediment and overlying water, with a strong emphasis on the influence of the redox potential (see Section 2.3.3), found pH values around 6 near the sediment in Esthwaite Water (English Lake District). In Ca-rich lakes CO_2 will dissolve the $CaCO_3$ which is either already present in the sediment, or has sedimented after its biogenic formation or abiogenic arrival in the epilimnion. Thus, the alkalinity will increase:

$$CaCO_3 + CO_2 + H_2O \longrightarrow Ca^{2+} + 2HCO_3^-. \tag{2.16}$$

Not necessarily two molecules of HCO_3^- are formed for every molecule of CO_2, because the pH shift must also be taken into account (see Section 2.2.2), while considerable amounts of free CO_2 will remain in solution. In the Vechten sandpit an alkalinity value in the hypolimnion of about $6\,\mathrm{mmol}\,\ell^{-1}$ was found at pH = 7.2, while the epilimnion contained $2\,\mathrm{mmol}\,\ell^{-1}$ at pH = 8. As at this pH the $(CO_2)_{\mathrm{free}}$ will be 20% of the $(CO_2)_{\mathrm{total}}$, it will have a concentration of $1.2\,\mathrm{mmol}\,\ell^{-1} = 52\,\mathrm{mg}\,\ell^{-1}$. Therefore, the water contains a hundred times more free CO_2 than water in equilibrium with air. Few data are available on the vertical distribution of Ca^{2+}, HCO_3^- and pH in hypolimnia. Mortimer (loc. cit.) studied exchange processes, not only between sediment and overlying water. He measured, therefore, the maximum concentrations just above the mud *in situ*, but also in experimental tanks. He found that in the experimental tanks more HCO_3^- accumulated in anaerobic than in aerobic conditions. The explanation of this phenomenon is that the aerobic tank was in equilibrium with air, so that the CO_2 produced could escape, but the anaerobic tank was sealed and the pH decreased.

Quantitative experiments in which redox potentials are changed independently of the pH and the CO_2/HCO_3^- system are clearly necessary.

As long as these processes continue to be so imperfectly understood quantitatively, it is probably more difficult than is often thought to calculate the respiratory quotient of lakes from the changes in O_2 and CO_2 concentrations in the hypolimnion. In Ca-rich waters $CaCO_3$ may precipitate, as the pH rises in the epilimnion during photosynthesis. The precipitate will sink until it reaches the lake bottom where pH values are lower resulting from CO_2 release, so that it will dissolve again. This process may take place near the mud and cannot be distinguished from direct dissolution of $CaCO_3$ from the mud. As $CaCO_3$ precipitates may adsorb phosphate, the Ca cycle will also influence that of phosphate (see Section 3.2.3). Chemistry and movement of Fe through the sediment/water interface is further discussed by Davison (1985).

2.3.3 THE REDOX SEQUENCE

The remaining sinking dead algal material enters the sediment, where bacteriological and chemical processes will continue to use O_2 and produce CO_2, respectively called Bacteriological Oxygen Demand (BOD) and Chemical Oxygen Demand (COD). Several authors have tried to differentiate between COD and BOD of sediment, e.g., by adding poisons, but the results remain dubious as many poisons will be inactivated by adsorption onto the sediment. (See Golterman, 1984). Furthermore, the production of reduced compounds (Fe^{2+}, H_2S, organic matter) may also decrease when the bacteria are killed, supplying less substrate for the COD.

After O_2 depletion, NO_3^- is the next compound to be used by org-C as an electron acceptor; the process will be discussed in Section 4.4. After depletion of NO_3^- the next electron acceptor is FeOOH in the sediment. Einsele (see Section 3.8.2) suggested that H_2S was the reducing agent for this process, Mortimer (1942) suggested that organic matter plays a role as well. Jones *et al.* (1983, 1984) demonstrated that a chemo-autotrophic* iron reducing bacterium (probably of the genus *Vibrio*) can reduce FeOOH, although the reduction rate of $FeCl_3$ was much higher. Physical contact between the bacteria and the Fe-particles was necessary, although about 30% of the activity was associated with extracellular compounds (therefore a chemical process). As FeOOH occurs in the sediment at the lake bottom, the Fe^{2+} formed will first appear in the interstitial water, the highest concentrations occurring in the top layer of the sediment, from where it diffuses into the overlying water layers. Considerable concentrations can be found: Mortimer (1942) found 20–30 mg ℓ^{-1} in Esthwaite water, Golterman (1976) reported about 5 mg ℓ^{-1} in the hypolimnion of Lake Vechten (the Netherlands), a hard water, stratifying sand pit.

Before FeOOH reduction, reduction of Mn compounds will take place, which will also escape through the interstitial water into the deeper layers of the hypolimnion (Davison, 1985). As no complexes or compounds of Mn and P are known, this process will no further be discussed here.

The presence of NO_3^- suppresses the FeOOH reducing activity and will therefore keep the P-adsorbing capacity of the sediment intact. (See Section 3.2.1 and Section 4.8.2.) Davison & Heaney (1978), Verdouw & Dekkers (1980) and Davison *et al.* (1982) studied the reduction of FeOOH further and showed that the reduction product, Fe^{2+}, does not form complexes with organic compounds, but may fix phosphate as vivianite.

Mortimer's results are summarized in Table 2.8. The lower potentials in the given range are those below which Mortimer could not detect the oxidized phase. For understanding redox potentials and their limitations, see Section 5.2. Not all arrows in Table 2.8 are chemical reactions: NO_2^- is not reduced to NH_3; NO_2^- may derive from both denitrification and NH_3 oxidation because some O_2 is still present. But denitrification of the NO_3^- released from the sediment may also produce NO_2^-. The higher potentials in Table 2.8 are those at which the oxidized form will begin to disappear, while the lower ones in the given range are those below which Mortimer could not detect the oxidized phase. The reduction reaction may, however, have taken place at a somewhat lower potential than was measured. These potentials are not real thermodynamic redox potentials, but just readings on the electrodes, which are the

TABLE 2.8: Apparent redox potentials at which reduction reactions occur in the mud, and prevalent O_2 concentrations (after Mortimer, 1941, 1942).

Reduction reactions	Apparent redox potential (mV)	Prevalent O_2 concentration (mg ℓ^{-1})
$NO_3^- \longrightarrow NO_2^-$	450–400	4
$NO_2^- \longrightarrow NH_3$	400–350	0.4
$Fe^{3+} \longrightarrow Fe^{2+}$	300–200	0.1
$SO_4^{2-} \longrightarrow S^{2-}$	100– 60	0.0

results of a mixture of many compounds, the oxidation/reduction reactions of which are not thermodynamically reversible.

The next electron acceptor is SO_4^{2-} which, via S, will be reduced to H_2S, mostly by *Desulfovibrio desulfuricans*. It is not clear whether this process is concomitant with the Fe^{3+} reduction, although Mortimer's sequence of events certainly suggests so.

Little is known of the nature of the organic matter used. In laboratory and model studies lactate is often used for the SO_4^{2-} reduction, which then becomes CH_3COOH as many strains of *Desulfovibrio desulfuricans* cannot metabolize CH_3COOH. Van Gemerden (1967) showed that dead cell walls of *Scenedesmus* and *Chromatium*, more likely substrates than lactate, can be used by pure cultures of *Desulfovibrio*. In older, over-simplified versions of these events the CH_3COOH, produced by the reduction of SO_4^{2-} by lactate, is finally converted into CH_4; methanogenesis and SO_4^{2-} reduction were seen as a sequence of reactions and SO_4^{2-} was often seen as an inhibitor of methanogenesis. (For a review, see Winfrey & Ward, 1983.) Another possibility is that the two processes are competitive, and that owing to a lower K_s for the SO_4^{2-} reduction this process is much faster. (See Figure 2.6.) This interaction between methanogenesis and SO_4^{2-} reduction remains a much discussed topic. Kristjansson *et al.* (1982) found that the different halfsaturation constants (1 μM for SO_4^{2-} reducing bacteria and 6 μM for the methanogenic bacteria) normally cause competition to be in favour of the SO_4^{2-} reducers. Senior *et al.* (1982) showed that the SO_4^{2-} reduction and methanogenesis are not separated in space or time, but occur within the same layer and at the same moment. The lower CH_4 production rate was explained by the lower halfsaturation constant. Oremland & Polcin (1982) demonstrated that SO_4^{2-} retarded methanogenesis when H_2 or CH_3COOH, but not when CH_3OH was the substrate. SO_4^{2-} reduction was stimulated by CH_3COOH and H_2 and by the combination. This suggests that SO_4^{2-} reducing bacteria will outdo methanogens in the competition for H_2 or CH_3COOH, but not for compounds such as CH_3OH. An incorrect interpretation of these events led to sulfide being considered an inhibitor of methanogenesis, instead of the concomitant SO_4^{2-} consumption. Winfrey & Ward (loc. sup.) found CH_4 production at a rather slow rate in all sediments of an intertidal marsh, and SO_4^{2-} reduction rates 100–1000-fold higher. When SO_4^{2-} was depleted CH_4 production became much higher. In some other studies CH_3OH and even CH_3NH_2 (methylamine) were used as a substrate for CH_4 production, but the presence of such compounds in natural sediment was not demonstrated.

Sequence of events

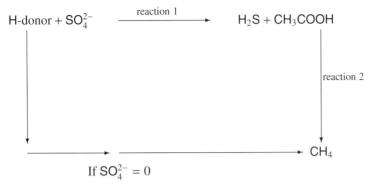

Competition

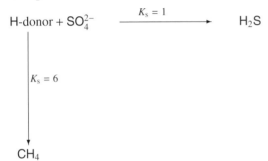

FIGURE 2.6: Relations between methanogenesis and sulphate reduction. Top: Sequence of two reactions. Bottom: Competition between the two processes, controlled by their K_s.

It is remarkable how little attention has been given to the influence of the temperature on the competition between methanogens and SO_4^{2-}-reducers. Zeikus & Winfrey (1976) demonstrated a severe limitation of methanogenesis by temperature, as the temperature for the methanogenetic bacteria was about 35–40 °C. Rates were considerably higher in sediment from 18 m water depth than from shallower waters, probably reflecting more anoxic conditions, and were higher in May than in January, reflecting the presence of more organic matter. But in all cases the temperature influence was high: in the January sediment, at 20 °C incubation, the rate was about 0.02–0.01 times the optimal rate at 40 °C and in May, at 10 °C incubation, it was 0.15–0.03 times the optimal rate. Abdollahi & Nedwell (1979) studied the influence of temperature on *Desulfovibrio* growth rate in sediment and found an optimal temperature of 35 ± 2 °C, with no seasonal influence in the field suggesting that other factors, probably the reducing capacity, were important. A Q_{10} of 3.5 was found which does not favour the SO_4^{2-}-reducers at hypolimnetic temperatures.

A factor which is often overlooked in the study of this competition is the fact that, concomitantly with SO_4^{2-} reduction, SO_4^{2-} production may occur. Urban *et al.* (1994) showed that reduction rates for SO_4^{2-} in cores (using $^{35}SO_4^{2-}$) of the soft water Little Rock Lake (Wiscon-

TABLE 2.9: Usage of [H]-donor produced by primary production.

Lake	Vechten		Blelham Tarn
	% of primary production	Idem in % of input into hypolimnion	
O_2 epilimnion	80 ± 20	–	–
O_2 hypolimnion	30	60	50
SO_4^{2-}	15	30	few
NO_3^-	n.d.	n.d.	15–25
Fe^{2+}	small	few	few
CH_4	small	few	15–25
Data from	Golterman, 1976, 1984		Jones & Simon, 1980. Jones, 1982

sin, USA) were higher than the calculated diffusive flux into the sediment suggested, which was explained by oxidation of reduced S-species. Nonlinear rates of SO_4^{2-} reduction and calculated turnover times of S^{2-}-pools supported the hypothesis that S^{2-} oxidation was nearly as rapid as SO_4^{2-} reduction. Laboratory measurements of the kinetics gave half-saturation values of 20–30 μmol ℓ^{-1} and a Q_{10} of 2.6.

Jones (1985) and Jones & Simon (1985) demonstrated that in cores the CH_3COOH pathway to CH_4 is less important than the reaction $H_2 + CO_2 \longrightarrow CH_4$. In older work CH_3COOH was wrongly suggested to be the major pathway, because the H_2 present in the original sediment was flushed out by He or N_2 to keep samples anoxic.

It is the low SO_4^{2-} concentration of the English Lake District waters that tends to keep the value for SO_4^{2-} reduction lower than in SO_4^{2-}-rich waters such as in Lake Vechten. SO_4^{2-} reduction and the competition with methanogenesis are important in relation to the P- and N-processes in sediment because of the formation of FeS. FeS plays a role in P-adsorption (see Section 3.8.4) and as a reducing agent in denitrification (see Section 4.4). Strong methanogenesis may use so much of the limited amount of org-C that only small quantities of FeS will be formed. Hines *et al.* (1989) analysed SO_4^{2-} reduction in sediment of a salt marsh (Chapman's marsh, New England) and demonstrated the importance of the vegetation. Above ground, elongation of *Spartina alterniflora* induced a fivefold increase in the reduction rate ($> 2,5 \mu$mol ml^{-1} d^{-1}), but a fourfold decrease upon flowering. It was suggested that these changes were induced by org-C_{diss} release during active growth. FeS concentrations were highest in the soils inhabited by the tallest plants. Reduction rates were similar in soils with *S. patens*, but did not increase when it was elongating. It is clear from this kind of studies that more attention should be given to the nature of the org-C which can serve as [H]-donor for the reduction of NO_3^- and SO_4^{2-} in anoxic sediment.

The relative usage of the [H]-donor from primary production is schematized in Table 2.9 for two comparable lakes. In Table 2.10 data are assembled for a hypothetical lake with a hypolimnion of 5–10 m and a primary production of $C = 100$ g m^{-2} y^{-1}. It follows from these calculations that of the Tot-FeOOH only 1–2% can be reduced by the dead organic matter

TABLE 2.10: Measured and calculated data for the iron reduction in hypolimnion of Lake 'Hypothesis'.

Measured data		
Hypolimnion depth	5–10 m	
Primary production	100 g m^{-2} y^{-1} of C	= 25–35 meq m^{-2} y^{-1}
Tot-FeOOH (in sediment)	500 g m^{-2}	
Fe^{2+} dissolved	5 g m^{-3}	= 25–50 g m^{-2}
Calculations		
2–3% of primary production	1 eq. m^{-2} y^{-1}	= 25–56 g m^{-2}
1% of Tot-FeOOH	50 g m^{-2}	

derived from primary production.[2] There are only a few lakes for which this distribution is known; sulphate is a crucial factor. Its concentration is low in the English Lake District, and relatively high in Lake Vechten, which explains the difference between these two situations, which resemble each other very much in other respects. See Table 2.9. A generalized scheme of [H] usage is given in Figure 2.7.

SO_4^{2-} reduction was studied by Hadas & Pinkas (1992) in Lake Kinneret (Israel). Using $Na_2{}^{35}SO_4$ the authors in the first place showed a decrease from 0.5 mMol ℓ^{-1} to 0.2 mMol ℓ^{-1} in the hypolimnion and a nearly quantitative reduction to H_2S. During stratification low values were found in the interstitial water, although the SO_4^{2-} was never depleted, probably because the quantity of org-C was limiting. The SO_4^{2-} reduction rate was low before overturn (12 nmol cm^{-3} d^{-1}) and increased to 1670 nmol cm^{-3} d^{-1} in the upper few millimeters of the sediment layer with another peak of 800 nmol cm^{-3} d^{-1} at 2 cm. SO_4^{2-} reduction capacity was also measured with the enzyme aryl-sulphatase. Maximal activity (per g ww.) was 670 nmol g^{-1} h^{-1} or lower in February, which was related to the higher SO_4^{2-} and lower H_2S concentrations available. In waters with high SO_4^{2-} concentrations the S oxidation and reduction processes may play a quantitatively important role.

The reduction of SO_4^{2-} will not only form H_2S; S^0 may be formed as well. In freshwater sediments the formation of pyrite, FeS_2, seems unlikely – there are no studies on this subject. FeS_2 can be formed in the marine environment. Rozan et al. (2002) described seasonal variations of the FeS_2 concentration with depth in sediment from the Rehoboth Bay (Delaware, USA), the only water source of which is Atlantic Ocean water. Pyritization was supposed to be caused by the reaction between FeS and H_2S, but a bacterial role was not established. FeS_2 seems to act in the P-cycle only by removing FeOOH, decreasing the adsorption capacity, while its action in the N-cycle is doubtful. Further information on pyrite formation and oxidation can be found in Postma (1982, 1983).

King (1988) showed that SO_4^{2-} reduction was the most important oxidizing process in muds of the salt marsh 'Belle Baruch' (South Carolina, USA), and this may well be true more generally. In Belle Baruch SO_4^{2-} reduction decreased with depth and was at all times greater in the short-form sites of Spartina alterniflora than in the tall-form sites, showing the influence of vegetation on oxido-reduction processes, probably by root-exudates. Dissolved

---

[2]This calculation is used in Section 3.8.4.4.

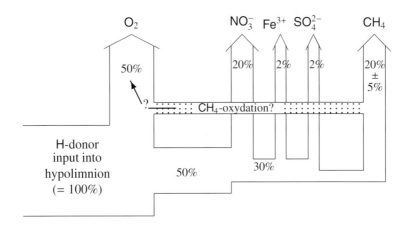

FIGURE 2.7: H-donor usage in hypolimnia of stratified lakes.

sulfides and pH varied seasonally in the short- but not in the tall-form sites, while pyrite (FeS_2) concentration was greater in the tall-form sites. As pyrite is not so easily formed in these environments it may be suggested that the presence of pyrite selected the form of *Spartina* present.

CH_4 production is an important process in wetlands and can be influenced by drying and wetting (see Section 4.4.4). As it is an anaerobic process, it will be stimulated by water logging, but the temporal lowering of the water level may stimulate methane-oxidizing bacteria and may cause oxidation of reduced compounds such as FeS. If NO_3^- is still present, this will also prevent CH_4 production.

Mountfort & Asher (1981) discussed the competition between methanogenesis and SO_4^{2-} reduction in an intertidal sediment in New Zealand. They mentioned a SO_4^{2-} reducing bacterium which can produce acetate from incomplete oxidation of long-chain fatty acids and propionate as well. Recent findings, that other SO_4^{2-} reducing bacteria may use acetate as H-donor, make it clear that no simple sequence of reactions is involved.

Boon *et al.* (1995, 1997) found peak CH_4 emissions of $1.7 \pm 0.05\,\mathrm{mmol\,m^{-2}\,h^{-1}}$ in an Australian wetland even at redox potentials between 176 and 243 mV. (The authors believe that the electrodes were not poisoned as they gave correct measures in a buffer solution. This is, however, not necessarily true (see Section 5.2).) At these peak emissions temporary wetlands reached the same levels as permanent wetlands. The authors noted that the lag phase in CH_4 production after re-flooding was shorter in the wetland than in rice-fields, where lag phases of several weeks were reported, a most unlikely event.

Mitchell & Baldwin (1999) studied the relation between methanogenesis and denitrification in temporary wetlands, i.e., aquatic ecosystems exposed to periods of desiccation. It is an important problem for the management of these wetlands, and of the large reservoirs for flood control or electricity production, which often have an appreciable amount of their sediment exposed to the air for a considerable period, and both processes may be important

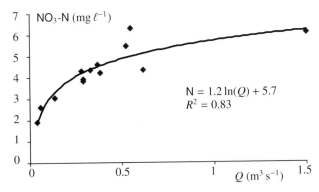

FIGURE 2.8: Nitrate concentration as a function of flow rate (Q). (Data calculated from Casey & Newton, 1972.)

as they lead to C and N losses. They used anaerobic mud-slurries with additions of glucose, acetate and nitrate, and found that methanogenesis did not occur as long as NO_3^- was present in concentrations $> 2.25 \cdot 10^{-5}$ M ($= 300 \mu g \ell^{-1}$ of N). They found no conclusive evidence, neither from their own study, nor from the literature, whether this was due to competition or inhibition. Glucose does seem not to be a suitable substrate for anaerobic processes – acetate is a better choice. However, they first flushed their mud-slurries with Ar-gas, removing a possible H-donor (see Jones, 1982). From the point of view of energy processes involved, denitrification is strongly competitive, and inhibition seems unlikely. The authors furthermore showed that denitrification was in the first instance limited by NO_3^- and, following the addition of N, by org-C. Nitrogen processes were not affected by desiccation, and there was no flush of mineral N upon rewetting. Methanogens were affected by desiccation but recovered over time upon rewetting.

2.4 River chemistry

Chemical processes are, of course, the same in rivers and lakes. The $Ca/HCO_3^-/CO_3^{2-}$ system is the same, but the influence of kinetic aspects may be different. The greatest difference is caused by the influence of the flow rate on the chemical composition. In principle the concentration of compounds may change with flow rate; three different phenomena may be observed:

1) the concentration remains constant regardless of the flow rate;
2) the concentration decreases with increasing flow rate;
3) the concentration increases with increasing flow rate.

Ad 1 Two mechanisms can be identified to account for this phenomenon:

   a) The water reaches an equilibrium with the soils through which it is percolating. This may happen to some anions while rain is seeping through agricultural land or in certain springs.
   b) The concentration approaches its chemical saturation value. Typical examples are

the Ca^{2+} and HCO_3^- ions, the concentrations of which are fixed by the pH and the solubility product of the $CaCO_3$, but this is not always the case in hard waters. Casey (1969) and Casey & Newton (1972, 1973) studied ionic composition in a number of rivers in the hard water area of Dorset (UK). In the river Frome they found that the alkalinity decreased with increasing flow rate, i.e., Case 2). A mean calculated dilution factor 2.4 times the mean observed dilution factor was found, suggesting that the water was not saturated with respect to $CaCO_3$. Although this is to be expected, the saturation cannot be calculated as no pH values were given. Other streams in the same district did indeed show a higher alkalinity and a very constant Ca^{2+} concentration, demonstrating that different patterns may be found for the same ion. Through the Ca-system, o-P may also be rather constant. Examples are the o-P concentrations in the Rhine and Rhône (Golterman & Meyer, 1985), which did not depend on flow rate but on pH.

Ad 2 This is the case when there is a constant input, diluted by the water volume. In principle $L = C^*Q$, with L = the load, C = the concentration and Q = the water volume. The relation between C and Q must be a hyperbole, but if sufficient data points are not available, or the two variables are inaccurately measured, this will not be found; in some cases a linear qualitative approach was found to satisfy.

Ad 3 This is a rather particular case. The case for particulate matter has already been discussed in Section 1.2. Another example is the NO_3^- concentration. The increase may happen if heavy rains push underground water, rich in N-fertilizer, out into the river. A very typical case is given by Casey & Newton (1972) for the river South Winterbourne (Dorset, UK), where the NO_3^- concentration increased logarithmically from ≈ 1 mg ℓ^{-1} up to 6 mg ℓ^{-1}, while the flow increased from 0.1 to 1.6 m^3 s^{-1}. (See Figure 2.8.) In other rivers in the same district the NO_3^- showed quite constant levels. The phenomenon is difficult to quantify, and no general rules can be given.

Rivers transport large amounts of o-P and P_{part}, and it is often necessary to know whether certain stretches retain or export it. This requires detailed measurements up- and downstream from the stretch, and the practical difficulties are considerable. A careful study was carried out by Bowes & House (2001) on a rather small system (river Swale, UK; flow rate = 40–80 m^3 s^{-1}) using an incremental mass-balance approach. During 4 campaigns samples were taken every 3 hr and analyzed for o-P, Tot-P_{dis}, Tot-P and $(SiO_2)_{dis}$. It was shown that, even during flood, there was retention of P and Si, while in the three other campaigns there was a net export. Cumulative residual loads were calculated and showed that these mass residues showed consistent patterns under various river discharges. During low-flow there was retention of P_{part} and even o-P, probably by adsorption onto the bed-load, while during high flow there was mobilization and export. During overbank flooding, not often considered in these studies, there was a high retention of P_{part} due to deposition of P-rich sediment onto the floodplain. The $(SiO_2)_{dis}$ showed no consistent pattern in relation to river discharge. Release of $(SiO_2)_{dis}$ from pore water and retention on the floodplain soil played opposite roles.

Chapter 3

Sediment and the phosphate cycle

Part 1: Speciation, fractionation and bioavailability

Ce qui est simple est toujours faux
Ce qui ne l'est pas est inutilisable.
(P. Valérie, Mauvaises Pensées et Autres)

3.1   Introduction

Most suspended matter in lakes, either autochthonous (primary production, etc.) or allochthonous, will eventually sink into the sediment, while a small part leaves through the outlet.
This includes algal detritus. During the passage through the epilimnion a large part of the
organic material already re-mineralizes; there are few estimates of this part, but it may be between 60% and 80% (see Section 2.3.1). Along the downward movement chemical changes
occur, in both thermocline and hypolimnion, but ultimately some N- and P-constituents will
enter the sediment, where the major part of both will be bound; some org-N will be released
after conversion, either into NH_4^+ (ammonification) or, under anaerobic circumstances, for a
part into N_2 by coupled nitrification/denitrification (see Section 4.3.1).

Phosphates enter the sediment in inorganic and organic form. Org-P will do so because,
during the growing season, in the photic zone of lakes, by far the greatest part of the phosphate
is in the form of org-P, as inorganic phosphate entering lakes is readily available for algae and
is usually rapidly taken up. After algal death a large part of its phosphate becomes o-P again,
but some org-P will sediment. Inorganic phosphate will also be transported downwards, either
included in the suspended matter entering lakes as a result of erosion, or after adsorption onto
this material in the lake. This phosphate will be adsorbed mainly onto FeOOH or be bound
as a form of Ca phosphate (see Section 3.2.1 and Section 3.2.2). These inorganic compounds
also form inside the sediment where further mineralization processes continue to break down
org-P and form o-P.

Stabel (1986) demonstrated the two mechanisms in Lake Constance (Switzerland/Germany), where $CaCO_3$ precipitation was the major mechanism for P-sedimentation, besides sedimentation of org-P. Stabel (1984) calculated the sedimentary P-flux to be 0.011 times the flux of the particulate organic matter and concluded that, therefore, P is mainly precipitated as a constituent of the particulate organic matter. This conclusion is not necessarily true, as during periods of algal blooms biogenic $CaCO_3 \approx P$ will also be formed. The absolute quantification of these transport processes remains difficult; future fractionation of the sinking material will contribute the information needed.

P_{part} can even be an important vehicle for horizontal P-transport. Golterman & Meyer (1985) have shown that, in a hard water river like the Rhine, more than 50% of the Tot-P is in the form of P_{part}; Moutin et al. (1998) demonstrated the same for the river Rhône. In this case the quantity transported depends on the flow rate. For example, El Habr & Golterman (1987) found a linear correlation ($R^2 = 0.7$; $N = 28$) between P_{part} and flow rate in the Rhône and thus a nearly exponential correlation between flow rate and total quantity of P_{part} transported. It also plays a role in vertical transport by air. Herut et al. (1999) demonstrated that a considerable quantity of P adsorbed onto loess arrived in the Eastern Mediterranean Sea by atmospherical transport from the Negev and subsequently contributed to the downward transport into the sediment. During this transport, solubilization may take place, contributing to the P_{aa} in the Mediterranean sea, the chemistry of which was studied by Ridame & Guieu (2002) in the Western Basin. In one week, up to 14% of the P_{part} in the dust could dissolve in sea water.

Although phosphorus is encountered in nature in one oxidation valence only (+ V, derived from P_2O_5), its chemistry in soil, water and sediment is complicated and many uncertainties and controversies remain. As many insoluble complexes of o-P exist, with a large variety of metal ions and -hydroxides, with clay minerals, or as organic complexes, most P is present in the solid phase, and very little in solution. In soils, Fe-, Al-, Ca- and clay- phosphates are considered to be the main inorg-P pools (Parfitt, 1978, 1989; Goldberg & Sposito, 1984). In aquatic systems, $Fe(OOH) \approx P$ and $CaCO_3 \approx P$ seem to be the most important inorg-P pools (Golterman et al., 1977; Golterman, 1988, 1995c). Furthermore, a considerable quantity of phosphate is present as org-P. Just as inorg-P, most of org-P is present in the solid phase, and very little in solution. As general processes plant and microbial uptake have some influence on P adsorption onto sediment, but after the death of the vegetation most of the phosphate is rapidly re-mineralized; the microbial pool is only small, but it has a rapid turnover and becomes o-P again. Quantitatively adsorption of o-P by mineral compounds is several times more important. The adsorption of o-P onto particles is a surface phenomenon and depends therefore on the size and nature of the particles. As smaller particles have a relatively larger surface, P_{ads} will be mainly on the smaller particles (P_{ads} is expressed in mg g^{-1}). Qiu & Mc-Comb (2000), comparing seven wetlands (Swan Coastal Plain, S.W. Australia), found that the P_{sed} concentration was inversely correlated with the coarse material, but positively with the water-, org-C and silt concentrations. A small part of Tot-P_{sed} was in the form of humic-P, substantially higher in lakes with higher Tot-P_{sed} and surrounded by woodland catchments (see Section 3.4.2). The positive correlation between org-C (varying between 5 and 25%) and Tot-P_{sed} (varying between 200 and 1000 μg g^{-1}) was also found in another estuarine sys-

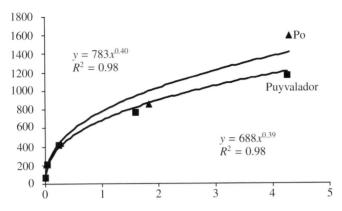

FIGURE 3.1: Phosphate adsorbed onto sediment against o-P_w. Sediment from the river Po (Italy), and from Lake Puyvalador (Pyrenees, France).

tem (McComb *et al.*, 1998). The within-year distribution was related to macrophyte biomass, which is a relatively rare phenomenon. The authors pointed out the great influence of algal biomass during low flow periods (35–52% was supposed to be bioavailable) on algal production in the following events.

When phosphate enters a waterbody it will be distributed over the different water and sediment compartments according to the following mass balance:

$$P_{in} = \Delta P_{sed} + P_{out} + \Delta(\text{o-}P)_w, \tag{3.1}$$

where: P_{in} = P entering the lake (g m^{-2} y^{-1}); P_{out} = P leaving the lake (g m^{-2} y^{-1}); ΔP_{sed} = P entering the sediment (g m^{-2} y^{-1}); $\Delta(\text{o-}P)_w$ = increase in P-concentration in water* depth (g m^{-2} y^{-1}). This equation is further elaborated in Section 3.10.1. As a result of P-accumulation in sediment, it is possible to define a P-retention time $t_p = P_{sed}/P_{in}$. In general the t_p increases with the water retention time, t_w ($t_w = V/Q = 1/\rho$, with ρ = the renewal rate), but no simple relation exists. This is further discussed in Section 3.10.1. In the past (O.E.C.D., 1982) it has been stated that the P-retention could not be quantified in reservoirs and played a minimal role because of the reservoirs' relatively short t_w. P-retention in reservoirs with short t_w may, however, be as large as in lakes. Garnier *et al.* (1999) showed that in three reservoirs in the Seine watershed (France) with residence times varying between 0.3–0.8 y, 60–80% of o-P, and 25–78% of Tot-P was retained. In tons, however, the Tot-P retention was 1.7 times the o-P retention, with most values $\approx$ 2.0 and with two negative exceptions. As no information about standard deviations was given, these data must be regarded with caution.

Most of the incoming P is in the form of inorg-P, although some org-P may arrive as well, either in the form of P_{alg} or adsorbed onto soil particles. Quantitatively, of the 4 compartments in Equation (3.1), by far the most of the Tot-P is found in the sediment and only traces are found in the water. The P entering the sediment compartment, P_{sed}, does not only come from decaying plant material; there is active chemical adsorption as well, either onto incoming or onto newly formed particles. Suspended particles and sediments have a large P-adsorption capacity due to their high concentrations of clay, silt and/or FeOOH. This can

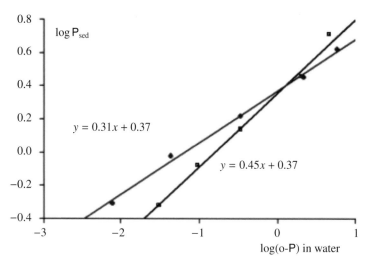

FIGURE 3.2: Log transformation of Figure 3.1.

be easily demonstrated in simple aquariums; two examples of such studies, in which sediment was shaken with water to which increasing quantities of P were added, are given in Figure 3.1 and Figure 3.2. Figure 3.2 is a mathematically correct logarithmic transformation of Figure 3.1, making it easier to find the constants of the adsorption equation. All sediments or suspended matter will adsorb phosphate and may release it when conditions change. The concept of sediment as a 'source' or 'sink' of phosphate is outdated. Sediment is not a P-source, as it first had to adsorb it, neither is it a P-sink, as the sediment is an integral part of the aquatic ecosystem, and the P_{sed} remains in this system and will be partly available for plant growth. The concept of 'internal loading' is not correct either; it is better to speak of 'recycling through the sediment compartment' or of 'P-release by sediment'. This quantity can become considerable. Knuuttila *et al.* (1994) found that in the shallow Finnish Lake Villikkalanjärvi, P-release from sediment could be as high as twice the external load, at least during part of the year. But it is essential to realize that the sediment originally obtained its phosphate from the external load.

Einsele (1936, 1937, 1938), Einsele & Vetter (1938) and Mortimer (1941, 1942), whose work will be discussed later (see Section 3.8.4), were the first to demonstrate the importance of sediment in the P-metabolism of lakes. A first *quantification* of the adsorption capacity was, however, made by Olsen (1964), who showed that the adsorption could be described by an adsorption equation (see Figure 3.3). In principle, two equations are often used to describe the adsorption. The first is the Freundlich one: $X_{ads} = A \cdot (X)^b$; the second, the Langmuir one, is:

$$X_{ads} = X_{max} \cdot \frac{X}{X + K_s},$$

in which X_{ads} = the adsorbed quantity of compound X; X_{max} = the maximal quantity that can be adsorbed; K_s = a constant, being the concentration at which 50% of the maximal quantity is adsorbed; A and b are constants.

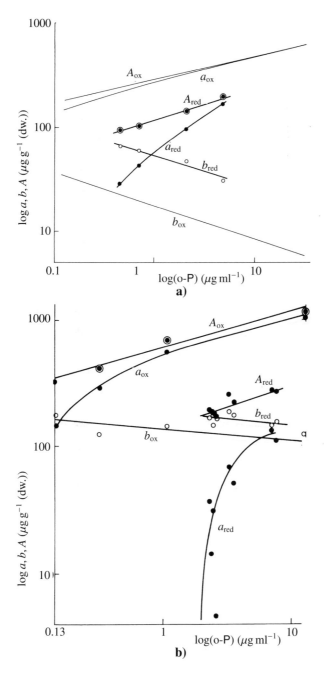

FIGURE 3.3: a) Phosphate adsorption and exchange for a reduced coarse sediment from 3 m in Furesø (Denmark). For comparison the results from experiments with the same sediment in oxidized state are indicated. For further details see the text. b) As a), but for a very fine-grained deep water sediment. From 35 m in Furesø. (Olsen, 1964, but redrawn from Golterman, 1975.)

TABLE 3.1: Comparison between the sum of P_{out}, P_{sed} and $\Delta(o\text{-}P)_w$ with P_{in}. P_{sed} calculated with Equation (3.2) using data as indicated in the text. Data on sediment load from Wagner (1972, 1976) and Olsen's adsorption constants for A and b of the Freundlich equation. Comparisons for mean values 1930–1940 and 1974. (From Golterman, 1980.)

Year	P_{out}	P_{sed} calculated as $L \cdot A \cdot \sqrt[3]{o\text{-}P_w}$	$\Delta(o\text{-}P)_w$	sum	P_{in}
1974	1.05	$6000 \cdot 0.62 \cdot (0.072)^{0.34} = 1.52$	1.0	**3.57**	**3.78**
1930–1940	0.13	$6000 \cdot 0.62 \cdot (0.008)^{0.34} = 0.72$	±0.1	**0.85**	**0.60**

Olsen chose, correctly, the Freundlich equation which for P-adsorption becomes:

$$P_{sed} = A \cdot (o\text{-}P_w)^b, \tag{3.2}$$

where P_{sed} = P concentration in sediment $(mg\,g^{-1})$; $o\text{-}P_w$ = o-P concentration in water $(mg\,\ell^{-1})$; A and b are constants. b is often found to be about 0.3–0.4.

Constant A reflects the adsorption capacity of sediment and is related to the Ca^{2+} and FeOOH concentrations of sediment and water. Equations such as depicted in Figure 3.1, Figure 3.2 and Figure 3.3 can easily be measured on isolated sediment samples, providing useful information of the P-adsorbing capacity of a given lake, viz., the constants A and b of the Freundlich equation. Olsen used a coarse shallow water sediment and a fine-grained one from deeper water (Furesø, Denmark), both under reduced and oxidized conditions. Equation (3.2) satisfactorily described the binding process for the 4 cases. Comparison of the two equations showed that the oxidized sediment adsorbed larger quantities than the reduced one; the deep, reduced sediment showed no adsorption below $P = 2\,mg\,\ell^{-1}$. The lack of adsorption under reduced conditions was probably due to FeS formed around the FeOOH particles, while the adsorption above $2\,mg\,\ell^{-1}$ was probably due to formation of $CaCO_3 \approx P$. Olsen showed that the P-exchange between sediment and water is a rapid process and that the P-uptake from the water by algae will be followed by P-release from the sediment with different rate constants for oxidized and reduced sediment.

The quantity of P_{sed} in Equation (3.1) cannot be calculated directly from P_{in}, as has been proposed in the past, but only from a study of the sediment itself and the equation for its adsorption capacity. In this way, Golterman (1980), using data from Wagner (1972, 1976) for the Lake of Constance, showed good agreement between the sum of P_{out}, P_{sed} and $\Delta(o\text{-}P)_w$ with P_{in}, calculating P_{sed} using Olsen's adsorption Equation (3.2) with his constants A and b for the sediment of Lake Constance, both for the period 1930–1940 and for 1974. (See Table 3.1.) The total sediment load $(6000\,g\,m^{-2}\,y^{-1})$ was taken from Müller (1966). Golterman (1984), in a review of this process based on more experimental work, strongly supported Olsen's proposal to describe the adsorption equilibrium with the Freundlich adsorption equation.

Due to the recent increase in P-loading, lake sediments have higher P-concentrations in the top layers than in the layers deposited before the period 1950–1960. Examples are, e.g., Edmondson (1974) for Lake Washington (Wash., USA) and Wagner (1972, 1976) for the

Lake of Constance; both examples have been discussed in Golterman, 1984. This increase can even be used to calculate the history of the P-loading quantitatively, if sufficiently precise P_{sed} data are available and the adsorption characteristics of the lake sediment are known.

3.2 Binding of inorganic phosphate onto sediment

FeOOH is one of the important compounds adsorbing o-P onto sediment and known as such for a long time already. It even plays a role in hard water systems, although there of course apatite or the $CaCO_3 \approx P$ system is important as well. The true chemical composition of $Fe(OOH) \approx P$ is not yet completely understood. The adsorption of o-P onto FeOOH is usually simplified as:

$$FeOOH + H_2PO_4^- \Longleftrightarrow Fe(O\text{-}HPO_4) + OH^-.$$

Syers & Curtin (1989) proposed a binuclear complex. This proposal was based on the infrared absorption studies of Parfitt *et al.* (1976). The infrared spectra were, however, made with dried samples and the drying will have changed the hydration degree. Its structure can schematically be depicted as follows:

It seems, however, likely that more complex hexahydrates such as $Fe(H_2O)_6(OOH)$ are involved. In the following the abbreviation $Fe(OOH) \approx P$ is used.

The 'apatite' compound is probably a mixture of $CaCO_3$ with apatite, $Ca_5(PO_4)_3.OH$. The small $Ca_5(PO_4)_3.OH$ crystals may be adsorbed onto the much larger $CaCO_3$ particles. The abbreviation $CaCO_3 \approx P$ is used.

Another way of describing the P-equilibrium between water and sediment is to use the EPC_0, i.e. the o-P concentration that causes no adsorption or release when sediment is suspended in o-P solutions. This is not a good concept, as the EPC_0 depends on factors like pH, presence of FeOOH or FeS, Ca^{2+} concentration and the history of P-loading. It is usually measured at a few concentrations, while often its value is obtained by linear interpolation; chemically this is not correct, because there is no evidence for a linear relation. House and Warwick (1999) demonstrated that the P-equilibrium between sediment and water controls P-concentration in the hard water river Swale (UK). They used the EPC_0 for kinetic purposes and analysed sorption results using the following equation:

$$Dn_a(t) = K_d c(t) - n_i, \tag{3.3}$$

where t = time, Dn_a = the change in the amount of o-P relative to the initial amount, K_d = the sorption constant and n_i = the amount sorbed when the concentration of o-P = 0. In this case the EPC_0 is used to quantify not the adsorption, but the kinetics of the process. The experimental adsorption of o-P onto FeOOH is discussed in Section 3.2.1 and onto the $CaCO_3$ system in Section 3.2.2. Furthermore, $Al(OH)_3$ is often suggested as an adsorbent in the literature, but rarely studied. The experimental evidence remains, however, weak and is usually based on (cor)relations between $Al(OH)_3$ and P_{sed}. Richardson (1985) discussed some of the mechanisms of P-retention capacity in different freshwater wetland types and demonstrated the large quantities that can be adsorbed. He showed wide quantitative differences in P-adsorption between the different soil types, in which he supposed extractable $Al(OH)_3$ to be the most important adsorbent, as it showed a higher correlation with a proposed 'phosphate adsorption index' than the Tot-Fe concentration did. This is probably caused by the fact that there are several forms of Fe in soils which are not active in phosphate adsorption. Richardson's equilibrium concentrations of o-P in the water ranging from 16–210 mg ℓ^{-1} are, however, unrealistically high and do not suggest a real adsorption mechanism; precipitation with $CaCO_3$ seems more likely. Richardson & Marshall (1986) demonstrated that chemical P-adsorption and uptake by organisms, mainly fungi and yeasts, were both responsible for the initial P-uptake by the peaty soil. P-uptake from fertilizer by sedge appeared to be negligible. Soil org-P was considered not to be available to plants, or to be released too slowly to contribute significantly to plant growth.

Pizarro *et al.* (1992) analysed sediments in the boundary area between fluvial and marine waters of the Rio de la Plata (Argentine). They found a good fit with the Langmuir equation between $Al(OH)_3$ and o-P ($R^2 = 0.92$) and a weaker one, $R^2 = 0.74$, with Tot-Fe. The Al correlation was, however, determined by one outlying, low, concentration which, if omitted, brings the R^2 down to 0.54. Moreover, one should not use Tot-Fe, but extractable FeOOH, as by far the largest part of Fe_{sed} is not involved in the adsorption.

In experimental studies with a marine clay, Lijklema (1980) and Danen-Louwerse *et al.* (1993) also suggested that $Al(OH)_3$ plays a role in P-adsorption onto sediment. These authors did, however, not measure the quantity of $Al(OH)_3$ present, nor did they give any evidence at all of the existence of the $Al(OH)_3 \approx P$ complex. Their extraction method (using NaOH and HCl and NH_4-oxalate-oxalic acid; pH = 3) does not distinguish between FeOOH bound or $Al(OH)_3$ bound P, while interference with org-P will occur (De Groot & Golterman, 1993).

Lijklema (1980) showed phosphate adsorption onto $Al(OH)_3$ in experiments *in vitro*, where high concentrations of $Al(OH)_3$ were used in limited volumes. Phosphate was more strongly adsorbed onto $Al(OH)_3$ than onto $Fe(OH)_3$ when Al-salts were added to an o-P solution, but far less when the o-P was added to $Al(OH)_3$ prepared in advance. These data are more relevant for P-elimination from sewage water than from lakes and cannot be used in a lake situation. Thermodynamically it seems unlikely that $Al(OH)_3$ accumulates in sediment, as it is relatively soluble ($pK_s = 10^{-8}$), while furthermore its solubility depends strongly on the pH; complex formation with other components will increase the minimal solubility considerably (Ringbom, 1963). Golterman (1995a) measured an Al-concentration of 55 μg ℓ^{-1} in an experimental tank with equilibrated hard water to which 1 g m^{-3} of Al^{3+} was added. De Haan *et al.* (1991) found concentrations of soluble (< 200 nm) Al between 10 and 50 μg ℓ^{-1}

in 5 humic rich Dutch lakes with pH values between 6.85 and 8.5 and suggested a linear correlation with org-C_{diss}. $Al(OH)_3$ does, therefore, most likely not play a role in the adsorption process in lakes, although in extremely acid waters it may do so (Ulrich, 1997), but this is, obviously, not a natural phenomenon.

The situation is different in soils, where $Al(OH)_3$ does play a role. Van Riemsdijk *et al.* (1977) studied the reaction between $Al(OH)_3$ and Al_2O_3 and showed that the reaction rate of the first compound slowed down after 1 day, and that of the second after 4 days, in both of which cases first order kinetics were obtained.

3.2.1 THE PROCESS OF PHOSPHATE ADSORPTION ONTO Fe(OOH)

In this section the chemical adsorption experiments, the influence of pH and Ca^{2+} concentration and the thermodynamical aspects will be treated. The importance of FeOOH as an adsorbent in sediment was understood long ago (Einsele, 1936; Mortimer, 1941, 1942; Olsen, 1964; Golterman, Viner & Lee, 1977; Lijklema, 1977, 1980). The equilibrium between P_{ads} and o-P_w can be described by Equation (3.2); some examples are given in Figure 3.4 (Golterman, 1995). In these laboratory experiments o-P was adsorbed onto suspensions of Fe(OOH) at different pH values and each curve was fitted with Equation (3.2). Some irregularities occurred, e.g., the curve at pH= 8.3. Later it was realized that this might be due to contamination with traces of Ca^{2+} or Mg^{2+}, the importance of which was not directly understood.

Lijklema (1977) was the first to emphasize the influence of the pH on the P-adsorption onto FeOOH; he used a logarithmic description taking into account the influence of the pH on the equilibrium concentrations:

$$P_{ads}/Fe = 0.298 - 0.031 \, pH + 0.201 \cdot (o\text{-}P)^{0.5}. \tag{3.4}$$

As this equation would give a positive Fe(OOH)$\approx$P concentration when no P is added, his data were later fitted by Golterman (1982a) on the Freundlich equation:

$$P_{ads}/Fe = A \cdot (o\text{-}P)^b. \tag{3.5}$$

This equation has a slightly lower correlation coefficient than Equation (3.4), but fits a chemical concept better.

A and b are constants under certain boundary conditions, while the constant A depends on the pH. A thermodynamic analysis of the adsorption process has been given by Stumm *et al.* (1980); one of the consequences for processes of adsorption onto sediment is that the adsorption of weak anions was shown to be highest around the pH value of pH = pK, in agreement with practical findings. However, the increasing solubility of FeOOH below pH= 5 also plays a role and the solubility diagram loses its value. Golterman (1994, 1995a) demonstrated in laboratory experiments that A does not depend on the pH linearly, but that the exponential curve:

$$A = 23600 \cdot 10^{-0.416 \, pH} \qquad (R^2 = 0.95; \quad N = 13) \tag{3.6}$$

fitted the data satisfactorily. Curve fitting showed b to fall in the range 0.25–0.6, but with b set at a value of 0.333 the R^2 value did not become significantly smaller than those obtained

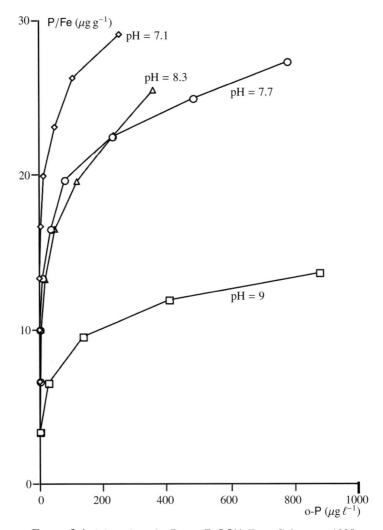

FIGURE 3.4: Adsorption of o-P onto FeOOH. From Golterman, 1995.

by best curve fitting. Furthermore, a theoretical calculation (Golterman, 1995b) showed that 0.33 is indeed a good approximation for the value of b. This means that the concentration in the sediment can be approached by the cubic root of the o-P concentration, a fact we will use below (see Section 3.2.4).

The constant A was found to depend not only on the pH, but also on the soluble Ca^{2+} concentration (Golterman, 1995a,c). The equation:

$$A = A_{max} \cdot C - (C - 1)e^{-Ca}$$

with Ca in mmol ℓ^{-1} and C a constant, was chosen as the descriptor, as the data suggested that the Ca^{2+} influence can be described by an asymptote. The maximal influence is obtained at $Ca^{2+} = 3$ mmol ℓ^{-1}. The experimental maximal value varied between 2.68 and 2.86 times the adsorption coefficient without Ca^{2+}. Influence of the electric double layer has been mentioned as the cause of this effect (Stumm & Morgan, 1981). The full Freundlich equation then becomes:

$$P_{ads} = 23600 \cdot \left(10^{-0.416\,pH}\right)\left(2.77 - 1.77\,e^{-Ca}\right) \cdot \sqrt[3]{\text{o-P}}, \qquad (3.7)$$

which holds true for a given FeOOH concentration in the sediment. ($10^{-0.416\,pH}$ can be written as $(H^+)^{0.416}$), with P_{ads} in μg per mg of Fe^{3+} and o-P in $\mu g\,\ell^{-1}$.

In the above-mentioned experiments the FeOOH suspension was made by adding NaOH to a $FeCl_3$ solution and carefully washing the precipitate with H_2O. The suspensions were used only after two weeks' ageing. Another way of preparing FeOOH is by oxidation of Fe^{2+} salts, which is what happens in a re-oxygenating hypolimnion. Mayer & Jarrel (1999) found that there was first an adsorption of a great quantity, a large part of which was later released probably owing to the growth of the FeOOH or Fe_2O_3 particles. Silica had a strong influence on the P-adsorption capacity, supposedly by changing the structure and surface chemistry of the Fe oxides or by competition with the P-anions. The molar P/Fe ratio (0.2) was, however, kept constant in these experiments and was quite high compared with natural waters. It was chosen rather for technical P-removal by Fe additions. The effect of Ca^{2+} on the P-adsorption is much stronger and may well overrule the silica effect, especially in natural waters.

The question has been raised why the empirical Freundlich equation gives such a good fit of the observed data, as there is no chemical or physical model behind it. The other equation, the Langmuir equation, is also often used. For the validity of this equation the assumption is needed that all adsorption sites are equal. Theoretically, Golterman (1995b) showed that Equation (3.2) is, at least, numerically correct if we suppose that for 1 molecule of P adsorbed two FeOOH molecules become inactive.

Let the concentrations in a theoretical adsorption experiment before adsorption be (M) = FeOOH and $(P^- + b)$ = o-P, of which a quantity b will be adsorbed. Let the adsorption equilibrium be:

$$\begin{array}{ccccccccc}
\text{FeOOH} & + & H_2PO_4^- & \Longleftrightarrow & \text{Fe(OOH)}{\approx}P + & OH^- & + & 2Fe_{inactive} & (3.8) \\
(M\text{-}3b) & & (P^-) & & (b) & (K_w/H^+) & & (2b) &
\end{array}$$

Then the concentrations after adsorption are as indicated between brackets. After substitution of these concentrations in the equilibrium equation we obtain:

$$K = (b)^3 \cdot \frac{OH^-}{P^-} \cdot (M - 3b), \qquad \text{with} \quad P^- = \frac{c}{1 + K_p/H^+}. \qquad (3.9)$$

Algebraic calculation yields a complex mathematical formula which, under certain conditions, indeed simplifies to Equation (3.2); a constant value for K is found. This argues strongly against the Langmuir adsorption equation, which demands a maximal amount of equal adsorption sites. The Langmuir equation, when applied to the chemical equilibrium, did not give a constant value for the equilibrium constant K, showing that chemical equilibrium and the Langmuir equation are incompatible. Furthermore, Miltenburg & Golterman (1998) found that the energy (enthalpy) of the adsorption of o-P onto FeOOH decreased with increasing amounts of P-adsorbed. At high Fe/P ratios ($\approx$ 60) this energy effect was 40–45 J mMol^{-1} which decreased to 0.5 J mMol^{-1} at Fe/P = 13. This is not in agreement with the supposition of equal adsorption sites. The decrease might be caused either by spherical hindrance or by shifts in the electron structure of the FeOOH particle. These low energy values indicate the adsorption character of the Fe(OOH)$\approx$P complex, as a coulombic binding mechanism would yield a higher energy.

In sediment without CaCO$_3$ the concentration of free FeOOH and the pH control the adsorption capacity of the FeOOH adsorption system through the constant A of the Freundlich adsorption equation. With CaCO$_3$ present the FeOOH represents only part of the adsorption system, but kinetically probably the most important one. The amount of Fe(OOH)$\approx$P fixed in sediment can be estimated by means of extraction with Ca-EDTA under reducing conditions (De Groot & Golterman, 1990; Golterman, 1996a). (See Section 3.4.1.) It is not yet known to what extent Fe-P-humic complexes are involved; the Ca-NTA or Ca-EDTA extracts are always dark brown and interactions seem possible. But the main characteristic of this compound is Fe(OOH)$\approx$P, while humic acid is likely to be (loosely?) attached to it. The Ca-NTA and Ca-EDTA extracts usually contain very little org-P, except in sediment rich in org-C as found, e.g., in the National Park Doñana (Spain, Díaz-Espejo, 1999). In eutrophic reservoirs in East Anglia (England), Redshaw et al. (1990) demonstrated a relation between P$_{\rightarrow alk}$ and org-Fe (i.e., pyrophosphate extractable) with P-adsorption. Comparing Fe-fractionation with that of P$_{sed}$ may produce some useful information, but the Fe-fractionation must be completely revised; influence of humic complexes must be taken into account and extractions should be repeated till depletion.

FeOOH constitutes a limited percentage ($\approx$ 5–10%) of the Tot-Fe in most (shallow) lakes and marshes, while under anoxic conditions an equal part of Tot-Fe may be in the form of FeS. Larger quantities of Fe (up to 90% of Tot-Fe) may be present in aluminosilicates. Shifts between FeOOH and FeS are likely to occur, depending on changes in redox and pH which influence P-adsorption.

Changes in the FeOOH/FeS system, which are further discussed in Section 3.8.4.2, are likely to have a more significant impact on the o-P adsorption in shallow lakes and marshes than in deeper lakes, where only a small part of the Fe will be in the form of FeS because of the limited amount of reducing power available (i.e., sinking detritus produced by phytoplankton) (Golterman, 1984). In winter the sudden high input of org-C from decaying macrophytes in marshes will produce a large reducing capacity in a few weeks, changing the situation more drastically than in deep lakes.

In an experimental study of P-exchange between sediment and water in freshwater and marine areas south of Stockholm (Sweden), Gunnars & Blomqvist (1997) have shown that there is a difference between fresh and brackish-marine environments and concluded that the P-flux into the sediment during shifts from anoxic to oxic conditions was mainly due to adsorption onto newly formed FeOOH. The freshwater system was characterized by high Fe/P ratios in the dissolved state after the preceding anoxic period. The lower ratios in the marine system were attributed to higher FeS concentrations in the marine systems, probably caused by higher SO$_4^{2-}$, but org-C concentrations providing the H-donor for the FeS formation may be supposed to have played a role as well. Neither the org-C data nor those of the extractable FeOOH concentrations of these sediments were given in this study. The question must be asked whether marine systems have less extractable FeOOH, a major agent for the P-adsorption.

Special attention is now often given to P-release when freshwater sediment enters a marine or brackish system. Here attention must not only be given to changing adsorption, but also to shifts in the pH and the Ca^{2+} and FeOOH concentrations. As Ca^{2+} plays a role as well, further examples are discussed in Section 3.2.3.

Slomp et al. (1998) have shown that in 4 different types of marine sediments from the North Sea the Freundlich equation also gives a good description of the adsorption of P onto these sediments. They attributed the P-adsorption to the presence of FeOOH and their b values of the Freundlich equation had a mean value of $1/2.9$, not far from the value mentioned above ($1/3$). In agreement with this, they found a strong correlation between the adsorption coefficient and NH_4-oxalate extractable Fe. They developed a kinetic model to describe the rate of exchange processes, the parameters of which were derived by curve fitting of experimental results. Pore water profiles were also measured and showed that adsorption limits the P-flux to the overlying water in most stations.

3.2.2 THE PROCESS OF PHOSPHATE PRECIPITATION WITH $CaCO_3$

Apatite, or $Ca_5(PO_4)_3.OH$, is ubiquitous in nature; it is found in igneous, metamorphic and sedimentary rocks (and in tooth and bone material). It is called 'insoluble', as its solubility in H_2O is $\approx 1\ mg\,\ell^{-1}$. Considerable quantities are found in fossil limestone watersheds. A second source is biogenic $Ca_5(PO_4)_3.OH$ formed in the water and co-precipitating with $CaCO_3$. The low solubility of $Ca_5(PO_4)_3.OH$ is known to control the solubility of o-P in water (Golterman, 1967, 1984). The exact value of the solubility product has remained a problem for a long time. Values between 10^{-30} and 10^{-60} have been proposed in the literature. A practical solution was given by Golterman and Meyer (1985), who determined the ionic product $(Ca)^5 \cdot (PO_4^{3-})^3 \cdot (OH)$ using 924 samples from the two hard water rivers Rhine and Rhône. They found a geometric mean value of 10^{-50} with a standard deviation of $10^{-0.3}$. It is now reasonable to use this 'apparent' solubility product for calculations of the solubility of o-P in hard water in equilibrium with the air, using the equation for the solubility product:

$$(Ca)^5 \cdot (PO_4^{3-})^3 \cdot (OH^-) = 10^{-50} \tag{3.10a}$$

or

$$(Ca^{2+})^5 \cdot \left[\frac{\text{o-P}}{1 + H^+/K_3 + (H^+)^2/K_2K_3}\right]^3 \cdot \frac{K_w}{H^+} = 10^{-50}, \tag{3.10b}$$

from which the maximal o-P concentration can be calculated, if Ca^{2+} and pH are known. A discussion on the precision of this calculation is given by Clymo and Golterman (1985). Work is still needed on two uncertainties: the influence of ion activity coefficients and the great uncertainty of the precise value of the K_3 of H_3PO_4. This constant needs to be measured with much greater precision than done so far and the temperature dependence must be established (see Golterman & Meyer, loc. cit.). Nearly the same value ($10^{-50.6}$) can be obtained from data on pore water Ca^{2+} and o-P concentrations of Mayer *et al.* (1999). From data of Löfgren & Ryding (1985) from interstitial waters of 9 eutrophic lakes in Sweden, a geometric mean value of $10^{-50.2}$ with a standard deviation of 10^{-1} can be calculated for the solubility product. The agreement with the value of Golterman & Meyer (loc. cit.) is excellent, but no information was given on which value of K_3 was used – while one point far from the mean value influenced the result strongly.

De Kanel & Morse (1978) studied the adsorption of o-P onto $CaCO_3$ (calcite and aragonite) and found an exponential decrease in available surface reaction sites or a linear increase in the activation energy associated with this adsorption. These results, although concerning a different chemical system, likewise show that the phosphate adsorption does not take place on equal sites. Spherical hindrance or shifts in electron structure may, again in this case, explain the phenomenon. De Jonge & Villerius (1989) demonstrated that P-adsorption onto (and release from) $CaCO_3$ (or apatite formation) also occurred in the Ems estuary (the Netherlands/Germany). A plot of $\log(P_{ads})$ against $\log(\text{o-P})$ gave straight lines depending on different factors such as salinity, pH and year. The function of $FeOOH$ was mentioned though not studied, and could well explain certain unexplained phenomena observed (see also Section 3.2.3).

The coprecipitation of $CaCO_3 \approx P$ with $CaCO_3$ can be used for lake management. Dittrich & Koschel (2002) used this coprecipitation as a tool to decrease the o-P concentration in hard water lakes. They treated the hypolimnion of Lake Schmaler Luzin (Germany), where $CaCO_3$ precipitation occurs naturally, with $Ca(OH)_2$ and found a decrease in the epilimnetic P-concentration. The idea of using $Ca(OH)_2$ to precipitate Ca^{2+} is often used in the preparation of drinking water and has the advantage that no Ca^{2+} is de facto added to the hypolimnion, according to the following reaction:

$$Ca(HCO_3)_2 + Ca(OH)_2 \rightleftharpoons 2CaCO_3 + 2H_2O.$$

The treatment gives no long-term solution, because the equilibrium situation is disturbed. It needs, therefore, to be repeated frequently if the P-input is not drastically reduced.

3.2.3 THE ADSORPTION AND PRECIPITATION PROCESSES TAKEN TOGETHER

In hard waters the two binding processes will occur at the same moment; even in soft waters this may happen if the pH and the o-P concentrations are high enough. The mathematical problem is to calculate the concentrations of o-P, $Fe(OOH) \approx P$ and $CaCO_3 \approx P$ as functions of a given concentration of total inorg-P and of the pH and the Ca^{2+} concentration in the water and the $FeOOH$ concentration in the sediment. A qualitative approach is given in Figure 3.5, where it is depicted how the different pools change under the influence of changes in pH: increasing pH causes a shift towards $CaCO_3 \approx P$ and a decrease in $Fe(OOH) \approx P$, and decreasing pH the opposite.

For quantitative work we have to calculate the concentrations of three unknown quantities, for which we have three equations, i.e., Equation (3.7), Equation (3.10a) and [Tot-P = o-P + $Fe(OOH) \approx P$ + $CaCO_3 \approx P$]. In a first diagram, incorrectly called 'phase diagram',

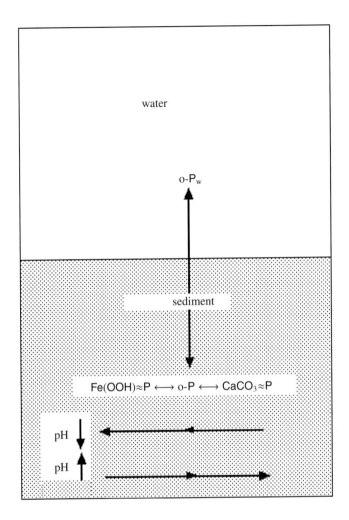

FIGURE 3.5: Shifts in the pools of o-P, $Fe(OOH) \approx P$ and $CaCO_3 \approx P$ under influence of pH changes.

Golterman (1988) depicted the solubility of o-P as a function of the pH, the FeOOH concentration in the sediment and the Ca^{2+} concentration in the water, but not the influence of Ca^{2+} on the P-adsorption onto FeOOH, as the adsorption studies were always carried out with FeOOH suspended in H_2O. Golterman (1995a,c) calculated the o-P, $Fe(OOH) \approx P$ and $CaCO_3 \approx P$ equilibria in a better solubility diagram and showed that NaCl, $MgCl_2$ and $CaCl_2$, compounds normally occurring in freshwater, largely increase the phosphate adsorption onto FeOOH. When naturally occurring concentrations of these cations were added to a suspension of $Fe(OOH) \approx P$ in equilibrium with o-P, the amount of P remaining in solution decreased by > 50%. With NaCl and $MgCl_2$ this does not cause further problems as their influence can be quantified; with $CaCl_2$ the problem was that it was difficult to distinguish the influence of Ca^{2+} on the adsorption process from the $CaCO_3 \approx P$ precipitation itself: Ca^{2+} added to the adsorption mixture disappeared rapidly from the solution and it was difficult to

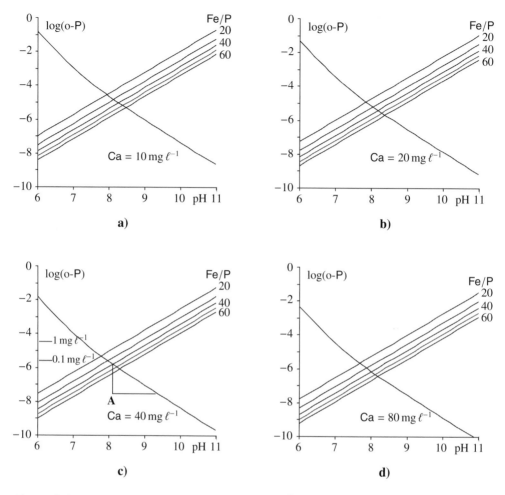

FIGURE 3.6: Solubility diagram for o-P as function of Ca^{2+} concentration in the water, FeOOH concentration in the sediment and pH. The influence of the Ca^{2+} concentration on the P-adsorption is taken into account. Ca = 10–80 mg ℓ^{-1}; FeOOH = 5.6 mmol ℓ^{-1}. (From Golterman, 1995.)

distinguish between apatite precipitation and adsorption onto FeOOH. Therefore, low values of Ca^{2+} and o-P had to be used, diminishing the precision. With Equation (3.7),

$$A = 23600 \cdot 10^{-0.416\,pH} \cdot \left(2.86 - 1.86\,e^{-Ca^{2+}}\right),$$

the solubility diagram of o-P was recalculated (see Figure 3.6a–d). These diagrams can be read like a city map. In the first place we see the equilibrium lines for the maximal solubility of o-P, drawn for different Ca^{2+} concentrations. These lines are the equilibrium concentrations between soluble o-P and solid $Ca_5(PO_4)_3.OH$: it is not possible to enter the area above these lines. So if the o-P concentration in a certain system was originally equal to point A (see Figure 3.6c), we can increase the o-P concentration up to this equilibrium line; a further addition of o-P will precipitate as $Ca_5(PO_4)_3.OH$. Furthermore lines are drawn for the $Fe(OOH)\approx P \Leftrightarrow$ o-P equilibria. These lines depend on the ratio of $FeOOH/Fe(OOH)\approx P$.

Each lake sediment is classified according to this ratio. When the o-P concentration increases from point A till the $Ca_5(PO_4)_3.OH$ equilibrium line, this concentration may be too low for a lake with, e.g., a sediment ratio $FeOOH/Fe(OOH)\approx P$ of 60. The $FeOOH/Fe(OOH)\approx P$ system will, therefore, release o-P till the concentration reaches the $Ca_5(PO_4)_3.OH$ equilibrium line. If on the other hand the $FeOOH/Fe(OOH)\approx P$ ratio was 20, this o-P concentration is too high for the $Ca_5(PO_4)_3.OH$ equilibrium, the $FeOOH/Fe(OOH)\approx P$ system will adsorb o-P and the $FeOOH/Fe(OOH)\approx P$ ratio will increase. In both cases the $FeOOH/Fe(OOH)\approx P$ system must have an o-P concentration equal to the $Ca_5(PO_4)_3.OH$ solubility graph. Each lake is, therefore, characterized by one equilibrium concentration for o-P. For each Ca^{2+} concentration in the water a different solubility diagram must be calculated with Equation (3.10b). The $CaCO_3$ concentration in the sediment is *not* important in this system. Ca^{2+} plays a double role, by both the apatite solubility limit and its influence on the $Fe(OOH)\approx P$ system. The diagram does, however, not yet take into account the presence of humic-$Fe(OOH)\approx P$ complexes. This complex may necessitate corrections of the solubility diagram, but the diagram may serve as a first approach to modelling the equilibrium of phosphate between sediment and overlying hard water. Preliminary experiments with Camargue sediment did not show deviations from the values predicted by the diagram (Golterman & De Groot, unpublished).

When lake sediment acidifies (e.g., by CO_2 production by mineralization), part of the $CaCO_3\approx P$ will be solubilized, but if $FeOOH$ is present, it will be readsorbed onto this compound which has a stronger adsorption at lower pH values (Equation (3.6) and solubility diagram Figure 3.6).

The solubility diagram describes the situation at a certain moment for a fixed quantity of inorg-P. For the description of the development with time a different approach is needed. Thus Golterman (1998), using chemical equilibrium equations, calculated the distribution of the two solid inorg-P compounds, $Fe(OOH)\approx P$ and $CaCO_3\approx P$, as a function of time, from the accumulated P-quantity, the water depth, the pH, the Ca^{2+} concentration in the water, the $Fe(OOH)$ concentration in the sediment and the maximal binding capacity of the sediment. In order to make these calculations possible the sediment was, virtually, divided in 1 cm layers, in which the P-accumulation occurred. When the o-P begins to enter the top layer, first most o-P is adsorbed onto the $Fe(OOH)$, while o-P increases in equilibrium with the $Fe(OOH)\approx P$ formed, therefore, very slowly. This concentration causes further downward diffusion. When in the top 1 cm layer the o-P increases sufficiently, the precipitation of $CaCO_3\approx P$ starts, till the moment that the maximal binding capacity is reached. At that moment o-P and $Fe(OOH)\approx P$ can increase further, followed by increased diffusion into the deeper layers, where the same sequence of events takes place. (See Figure 3.7.)

Gunnars *et al.* (2002) studied co-precipitation of Ca^{2+} and o-P with $FeOOH$, when Fe^{2+} was left to be oxidized and form $FeOOH$ by hydrolysis in freshwater and brackish seawater. They noted a slower rate in freshwater, a half-life time, $t_{1/2}$, of 7 hr against 20 min for the brackish water. On the time scale of lake events this is nearly immediate, as in 1 day 90% will have been formed. The amount of o-P co-precipitated depended on the initial Fe/P ratio, but only one point fell in the range of natural Fe/P ratios. Ca co-precipitation followed the same mechanism: the Fe/Ca ratio in the particles decreased from 14 when the initial Fe/Ca was near 0.2, to about 4 when the initial Fe/Ca ratio was < 0.05. Increasing salinity enhanced

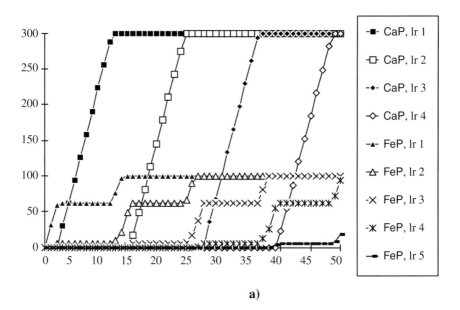

a)

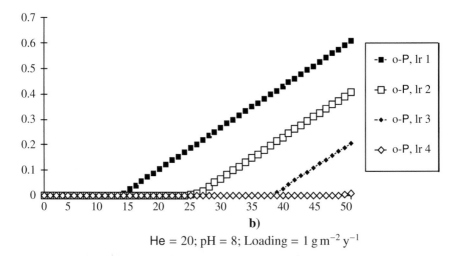

b)

He = 20; pH = 8; Loading = 1 g m^{-2} y^{-1}

FIGURE 3.7: Concentrations of o-P, CaCO$_3$≈P and Fe(OOH)≈P (mmol m^{-2}) in subsequent sediment layers in a 2 m deep lake as function of accumulative P-loading (g m^{-2}). pH= 8; Ca^{2+} = 40 mg ℓ^{-1}. Note the difference in scale.

the adsorption strongly. The authors explained the adsorption of Ca^{2+} onto the particles by surface adsorption processes, but the Fe/Ca seems rather high for such a process. These findings are more related to events in the hypolimnion, at autumn turnover when Fe^{2+} is converted into $FeOOH$, than to $FeOOH$ in sediment in constant contact with Ca^{2+} and o-P solutions, where the particles already exist. A sequential fractionation with Ca-EDTA and Na_2-EDTA may give much information on the composition of the precipitate.

In shallow lake systems the equilibrium between o-P and $CaCO_3 \approx P$ and $Fe(OOH) \approx P$ rapidly leads to an equilibrium between o-P in water and sediment, which causes the sediment to accumulate much P, explaining the so-called resilience of shallow lakes. Furthermore, the presence of Ca^{2+} in shallow overlying waters strongly limits the solubility of o-P, because the same P-loading gives a higher P-quantity per m^2 than in deeper waters. This causes an apatite precipitation for the concentrations above those controlled by Ca^{2+}, pH and temperature, which apatite is bioavailable owing to its fine grain size. Chemically, shallow lakes are not different from the deeper ones.

Two examples may demonstrate the use of these diagrams. The very eutrophic Lake Veluwemeer (the Netherlands) was flushed through with clearer, more alkaline water, the objective being that because of the higher pH the P_{sed} would be released from the sediment and washed out. o-P was in fact released from the $Fe(OOH) \approx P$ pool and appeared in the water, but precipitated as $Ca_5(PO_4)_3.OH$, because of the high pH. Actually the lake was loaded with the o-P from the flushing water; when the experiment came to an end the pH decreased again and the apatite was transformed back into $Fe(OOH) \approx P$.

In another lake (Groot Vogelenzang, the Netherlands) $FeOOH$ was injected into the lake bottom (Boers et al., 1992). Although the quantities were relatively large, they were small compared with the $FeOOH$ already present; the ratio $FeOOH/Fe(OOH) \approx P$ changed very little. The authors stated that there was an improvement, lasting three months, and attributed the subsequent increase in Tot-P and algal bloom to the hydrology of this lake. The improvement, however, occurred when the Tot-P was already in rapid decline (October–November) and the hydrology could never have changed the system back so rapidly. Re-equilibration between sediment $Fe(OOH) \approx P$ and o-P_w explains these events much better. The authors stated that the cost is low compared with dredging, but the negative results made it an expensive experiment. Knowledge of the solubility diagrams of these two lakes would have prevented large financial losses. There are a few reports in the literature in which the $Fe(OOH) \approx P$ and the $CaCO_3 \approx P$ systems are shown to be important together. Some of these concern estuaries, where greater changes in pH and salinity occur than in lakes, causing shifts between the two P-fractions. These examples will be discussed in Section 3.5.

Søndergaard et al. (1996) analysed sediments from 32 meso- to hypertrophic Danish lakes with a wide range of Tot-P_{sed} and, among other factors, of the $CaCO_3$ concentration in the sediment. They found strong correlations between Tot-P_{sed} and both external P-loading and Tot-Fe_{sed}, but not between Tot-P_{sed} and Ca_{sed}. This is obvious, as the o-P_w is controlled by the Ca^{2+} in the water and not by the $CaCO_3$ in the sediment, while a considerable part of the $P_{\rightarrow NaOH}$ will have been recovered in the $P_{\rightarrow HCl}$ pool because the Hieltjes & Lijklema method was used (see Section 3.4.1). In one of the lakes with an extremely high $Fe(OOH) \approx P$ concentration the $P_{\rightarrow HCl}$ dependence on depth strongly resembled that of the $P_{\rightarrow NaOH}$. $P_{\rightarrow NaOH}$

contributed $\approx 35\%$, $P_{\to HCl} \approx 20\%$ and the res-P (mainly org-P) $\approx 41\%$ to Tot-P_{sed}. Tot-P_{sed} and res-P decreased with depth, the first from ≈ 3 to ≈ 1 mg g^{-1} (dw.). Relative to Tot-P_{sed}, res-P increased with depth in the sediment to 63%, while $P_{\to NaOH}$ remained 21% at greater depths. This pool may consist of org-P_{sed} or $Fe(OOH) \approx P$, as the authors suggested.

Transfer between different P-fractions will also occur when river sediment enters a lake. Selig & Schlungbaum (2002), analysing P-transport in the Warnow river (North Germany), three lakes (Lakes Barnin, Bützow and Klein Pritzer) and an impoundment in its watershed, and its tributaries, noticed, in the first place, a shift in grain size from 0.5–0.6 mm in the river to ≈ 0.1 mm in the lake. This implies that the coarse sediment remained as bed load in the lower parts of the river. Fine sediment with a high P-concentration accumulated in the lake and impoundment sediments. In the upstream sediments $P_{\to BD}$ accounted for $> 60\%$ and $P_{\to HCl}$ for $> 20\%$ of Tot-P, with $P_{\to NaOH}$ accounting for traces only. In the impoundment and lake sediments $P_{\to BD}$ was lower, while $P_{\to NaOH}$ (especially the o-$P_{\to NaOH}$) and $P_{\to HCl}$ became dominant. The nr-$P_{\to NaOH}$ also increased, which the authors assumed to be humic bound P. In this agriculture-loaded system one might expect to find phytate as well. It is difficult to relate these operationally defined fractions to chemical forms as the Psenner scheme of P-fractionation was used. The shifts from $P_{\to BD}$ to $P_{\to HCl}$ may, e.g., be caused by changes in the $CaCO_3 \approx P$ extractability, while the $P_{\to NaOH}$ will consist of $Fe(OOH) \approx P$ and/or org-P.

Depending on the chemical composition of P_{sed}, a large part of it is available to phytoplankton. (See Section 3.6.) As bioavailability is strongly related to the chemical nature of the phosphate compounds, we developed a fractionation scheme for sediment phosphate. (See Section 3.4.1.)

3.2.4   PRACTICAL CONSEQUENCES OF P-BINDING ONTO SEDIMENT

The presence of P_{sed} has consequences for calculating what will happen whenever the P-input is stopped or decreased, especially for shallow waters. In these waters, where the o-P concentration is in equilibrium with the o-P concentrations, Equation (3.1) can be written as:

$$P_{ads} + \text{o-P} = n \cdot L$$

or

$$A \cdot (\text{o-P})^b + \text{o-P} = n \cdot L, \tag{3.11}$$

where L = the net annual loading (g m^{-2} y^{-1}) and n = the number of years.

As for natural sediment b is very often near 0.33, as in Equation (3.5), we can rewrite Equation (3.11) as

$$\sqrt[3]{\text{o-P}} + \text{o-P} - n \cdot L = 0. \tag{3.12}$$

This equation has only one real root, which can be calculated (see Figure 3.8). In this Figure it is shown how at first phosphate mainly enters the sediment while the concentration in the water only increases very slowly; later on it increases rapidly. The situation is not different for other values of b (e.g., between 0.3 and 0.45), but the calculations become difficult.

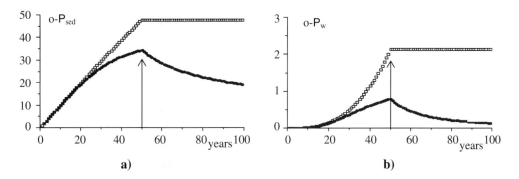

FIGURE 3.8: Changes over time of o-P_{sed} (a) and o-P_w (b) in a shallow mud/water system with a flushing rate of $1\,y^{-1}$ (lower curve) and without flushing (upper curve). The arrow indicates the moment when the loading is reduced to 0.

In Figure 3.8, it is furthermore shown how the P-concentration will decrease, but only very slowly, when the P-input is eventually stopped: owing to its buffering capacity the sediment will for a long time continue to release o-P into the water, which process may be slower than the building up stage. (For further details see Golterman, 1991a.) In this way we can visualize the principle of the influence of the phosphate binding capacity on the phosphate transport through wetlands and its final accumulation there, but a precise quantification is, unfortunately, not yet possible. A preliminary example is the P-flux through ricefields and wetlands in the Camargue (France). Using the chemical characteristics of the sediment, the relatively well-known amounts of P-input via fertilizers, and the P-binding processes, Golterman (1999) calculated the amount of P arriving in the central lake of the Camargue, the Vaccarès, which is a nature reserve. In the Camargue, ricefields are the major agricultural activity and receive much P-fertilizer and water, which are finally exported towards the Vaccarès. The amount of P_{sed} that accumulated in the sediment of the Vaccarès was calculated with an imprecision of a factor 2, but this uncertainty was due for a great deal to the lack of data from the pre-fertilizer period and the lack of precision of the hydrological data. The study was, however, relatively easy as the fertilizer input was the major source of the P-input and was sufficiently quantifiable.

In the same way slow release of P_{sed} delays the improvement of the water quality of large lakes when P-input has been stopped. Here the more complicated water movement (formation of epi- and hypolimnion, with different water retention time) makes calculations more difficult. Great lakes, like the lakes of Constance and Geneva, often improved more slowly than had been predicted.

3.3 Organic phosphates

Besides the inorganic phosphate in sediment, organic phosphates are also found. In most lakes these compounds are the constituents of sedimenting dead phytoplankton; they may represent the major input of P_{sed}. Although a large part of the phytoplankton-P will already have been lost along the way towards the sediment, a small part is not easily biodegradable (sometimes called refractory) and becomes P_{sed}. The results of extraction and identification of org-P_{sed} compounds have so far been dubious; their identification depends entirely on the kind of extraction used. With the NTA and EDTA extractions usually only small amounts of org-P are extracted. (Only in sediment rich in organic matter, such as is found in the National Park Doñana (Spain), somewhat larger amounts of org-P were measured in these extracts.) As it is unlikely that any inorg-P remains in the pellet after extractions of Fe and Ca bound phosphate, it is usually understood that the P remaining after EDTA (or NTA) extractions is org-P. In the past, attempts to extract these compounds failed, as, after the necessary hydrolysis, the o-P formed was rapidly adsorbed onto the FeOOH, while the NaOH and the HCl, used to remove the inorg-P first, also hydrolysed a substantial part of the org-P_{sed}. Since in the present extraction schemes (see Section 3.4), using chelators as first extractants, FeOOH and $CaCO_3$ are extracted as well, this danger is no longer present. In this way De Groot (1990) and De Groot & Golterman (1993) were able to analyse the org-P compounds further. It was found that a quantity, sometimes small, sometimes large, of the org-P_{sed} could be extracted with strong acids (e.g., 0.5 M H_2SO_4), but over a relatively short period, 30 min. This pool, originally called Acid Soluble Organic Phosphate (ASOP) but now called org-$P_{\rightarrow ac}$, appeared to be biologically active. (See Section 3.5.)

Lastly, the org-$P_{\rightarrow ac}$ *may* contain small amounts of o-P not extracted by the EDTA extractants – it is also a safety extraction before the analysis of the much more important pool of org-$P_{\rightarrow alk}$ which is the org-P remaining in the pellet after the acid extraction; it can be extracted with hot 2 M NaOH. De Groot & Golterman (1993) demonstrated with the enzyme 'phytase 2' that a large part of this pool can be hydrolysed to o-P, while alkaline phosphatase did not release o-P. This strongly suggests that the org-P was phytate, an inositol-hexaphosphate. Experiments with test compounds showed that it is probably Fe_4 phytate. This compound could be prepared by adsorbing phytate onto FeOOH and was shown not to be soluble in HCl, unlike Ca- and Mg-phytate. Phytate appeared to be more strongly adsorbed onto FeOOH than o-P, which it could even replace. Furthermore, Fe_4 phytate prepared in the laboratory was extracted neither by Ca-NTA nor by 0.5 M HCl, and only partly by EDTA, but it was extracted by hot 2 M NaOH. FeOOH with adsorbed phytate appeared to be protected against an attack with acid or chelating compounds, while free FeOOH was dissolved with H^+. A white precipitate then formed with a ratio of 4.3 mol Fe per 6 mol o-P. This presents strong, but not yet conclusive, evidence of the presence of humic bound phosphate and Fe-phytate in the org-P fraction of sediment.

Further work on the identification of the organic part of these molecules is needed, together with analyses of other sediments, viz., from deep lakes. Masahiro Suzumura & Akiyoshi Kamatani (1993) isolated and determined phytate in sediment from Tokyo Bay, using anion-exchange chromatography with nuclear magnetic resonance. They found concentrations of about $4 \mu g \, g^{-1}$ and suggested that phytate is not a significant component of org-P in marine sediment.

In the past the question has been raised how org-P in sediment could occur in such large amounts, most of the org-C compounds being biodegradable. The presence of humic bound and inositol phosphate explains this presence, as both compounds form complexes which are highly resistant to bacterial attacks and are moreover strongly adsorbed onto FeOOH rendering them nearly insoluble.

3.4 Fractionation of sediment bound phosphate

3.4.1 FRACTIONATION OF INORGANIC P-COMPOUNDS

The first fractionation scheme of P-compounds in soils is that of Chang & Jackson (1957; see also Jackson, 1958) using NaOH and HCl, while later a NH_4F extraction step was added. Fluoride was later omitted as it may induce the formation of fluor-apatite. Olsen *et al.* (1954) introduced an extraction of P_{sed} with 0.5 M $NaHCO_3$, which is still often used in soil science, but rarely in limnology. It will extract part of the $Fe(OOH){\approx}P$, and probably some org-P; a second extraction will again yield a considerable amount of extractable P. The Chang & Jackson procedure was slightly modified by Hieltjes & Lijklema (1980; extraction 'H&L'). They introduced a first step with NH_4Cl, which is supposed to extract the 'loosely bound P'. The chemical identity of this fraction is not clear, it is probably o-P from the interstitial water which is usually only a few percent of Tot-P. It will also include some apatite. At pH = 7 of this extractant, about 3 mg ℓ^{-1} of P, may be dissolved. Thus, if an extractant volume of 50 ml containing 500 mg sediment is used, this will mean a solubilization of 150 μg per 500 mg sediment or 300 $\mu g\, g^{-1}$, which is a considerable quantity of Tot-P. Whether this quantity will be attained depends on the grain size of the apatite and the duration of the NH_4Cl extraction, which is usually not mentioned. Pettersson & Istvanovics (1988) presented evidence for this error and showed that repeated extractions with NH_4Cl do indeed extract more and more P. Hieltjes & Lijklema (1980) tested neither the influence of the duration of the NaOH extractions (17 hr) nor that of the concentration of the NaOH and no checks were made by estimating the recovery of added standard compounds. It is not clear why the shaking time for the extraction of $P_{\rightarrow alk}$ is always set at 17 hr, as the P from the $Fe(OOH){\approx}P$ complex will be dissolved in a much shorter time – except that 17 hr is convenient for overnight experimenting, but the long duration carries the risk of hydrolysis of org-P_{sed}.

Williams *et al.* (1976, 1980) used an extraction method with Na-citrate plus dithionite plus bicarbonate ('CDB'), at pH = 7–8, followed by NaOH. The CDB is supposed to extract the $Fe(OOH){\approx}P$ which leaves only org-P for the NaOH step. A subsequent extraction with HCl is supposed to extract $CaCO_3{\approx}P$. The Williams extraction was later changed, omitting the CDB step, by the Geneva geochemists group (Favarger, pers. commun.) and the C.C.I.W. group (Manning, pers. commun.), but out of deference the name of the procedure was not changed. The reason for this change was the assumption that $P_{\rightarrow alk}$ reflects the P_{aa} better than $P_{\rightarrow CDB}$. Pardo *et al.* (1999) modified the Williams method by adding the measurement of org-P. This was done by calculating the difference between Tot-P and inorg-P, the latter measured by a single extraction with 1 M HCl. It was, however, not shown that this quantity of inorg-P equalled the sum of the inorg-$P_{\rightarrow alk}$ and inorg-$P_{\rightarrow HCl}$, which it ought to, but such is never the case. The method was invalidated by De Vicente *et al.* (2002) who showed in a hard water, eutrophic lagoon (Albufera de Adra, SE Spain) that part of the $Fe(OOH){\approx}P$ extracted

TABLE 3.2: Comparison of extractions of $Fe(OOH) \approx P$ and $CaCO_3 \approx P$ with Ca-NTA (plus dithionite) and Na_2-EDTA extractants with a NaOH extraction. Added: $Fe(OOH) \approx P$ = 14.6 mg; $CaCO_3 \approx P$ = 11.7 mg. (The two $CaCO_3 \approx P$ suspensions contained small quantities of o-P.) (Data from Golterman & Booman, 1988.)

Suspension	Ca-NTA extraction (1/2 hr)	Na_2-EDTA extraction (1–2 hr)	0.1 M NaOH extraction (17 hr)
$Fe(OOH) \approx P$ (P = 14.6 mg)	14.9 (= 102%)	1.5 (= 10%; 2 days)	11.2–11.5 (76–79%)
$CaCO_3 \approx P$ (P = 11.7 mg)	0.0	10.7–10.8	
id. (containing 0.2 mg o-P)	n. d.	12.1	
id. (containing 2.0 mg o-P)	2.4	10.1	
Mixture of 15 mg $Fe(OOH) \approx P$ + 25 mg of $CaCO_3 \approx P$	14.5 15.2 16.5	24 25 26.1	

with NaOH reprecipitated as $CaCO_3 \approx P$.

For a long time it was not realized that NaOH does the same as NH_4F, inducing the formation of hydroxyl-apatite instead of fluor-apatite (Golterman & Booman, 1988; De Groot & Golterman, 1990). These authors furthermore did not only show that apatite formation leads to wrong results, but also that NaOH hydrolyses a certain quantity of the org-P, the quantity depending on the extraction duration. From Table 3.2 it can be seen that $Fe(OOH) \approx P$ and $CaCO_3 \approx P$ are recovered with Ca-NTA and Na_2-EDTA respectively, but that NaOH extracts only 80% of the $Fe(OOH) \approx P$; probably 2 extractions are needed. Table 3.3 shows that NaOH extracts more than just $Fe(OOH) \approx P$, and again that $CaCO_3 \approx P$ is formed. Several more modifications of the NaOH/HCl extractions have been published, but will not be discussed here, as most present the same problems as outlined in Golterman & Booman (1988) and De Groot & Golterman (1990). NaOH extractions were once more invalidated by Golterman (1996a), Golterman et al. (1998) and by Romero-Gonzalez et al. (2001). All these authors demonstrated that the duration of the extraction and the concentration of the NaOH always influence the quantities of P extracted and that, therefore, NaOH does not extract a specific fraction. (See Figure 3.9 and Figure 3.10.)

Romero-Gonzalez et al. (2001), using sediment from a soft water river system, the Catatumbo River (Venezuela), showed that 1.0 M NaOH extracted more phosphate than 0.1 M NaOH, while with increasing time the $P_{\rightarrow NaOH}$ decreased 4–5 fold. From Figure 3.8 and Figure 3.9 it can be seen that the concentration of $P_{\rightarrow NaOH}$ increased with time in some cases, but decreased in others. The decrease is caused by re-adsorption onto $CaCO_3$, while the increase is due to the hydrolysis of the org-P_{sed}. Decrease or increase, therefore, depend on the relative concentration of these compounds – even in the case of the soft water in the Catatumbo river the authors suggest that the decrease with time is due to the formation of Ca- or Mg – P complexes. The occurrence of these processes can be demonstrated by comparing results with Hieltjes & Lijklema extraction with those of Golterman & Booman's. See Table 3.3.

TABLE 3.3: Amounts of Fe(OOH)≈P, CaCO$_3$≈P, org-P and their sum (ΣP) in 3 sediments from Camargue marshes and 2 sediments from Lake Balaton (P in μg g^{-1} dw.). From De Groot & Golterman, 1990.

Marsh	Relongue I	Relongue II	Ricefield I	Ricefield II	id +162	Balaton T	Balaton K
Fe(OOH)≈P (G & B)	228	147	78	82	190	30	102
P$_{→NaOH}$ (H&L)	20	47	12.5	5.3	3.5	16	42
CaCO$_3$≈P (G&B)	245	230	271	303	350	112	216
P$_{→HCl}$ (H&L)	539	410	423	421	575	268	346
org-P (G &B)	267	322	373			280	297
org-P (H&L)	177	280	163			121	171
ΣP (G&B)	740	699	722			420	615
ΣP (H&L)	736	737	599			406	559
Tot-P	709	747	746				433	627

The hydrolysis of org-P was already mentioned by Boström *et al.* (1988), but as these authors presented no chemical evidence, it never received much attention.

Pettersson *et al.* (1988) gave an overview of different operational[1] chemical extractions, without comparison between these methods, while factors like duration, number of extractions, and influence of the NaOH concentration were not examined. These authors did, however, state that these methods cannot be used for identification of discrete P-compounds. Extractions with chelators were only mentioned, but not critically analysed.

Psenner *et al.* (1985) and Psenner & Puckso (1988) re-introduced a first step with a reducing agent, i.e., dithionite, in a NaHCO$_3$ solution without citric acid, which was then followed by NaOH. Dithionite/bicarbonate ('BD') extracts Fe(OOH)≈P, but not all the FeOOH itself, rendering subsequent steps uncertain, because of the possibility of re-adsorption. The NaOH step is supposed to extract org-P$_{sed}$ or non-reactive-P$_{sed}$, but the problem of the hydrolysis of org-P$_{sed}$ was not fully appreciated. The next steps were extractions with HCl, and subsequently with hot NaOH. The hot NaOH extracted only 10% of the first (cold) NaOH step. It is not clear to what extent the FeOOH itself is removed, which is essential for further analysis. Two major points were not addressed: dithionite/bicarbonate probably does not extract all Fe(OOH)≈P in one extraction and the recovery after addition of known compounds was

---

[1]i.e., extractions defined by the extraction method, and not by the target compound.

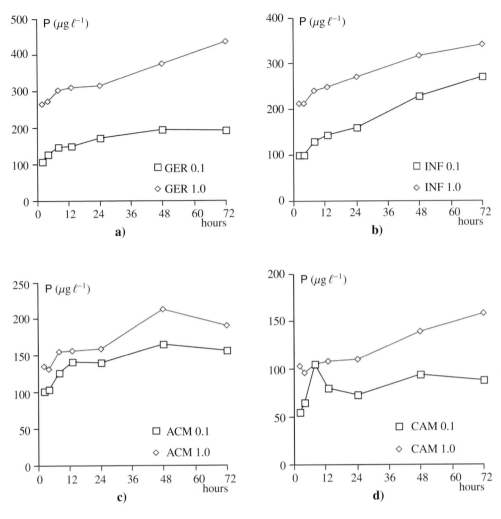

FIGURE 3.9: Influence of the duration of extraction and the concentration of NaOH on the amount of P-extracted. Two samples rich in org-C, but poor in $CaCO_3$ from the National Park Doñana (Infraqueable and Acebuche de Matalascañas, 'INF' and 'ACM') and one rich in org-C with medium Ca from a nearby reservoir, El Gergal ('GER'), and one poor in org-C, but rich in $CaCO_3$ from the Camargue ('CAM'). From Golterman *et al.*, 1998.

not studied.

Using three different marine sediments on the Danish coast, Jensen & Thamdrup (1993) further examined the 'BD'-reagent. They found a highly significant correlation between $P_{\rightarrow BD}$ and Fe extracted with oxalate/HCl, which they considered to be an indication of the fact that 'BD' indeed extracts $Fe(OOH) \approx P$. In two samples the ratio Fe/P was 8–11 and in the third one 17 (mol / mol). One may wonder why this is an argument for 'BD' as the extractant, because with a certain amount of FeOOH in a sediment it will adsorb increasing amounts of o-P when that enters the sediment. The authors distinguished crystalline and amorphous

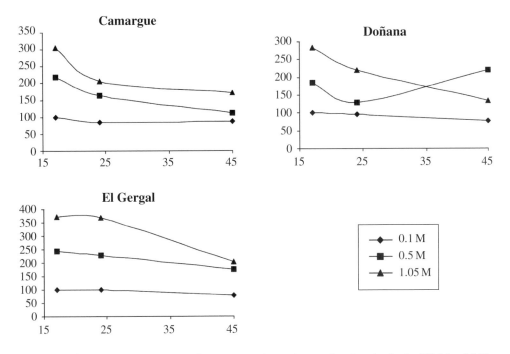

FIGURE 3.10: The influence of the NaOH concentration and extraction duration in the 'Hieltjes & Lijklema' extraction scheme (17 hr extraction with 0.1 M NaOH). Data from sediments from the Camargue, National Park Doñana and El Gergal. Results in $\mu g\, g^{-1}$.

FeOOH and discussed the different extractants for these two forms. They mentioned a difference between the anaerobic oxalate and the 'CDB' extractions which both extract both forms of FeOOH, while NH$_2$.OH.HCl only extracted the amorphous FeOOH. This difference probably depends on the concentration of the extractant and the duration of the extraction. Golterman (unpubl.), using a calcareous sediment with about 0.6% extractable FeOOH, found no difference between NH$_2$.OH, ascorbic acid and dithionite as reductants. Jensen & Thamdrup (loc. cit.) found NH$_4$Cl to extract more Ca than 0.46 M NaCl (as mentioned above); therefore NH$_4$Cl was replaced by NaCl. After freezing of the samples P$_{\rightarrow BD}$ was overestimated owing to hydrolysis of org-P. The authors stated that 'BD' extracted all P$_{\rightarrow BD}$ in one single extraction, but after the first extraction two 'BD' washes took place followed by one NaCl wash. This is as a matter of fact equal to three 'BD' extractions. It seems more logical to estimate the FeOOH in the Ca-EDTA extract, together with the o-P in the same extractant.

Burke *et al.* (1989) used a series of extractants on 10 surface sediments from shallow bays of Lake Manquarie (New South Wales) to study P-fractionation. They showed that the quantities extracted increased with the extraction time in a 0.5 M NaHCO$_3$ solution adjusted at pH= 8.5. Furthermore, they showed a strong dependence of the quantities extracted on the ratio sediment/extractant-volume, but did not repeat the extractions; it is therefore not certain that the extractions were complete.

In none of the above mentioned methods was it checked whether a second

extraction extracts any more of the target compound. We have shown that with NaOH this is certainly the case.

For several reasons Golterman (1976b) tried a different approach, i.e., using chelators specific for certain compounds. Compounds like NTA and EDTA will solubilize Ca and Fe compounds by complexation, and this can be done at a pH near that of the sediment in question. NTA complexed in advance with Ca^{2+} will not be able to attack Ca-salts; this extractant becomes specific for $Fe(OOH) \approx P$:

$$[CaCO_3 \approx P]_{solid} + [Fe(OOH) \approx P]_{solid} + NTA \longrightarrow Ca\text{-}NTA + Fe\text{-}NTA + o\text{-}P \quad (3.13a)$$

but

$$[CaCO_3 \approx P]_{solid} + [Fe(OOH) \approx P]_{solid} + Ca\text{-}EDTA \longrightarrow \quad (3.13b)$$
$$[CaCO_3 \approx P]_{solid} + Fe\text{-}EDTA + o\text{-}P.$$

In theory EDTA should work more quickly and efficiently than NTA, resulting in a stabler complex. This appeared, however, not to be the case, NTA giving better results. Nevertheless the extraction was time-consuming, the kinetics were slow, probably owing to the reaction rate of the formation of the Fe-chelator complex, which is even slower with EDTA. As the reduction of FeOOH by reducing compounds is much quicker, Golterman (1982b) tried extractions with Ca-NTA plus ascorbic acid or NH_2OH. These reducing agents are only active at a slightly acid pH, therefore the extractions were made at pH = 5. The results were promising; o-P added to a Camargue soil sample was quantitatively recovered as $Fe(OOH) \approx P$ or $CaCO_3 \approx P$ and, because of the short duration of the extractions, no apatite was solubilized in the first step. Ascorbic acid has the disadvantage that after some time a precipitate appears with the Ca from the Ca-NTA, while $NH_2OH.HCl$ does not work at pH > 5. Therefore Golterman & Booman (1988) also tried dithionite ($Na_2S_2O_2$) as the reducing agent, which is the only compound reducing at a pH = 8. The results were satisfactory: artificially prepared suspensions of $Fe(OOH) \approx P$ and $CaCO_3 \approx P$ were quantitatively recovered, even when added to Camargue soils. (See Table 3.2.) As NTA combined with dithionite was not a better extractant than EDTA with dithionite and as the extracting capacity could be increased with EDTA, Golterman (1996a) later changed the Ca-NTA (0.02 M) to Ca-EDTA (0.05 M). The extractions are always repeated till the target compounds are depleted. This is needed as reactions (3.13a) and (3.13b) are equilibrium processes and will not extract 100% in one step. (This is the only method in which this repetition is done.)

Ruttenberg (1992) developed the sediment extraction ('SEDEX') sequence, using first two extractions with $MgCl_2$, then an extraction with citrate/dithionite/bicarbonate, then a Na-acetate/acetic acid buffer (pH = 4) extraction and finally an H_2SO_4 extraction. Between the different steps washings with $MgCl_2$ were added. Na-acetate is supposed to extract only the biogenic $CaCO_3 \approx P$, and H_2SO_4 the detrital* (boulder) apatite, $Ca_5(PO_4)_3OH$. Golterman (1996a) found, however, that EDTA extracted a larger quantity of $CaCO_3 \approx P$ from Camargue sediment than Na-acetate plus H_2SO_4. The difference was $423\,\mu g\,g^{-1}$ against $330\,\mu g\,g^{-1}$ (with a Tot-P = 0.97 mg g^{-1}); at the moment this difference is not understood. Theoretically it seems probable that the citrate attacks some $CaCO_3 \approx P$ as well; Ruttenberg's Figure 5 demonstrates that this is indeed the case; the explanation is the same as for the extraction of apatite with NH_4Cl (see above, page 73). In the case of eroded apatite this quantity will be small, although depending on grain size, but it may be important with biogenic $CaCO_3 \approx P$. This extraction of $CaCO_3 \approx P$ is prevented in the EDTA method by using Ca-EDTA as the first step.

A modified SEDEX extraction was used by Baldwin (1996b). Because the first extractant in the SEDEX procedure ('CDB') extracts both $Fe(OOH) \approx P$ and some org-P_{sed}, Baldwin added extractions with $NaHCO_3$ before the 'CDB'. He showed that 3–4 extractions are often necessary to remove 95% of this fraction and criticized the fact that most extractions in the literature are only done once (which is true, except for the EDTA method). However, it is not clear what is extracted with the $NaHCO_3$, as both $Fe(OOH) \approx P$ and humic-P will be extracted. Baldwin indeed found a strong correlation between the extracted o-P and humic compounds, measured by the O.D. at 250 nm. The problem is therefore the same as for the CDB (and EDTA) extracts. After the $NaHCO_3$ and 'CDB' extractions, the next extraction with NaOH was carried out, which apparently removed only org-P. Baldwin showed that the $MgCl_2$ of the SEDEX method removed only 2% of Tot-P, which is probably the interstitial o-P. $NaHCO_3$ removed between 20 and 60% of Tot-P_{sed}, while the extracted P was usually in the molybdate reactive form, i.e., o-P. The next NaOH step removed only 5–30% of Tot-P_{sed}. The combination of $NaHCO_3$ and NaOH extractions does not seem logical. Baldwin also found a good correlation between Fe and P extracted by 'CDB', and repeated the notion that CDB may extract Ca bound P as well. Baldwin presented a ^{31}P NMR spectrum with a peak at -13.9 p. p. m. which was assigned to polyphosphate. As no details of the apparatus were given, and no other phosphate compounds were tested except o-P, the peak may as well be due to other P-compounds.

Jensen *et al.* (1998) combined the Psenner and SEDEX methods by using the 'BD' reagent without citrate, but the next acid step was split into a Na acetate/acetic acid and an HCl extraction, in order to distinguish between detrital and biogenic apatite. Between the 'BD' step and the Na-acetate step an extraction was carried out with 0.1 M NaOH (18 hr). It is not clear what the target compounds for this extractant are (probably mostly org-P), but a considerable part of the extracted o-P will re-precipitate as $CaCO_3 \approx P$. These authors also measured the amount of Fe extracted with the 'BD'. This amount will, however, depend on the time the extract is left in contact with air, as the $Fe(OOH)$ extracted as Fe^{2+} will rapidly precipitate again, certainly if aeration is used to remove the dithionite. In the pooled acid extractions more Fe was measured which was believed to originate from FeS or FeS_2, but the presence of these compounds was not demonstrated, neither was the apparent redox potential measured. The results were used to study the availability of P_{sed} for seagrass beds in Bermuda (Atlantic Ocean, UK). (See Section 3.6.)

Anderson & Delaney (2000) simplified the SEDEX procedure by combining the first and second steps, arguing that the distinction between the P extracted in the first step (so called 'loosely bound', whatever that may mean) and the P associated with the metal oxides is arbitrary, as most P is adsorbed onto the metal oxide surfaces. This combination is also applied in the EDTA method. Furthermore, Anderson & Delaney replaced the traditional spectrophotometric analysis by an automated flow injection analysis, considerably reducing the time needed.

There are several more fractionation methods. For example, House *et al.* (1995) used the iron-oxide stripping method, which might be promising as a fractionation method, with the difficulty that several extractions may be necessary, as tailing (i.e., each next fraction still containing a considerable quantity) will occur. A similar approach has been made by Huettl *et al.* (1979). They loaded a cation-exchange column with $Al(OH)_3$, measured the amount of P-adsorbed, and compared this with P_{aa}. The authors tested 5 slightly acid soils, without $CaCO_3 \approx P$, in which the Al-column extractable-P was measured, while the P_{aa} was supposed to be equal to the quantity of P found by extraction with NaOH before and after cell growth (the duration is not mentioned), but the cell growth itself was not measured. Although the amount extracted still increased with time after 100 hr, the authors found an excellent correlation between P extracted on the column and the supposed P_{aa}. The experiment, therefore, only shows that, in these acid soils, the column-extrable P is more or less equal to $P_{\rightarrow NaOH}$.

When using these techniques several possible sources of errors must first be identified: incomplete transfer of P to the exchanger, slow dissolution of P_{part}, incomplete retrieval from the exchanger, etc. (Waller & Pickering, 1992). The conditions of these extractions must, therefore, be carefully checked – which is not often done. Uusitalo & Markku Yli-Halla (1999) pointed out another error in both the Fe-strip method and the anion exchange method. They showed that adsorption of small soil particles may occur and suggested that the strips and beads be enclosed in nylon mesh bags. Al and Si determinations may be used to check for the presence of adsorbed particulate material. They furthermore showed that repeated extractions once more yielded phosphate. Overviewing these two methods it must be concluded that they are not yet sufficiently defined, and that it is not clear at all which P-fractions (chemical as well as biological) are extracted.

In future the extractions with chelators can be further improved, e.g., by the use of chelators specific for Ca^{2+} or Fe^{3+}. First attempts were promising, but this method was pursued no further, as in those days the specificity depended on the pH, which must be adjusted to the pH of the sediment and cannot be set at the optimal pH of the separation.

Two papers compared different extraction procedures. Using National Park Doñana sediment, synthetic compounds and unicellular algae, Jáuregui & García Sanchez (1993) compared 4 fractionation schemes: the Williams *et al.* (1971, 1976), the 'H&L' (Hieltjes & Lijklema, 1980), the 'GB' (Golterman & Booman, 1988) and a modification of the latter, using citric acid instead of the Ca-NTA plus Na_2-EDTA extractions. This modification can only be used if no distinction is needed between $Fe(OOH) \approx P$ and $CaCO_3 \approx P$. They concluded that in that case chelators, like citric acid, are good tools for the study of P-fractionation; citric acid can be used for P-fractionation when a N-fractionation is also needed. Barbanti *et al.* (1994) compared 7 different extraction schemes, both theoretically and practically, and concluded that chelators gave the best results. They found that in the SEDEX method re-adsorption was significant, but that the GB method gave an incomplete extraction of detrital apatite. In this respect the SEDEX method is superior to the (old) NTA/EDTA method. In the present EDTA method this apatite will of course appear as org-$P_{\rightarrow ac}$ (ASOP, Acid Soluble org-P), as an acid extraction follows the EDTA extractions (De Groot, 1990).

In the cold NaOH extracts from certain sediments some phosphate appears which does not react with the blue/molybdate method. It has been suggested that this might be a polyphosphate (Hupfer *et al.*, 1995), but the evidence is rather weak and it is more likely that phytate and/or humic-P form this pool (Golterman *et al.*, 1998). See below, page 81.

3.4.2 FRACTIONATION OF ORGANIC P-COMPOUNDS

The fractionation of org-P$_{sed}$ has been much less developed than that of inorg-P$_{sed}$. Several authors (among which myself) have used enzymatic methods to identify specific target compounds. However, if FeOOH and/or CaCO$_3$ are not removed from the sediment in a preceding step, these two compounds will adsorb the liberated o-P, so that the concentration of the target compound will be largely underestimated.

De Groot (1990) made a first improvement by trying to analyse the org-P after the complete removal of the inorganic components. He used HCl as the third extractant (org-P$_{\to ac}$), after the NTA/EDTA extractions, followed by hot NaOH. We do not know what the org-P$_{\to ac}$ is. It may be traces left over from the EDTA extractions, it may be a small quantity of fulvic-P; we found a constant ratio Fe/P in this extract. The total quantities are, however, always small. In the hot NaOH extract, de Groot & Golterman (1993) identified humic-P by precipitation and phytate by using phytase. A difficulty with this enzyme is that humic acid inhibits its activity, causing an underestimation. For future work it will be necessary to purify the hot NaOH extract first, e.g., by a mild oxidation with H$_2$O$_2$, or adsorption onto and re-dissolution from freshly added FeOOH. ^{31}P-NMR spectroscopy, a method with which a detailed structure and configuration of molecules can be determined rather than their concentration, is likely to be a possible way of analysing concentrated extracts, but only after purification (Golterman, 2001).

Nissenbaum (1979) found that marine humic acids contained 0.1–0.2% of P, with C/P ratios of 300–400, while marine fulvic acids contained 0.4–0.8% of P and C/P ratios of 80–100. The humic-P complexes were extracted from marine sediment with a high deposition rate, i.e., in early stages of diagenesis, and were purified by repeated acid precipitation. As inorg-P is then soluble, there is little doubt concerning the humic-P complex character. (It is not clear how the fulvic-P complexes were purified or what definition for 'fulvic' was used. The article suggests that fulvic acids are precursors of humic acids.) He calculated that humic-P may account for 20–50% of org-P$_{sed}$ and mentioned that the chemical speciation is unknown, but that it is probably bound organically. The humic-P complex is suggested to be a precursor for apatite during later stages of apatite diagenesis.

Although the presence of humic material in the hot NaOH is evident, the complex of humic material and phosphate has not yet been proved. Qiu & McComb (2000) extracted humic matter with 0.1 M NaOH, precipitated the humic acid with HCl (at pH = 1), redissolved the precipitate in NaOH and precipitated it again with HCl. Silicates were then removed by 0.3 M HF and org-P determined after destruction by HClO$_4$. They considered humic-P to be more or less identical to org-P and showed that a large part of the P$_{\to NaOH}$ consists of humic-P, again demonstrating the errors in the Hieltjes & Lijklema extraction.

Studying sediment rich in org-C from a Danish wetland (org-C = 20%) and a Danish lake

(Lake Kvie; org-C $\approx$ 10–15%), Paludan & Jensen (1995) used a modified Psenner extraction scheme with H_2O, 'BD', cold 0.1 M NaOH and 0.5 M HCl respectively, while the 'BD' and NaOH extractions were followed by an extra wash with the same fluids, which were not analysed separately. In the NaOH fraction separation was made between humic-P and the rest by precipitating the humic-P with H_2SO_4 (pH = 1). The precipitate contained $Al(OH)_3$ and FeOOH. The soluble P was supposedly attached to FeOOH or $Al(OH)_3$. $Al(OH)_3$ and FeOOH were separately extracted with ammonium oxalate and HCl, respectively. The (constant) ratio between Al and $P_{\rightarrow NaOH}$ being higher than that between Fe and P, the authors suggested the existence of a humic$\approx Al(OH)_3 \approx P$ complex. Real evidence can, however, only be obtained by an extraction with, e.g., Ca-EDTA *without* dithionite, eventually followed by one with dithionite. Only in this way the two complexes, $Al(OH)_3 \approx P$ and FeOOH$\approx P$, with or without humic matter, can be shown to co-exist – an unlikely situation.

Finally, it must be noted that both the Ca-EDTA and the Na_2-EDTA extracts are always dark brown; probably most of the humic material is already extracted with these chelators. As the Fe-humic complex is destroyed by EDTA, the evidence for the existence of this complex is destroyed as well. Serrano *et al.* (2000) reported the presence in the EDTA extracts of a brown compound together with P_{org} extractable with butanol. The concentration of the org-$P_{\rightarrow EDTA}$ extractable with butanol varied between P = 30 $\mu g\,g^{-1}$ and 50 $\mu g\,g^{-1}$ in temporary ponds of National Park Doñana. It is likely that some fulvic and humic/Fe/P complexes are already extracted with the EDTA. Part of the o-P found in the EDTA extracts may thus derive from these complexes.

De Haan *et al.* (1990) established direct evidence for the existence of humic/Fe/P complexes by using Sephadex G-100 gel filtration and ^{55}Fe plus ^{32}P labelling. Both isotopes were included in molecular particles of 10–20,000 MW and were eluted in the same elution volume, but while 20% of ^{55}Fe was included in this complex after 1 minute the ^{32}P-inclusion took several hours.

Golterman *et al.* (1998) published a fractionation scheme with which a separation between polyphosphate, phytate and humic bound-P may be obtained, using cold and hot TCA extractions in order to demonstrate the presence of polyphosphates. (See Section 5.1.7, Section 5.1.8 and Figure 3.22.)

3.5   Variability of and transfer between P-fractions

Mesnage & Picot (1995), studying sediment in a Mediterranean lagoon (Languedoc, France) found a wide variation in the different P-fractions, including org-$P_{\rightarrow ac}$, over time and as a function of depth, and a significant difference between sediment under or beside oyster banks. Moutin *et al.* (1993) studied the variations of o-P_w and the different P-fractions in two Mediterranean coastal lagoons (Languedoc, France). The org-P_{sed} varied, which could, however, not be linked to known processes or events. They found gradients for P_{sed} and Fe(OOH)$\approx P$ in the sediment and a decrease in Fe(OOH)$\approx P$ during summer and autumn. This decrease was concomitant with an increase in pH and compensated by an increase in $CaCO_3 \approx P$, in agreement with the solubility diagram. Ca^{2+} appeared to be the regulator of o-P_w concentrations. The increase in Fe(OOH)≈P was strongly correlated with an increase in FeOOH at all sampling sites. Calculations showed that under anoxic conditions during summer about $1400\,\mu g\,m^{-2}$ of Fe disappeared from the sediment by reduction, in agreement with the assumption mentioned for Lake Hypothesis in Section 2.3.3 and Section 3.8.4.4; only a small part of the primary production appeared to be indeed available for Fe^{3+} reduction. The authors also showed a strong increase in P_{sed} during summer/autumn, which might be explained by redistribution, by dying of populations of Ulva and precipitation of $CaCO_3$≈P due to increasing pH.

De Groot & Fabre (1993) found that, when a Camargue marsh (France) was left to dry, org-$P_{\rightarrow ac}$ was converted into o-P, and might, therefore, become biologically active. Further work has suggested that this pool may be a humic or fulvic bound phosphate. In Camargue sediment a small quantity of Fe^{3+} always dissolved in the acid extract together with some phosphate, normally at a mol / mol ratio of 2–3. As the phosphate released at the same time becomes o-P, we have no evidence that the P was indeed bound to humic compounds, but it seems likely. Díaz-Espejo *et al.* (1999) studied the different P-fractions in the sediments of sandy, temporary ponds (pH = 5.2–5.3) in the National Park Doñana (Spain), before and directly after filling. The percentage of org-C in the fine sediment (< 0.1 mm) was high, around 15%, and did not change during the filling, while the Tot-Fe and FeOOH increased significantly. Besides the usual presence of Fe(OOH)≈P and $CaCO_3$≈P, org-P fractions were also found in the extracts with Ca-EDTA and Na_2-EDTA, which moreover contained high concentrations of humic compounds. The presence of $CaCO_3$≈P in a pond with a pH = 5 is remarkable, and it must be investigated whether this pool was hydrolysed from the org-P in the EDTA fractions. The sum of the two org-P fractions amounted to around $300\,\mu g\,g^{-1}$ and contributed about 70% to the sum of all fractions. org-$P_{\rightarrow ac}$ increased about 30% in both ponds during filling. Fe(OOH)≈P increased only in the pond where FeOOH increased. In an adsorption experiment both Fe(OOH)≈P and $CaCO_3$≈P increased.

Goedkoop & Pettersson (2000) followed the sedimentation of spring diatoms in Lake Erken (Sweden) and found that all P-forms showed considerable seasonal variations. In the sediment Tot-P contained 25% reactive $P_{\rightarrow NaOH}$, 17% non-reactive $P_{\rightarrow NaOH}$ and 25% $P_{\rightarrow HCl}$. During sedimentation, the NH_4Cl and NaOH extractable fractions increased from 420 to $540\,\mu g\,g^{-1}$ and from 115 to $186\,\mu g\,g^{-1}$, respectively, in three days. The authors concluded that non-reactive $P_{\rightarrow alk}$ reflects bacterial polyphosphate, but did not give a critical discussion of the analysis of this fraction. They did not analyse for phytate or humic bound phosphate.

Masahiro Suzumura & Akiyoshi Kamatani (1995a,b) pointed out that phytate is an important carrier of org-P in *river* sediment and found concentrations of up to $175\,\mu g\,g^{-1}$ in some Japanese soils; this phytate is carried by the river to the marine sediment. They found mineralization in the marine sediment especially under anoxic conditions: under anoxic conditions phytate mineralized completely in 40 days, while under oxic conditions about 50% remained after 60 days. This is not in agreement with the findings of De Groot & Golterman (loc. cit.) in the Camargue, but the difference may be explained by the presence of larger quantities of FeOOH found in Camargue sediment, onto which the phytate appeared to be strongly adsorbed. The difference between P-adsorption onto FeOOH in freshwater and marine systems has been noticed above (see Section 3.2.1). Together with the phytate-P humic-P is extracted. (For their separation see Section 3.4.2.) These compounds may constitute a large part of the org-P pool. Qiu & McComb (2000), studying productive wetlands of the Swan Coastal Plain (SW Australia), found that humic-P varied between 5 and 73% of P_{sed}. Their Table 3 shows, however, that the sum of the P-fractions is often rather different from Tot-P, while it should be equal (±5%) This difference may be attributed to the org-P in the NaOH extracts, and partly to the inclusion of P associated with FeOOH in the HCl extracts of the Hieltjes & Lijklema scheme (see Section 3.4.1). Further, stronger evidence of the presence of humic-P compounds could have been obtained by measuring the Fe and P-concentration in the peaks of the Gel Permeation Chromatograms.

Rivers seem to have much more variation in Tot-P values and in the different P-fractions than lakes, but only a few studies are available. Fabre (1992) found important differences in Tot-P concentrations and in fractional composition in sediment of the river Garonne (France). Samples were taken at different places, i.e., under shallow water, in the dried river bank and in the riparian forest, and at different dates. Tot-P varied between 0.55 and 2.07 mg g^{-1} depending on site and hydrological conditions. Fe(OOH)≈P varied between 190 and 875 μg g^{-1}, while CaCO$_3$≈P varied between 75 and 235 μg ℓ^{-1}, depending on sampling place, but not so much on sampling moment. As most of these sediments originate directly from erosion, these variations reflect the geochemistry of the watershed. Moutin *et al.* (1998) found the same phenomenon in the River Rhône (France). Tot-P varied between 200 and 1200 μg ℓ^{-1}, while Fe(OOH)≈P and CaCO$_3$≈P varied between 45 and 405 μg ℓ^{-1} and between 90 and 570 μg ℓ^{-1}, respectively. The relative composition changed much less; Fe(OOH)≈P varied between 22.5 and 34% and org-P varied between 35 and 19%, while CaCO$_3$≈P remained rather constant at about 45%. The source of P_{part} in this river is much more variable than that of the Garonne, because while a large part of the river water comes from the Lake of Geneva, and will therefore be strongly buffered, another large part comes from agricultural sources (i.e., the Saône); the relative quantities vary widely throughout the year.

Phosphate release during anoxic conditions in sediment will most likely come from these organic pools (see Section 3.8.4.7). After mineralization by bacteria, the part laying on top of the sediment can diffuse into the overlying water without interaction with FeOOH, as this may be covered by a layer of FeS. Circumstantial evidence for this inhibition is given by Kleeberg (1997) in his studies of S-cycling and P-release in the eutrophic Scharmützel See (Germany). However, neither a possible release of org-P$_{sed}$ nor the dissolution of apatite was taken into account. The methodology of P- and S-fractionation (only some 20–30% of Tot-S was recovered) was not yet appropriate for this kind of problem.

Hupfer *et al.* (1995), using a modified Psenner scheme, showed shifts in P-fractions in settling seston and surface sediment. Sedimenting P-flux varied between 1 and 10 mg g^{-1} (dw.) in different seasons and the settling particles became enriched with reductant-soluble phosphate. $P_{→ac}$ and $P_{→alk}$ were low, org-P$_{sed}$ (measured as non-reactive $P_{→alk}$) showed a sharp increase around February. The dubious points of the Psenner scheme were, however, not addressed. Temporal differences in the composition of the org-P may have caused some of these changes. Only small quantities of CaCO$_3$≈P were found in the sedimenting material, but part of this compound may have been extracted with the dithionite/bicarbonate. Pettersson and Istvanovics (1988) presented some indirect evidence for this by comparing the scheme of

Hieltjes & Lijklema with that of Psenner, but did not demonstrate this by using well-defined compounds. Hupfer *et al.* (loc. cit.) gave some dubious arguments against the Pettersson & Istvanovics evidence. Another point of doubt is that the collected material remained 2 weeks in the traps, where increased bacterial activity – due to increased substrate concentration – must have changed the original composition. Low temperatures during winter may have caused the February maximum.

Penn & Auer (1997) analysed settling particles collected in sediment traps in the calcareous eutrophic Onondaga Lake (N.Y., USA). Traps were deployed in triplicate at depths of 5, 11 and 18 m, representing epi-, meta- and hypolimnion, during only 1–2 weeks, the short duration supposedly minimizing within-trap transfers. Seston dry weight sedimentation amounted to $8.9 \, g \, m^{-2} \, d^{-1}$ (st. dev. 3.9; $n = 90$) with high values in summer, and settling P to $22.9 \, mg \, m^{-2} \, d^{-1}$ (st. dev. 8.8; $n = 36$). They used the 'H&L' fractionation technique, but distinguished between reactive and non-reactive $P_{\rightarrow NaOH}$. Realizing that during the NH_4Cl extraction P is re-adsorbed onto $FeOOH$, they combined the $P_{\rightarrow NH_4Cl}$ and $P_{\rightarrow alk}$ and called this 'loosely bound P', a confusing name as it is already often used for $P_{\rightarrow NH_4Cl}$. The re-adsorption of $P_{\rightarrow alk}$ onto $CaCO_3$ was not taken into account. They found a strong increase in 'loosely bound P' during autum turn-over when dissolved Fe^{2+} mixed with surface waters and precipitated as $FeOOH$. $CaCO_3 \approx P$ was supposed to be terrigenous, while its formation in the lake together with precipitation of $CaCO_3$ was not considered. They estimated the labile (or exchangeable)-P (i.e., the loosely bound and extractable biogenic P-forms) of the settling particles to be $\approx 50\%$ of the Tot-P downward flux over the whole study period. The downward flux of labile P was $11 \, mg \, m^{-2} \, d^{-1}$ of P in July–November and under ice still $6 \, mg \, m^{-2} \, d^{-1}$. The Tot-P concentration of the seston increased from $1.5–2 \, mg \, g^{-1}$ dw. in July to $4.5 \, mg \, g^{-1}$ dw. in November, but an explanation for this phenomenon was not presented. The annual average downward flux of labile P was $9 \, mg \, m^{-2} \, d^{-1}$, while the sediment release rate was $10 \, mg \, m^{-2} \, d^{-1}$, indicating a steady state with respect to the external P-loading of the sediment. There was a close similarity between the fractional P-composition of seston collected at 18 m depth and that of the surface sediment, 'loosely bound P' being the largest fraction ($\approx 39\%$). This fraction, however, contained $FeOOH \approx P$ and org-P; transfer between these pools will remain unnoticed. Differences with older work on the same lake (Wodka *et al.*, 1985) can also be due to the different interpretation of the P-fractions, while in the former study the $CaCO_3 \approx P$ was considered to be biogenically formed in the epilimnion. The difference in the downward P-flux between the two studies is directly linked to management measures concerning P-retention taken in the watershed, indicating the high efficiency of these measures.

House *et al.* (1998) discussed the EPC_0 of mixtures of different sediments in relation to P-transport by rivers. They calculated a 'fractional distribution coefficient' by assuming a fractional contribution of the two components. They mentioned that for most sediments, the adsorption equation is linear at natural P concentrations (supposed to be $< 600 \, \mu g \, \ell^{-1}$) because of the linearity of both the Langmuir and Freundlich equations at low concentrations. However, the Freundlich equation is certainly not linear at these concentrations and the apatite solubility product limits the solubility of o-P to much lower concentrations. They showed that in some estuaries release, and in others uptake, of o-P takes place and suggested that competition with SO_4^{2-} for adsorption sites may be important, but present no evidence for this. It seems more likely that changes in the chemical composition of the sediments must be taken into account.

Bonetto *et al.* (1994) compared the P-composition of the Paraná River (Argentina) sediment with that of the surrounding marshes and found large shifts to a non-reactive P-fraction with a decrease in the dithionite/bicarbonate fraction, which could be identified no further. $P_{\rightarrow NaOH}$ and $CaCO_3 \approx P$ decreased, the latter because of the decrease in pH when the river sediment settled in the marsh.

Kleeberg (2002) analysed the P-flux from the epilimnion into the hypolimnion in the eutrophic Lake Scharmützel (NE Germany). Variable but large amounts of P sedimented, either with allochthonous particles, with diatoms (spring), or FeOOH (overturn); almost 60% of the P in the water column reached the sediment surface. org-P was the major constituent, while Ca and Fe components amounted to only < 6% and 12% respectively. Within the water column 16–18% of the total P-flux of the primary production was recycled and at the sediment surface $\approx$ 75%. Incorporation into recent sediment accounted for 10–23% of P indicating a long residence time for P and a high recycling in this sediment.

Several studies addressed the shifts between different P-fractions in freshwater/marine interfaces, often with a brackish water gradient.

Paludan & Morris (1999) analysed P-fractionation and distribution of P_{part} in intertidal marsh sediments at three sites along a salinity gradient in a river-dominated estuary changing into a marine-dominated salt marsh (Cooper River estuary, Sth. Carolina, USA). Fractionation was carried out in 5 steps:

1) H_2O (or NaCl solutions for the brackish or salt waters);
2) 0.11 dithionite/bicarbonate;
3) 0.1 M NaOH;
4) 0.5 M HCl;
5) combustion followed by extraction with hot 1.0 M HCl.

In all supernatants org-P was also measured as the difference between Tot-P and o-P in the fractions. Cold 0.1 M NaOH was supposed to extract org-P plus $Al(OH)_3 \approx P$, but the presence of $Al(OH)_3 \approx P$ was not confirmed. Humic-P was determined after precipitation with HCl (pH = 1). It was not shown whether the 'BD' extracted all $FeOOH \approx P$ in one step; a second 'BD' extraction might still have extracted some $Fe(OOH) \approx P$. Average Tot-P concentration in the upper m was $\approx 750 \, \mu g \, g^{-1}$ (dw.), with a st. dev. = $344 \, \mu g \, g^{-1}$. In the freshwater marsh the humic-P was the greatest fraction (24–51%), which decreased to 1–23% in the salt marsh. The sum of o-P was 15–40% of Tot-P in the freshwater marsh, and increased to 33–85% in the salt marsh. Most of inorg-P was present as $CaCO_3 \approx P$, 36–85% and 13–38% in two different salt marsh sediments, the second being more mature, but less under the influence of freshwater input. $Fe(OOH) \approx P$ was dominant in the surface of freshwater and brackish sediments, while $CaCO_3 \approx P$ dominated in the salt marshes. The authors suggested that these differences are the consequence of changes in salinity and concentrations of Fe and Ca; differences in pH, which varied largely over depth, must, however, also be taken into account. Only a few data on the presence of FeOOH are available, i.e., values between 750 and $3000 \, \mu g \, g^{-1}$. As a conclusion it seems likely that in estuaries the original, fluvial $Fe(OOH) \approx P$ and humic-P (better: org-P) will be converted to $CaCO_3 \approx P$. These authors suggested that the sum of all inorg-P and org-P extracted is available for macrophytes which seems unlikely and is not confirmed by data.

De Jonge & Villerius (1989) showed a partial release from estuarine sediment when the pH decreased and noted the importance of the amounts of $CaCO_3$ present. The influence of salinity changes was limited, which is in agreement with the fact that salinity has a stronger influence on the $Fe(OOH) \approx P$ than on the $CaCO_3 \approx P$ system.

Carignan & Vaithiyanathan (1999) compared P-adsorption characteristics in suspended sediment from 3 different rivers in Argentina: the Bermejo, the Paraguay and the Upper Paraná. They described P-adsorption by means of a linear adsorption equation, arguing that both the Langmuir and Freundlich equations are nearly linear at low concentrations, and used the Olsen and Sommers (1982) extraction for the different P-fractions [1: 0.1 M NaOH + 1 M NaCl followed by citrate-bicarbonate ('non-occluded'); 2: citrate+bicarbonate+dithionite at 85 °C for 15 min ('occluded'); 3: 1 M HCl ('calcium bound')]. The 'linear' adsorption coefficient ranged from $0.25\ell\,g^{-1}$ to $1.38\ell\,g^{-1}$, while the equilibrium o-P ranged from $0.17\,\mu mol\,\ell^{-1}$ to $2.92\,\mu mol\,\ell^{-1}$. At these concentrations, however, neither the Freundlich, nor the Langmuir adsorption equations are linear. P-release after 60 hr ranged from $7.4\,\mu g\,g^{-1}$ (Paraguay) to $25\,\mu g\,g^{-1}$ (Upper Paraná). These values increased considerably with a small decrease in pH. In the $CaCO_3$ containing Bermejo and Paraguay rivers, a pH decrease of 1–1.5 units caused a 10-fold increase in o-P and a 5–10-fold increase in 'desorbable' P. In the Upper Paraná a change in pH had the opposite effect, demonstrating the influence of $Fe(OOH) \approx P$. The decrease in pH also caused a dissolution of $CaCO_3 \approx P$, rendering this compound available. The shifts in P fractions caused by the pH shifts agree well with the general characteristics of the solubility diagram as outlined above and explain the o-P excess in these rivers and their flood plains, where such pH shifts do occur after sedimentation of the suspended solids. The chemical shifts were, however, not demonstrated experimentally.

Pizarro *et al.* (1992) noticed a relative decrease in P-adsorption in sediment of the Rio de la Plata when going from the fluvial to the marine environment. This decrease was explained by the higher pH of the estuary water, which is in agreement with the solubility diagram, but also by the more than double Fe-concentrations in the marine sediment.

Sundareshwar & Morris (1999) studied P-adsorption characteristics in intertidal marsh sediments along a salinity gradient in the Cooper river estuary (Sth. Carolina, USA). Results were fitted on the Freundlich equation after a 24 hr equilibration, which is not long enough to ensure adsorption onto $CaCO_3$. They used the following equation to calculate P-adsorption:

$$P_{ads} = A \cdot (\text{o-P})^b - Q,$$

with A and b as in Equation (3.2) and Q the amount of P adsorbed prior to any treatment. It is not clear why Q was subtracted from P_{ads}, as part of it was already adsorbed and is therefore included in the quantities of the Freundlich equation. Nevertheless, they found higher adsorption values in surface sediments than at 10–20 cm depth, and decreasing adsorption towards the saline waters. This was related to a 8.5 fold decrease in sediment particle surface in salt water sediment and possible shifts in the Fe and Al concentrations. In some sediments they found rather high values for b (approaching 1) and rather low for A, which may be due to the subtraction of Q. If Q is large in comparison to the experimentally adsorbed P, the error becomes serious. In the four above mentioned studies evidence for the observed or supposedly occurring processes could have been re-inforced by a better chemical P-fractionation and an estimation of extractable FeOOH.

Serrano $et\ al.$ (2003) studied the variability of P_{sed} in time and space in the sediment of a temporary pond in the National Park Doñana at three different sites (open water, littoral and flood plain) over a period of 3 years. Fine sediments were rich in organic matter (9–25%) and contained P-concentrations between 182 and 655 mg kg^{-1}. Using the EDTA fractionation technique it was shown that org-P_{sed} was 64–94% of Tot-P_{sed}. Flood plain sediment was poorer in Fe(OOH)≈P but richer in org-P_{sed} than the open water site. Organic matter increased significantly in the sediment of the open water site at the end of each dry season, while it decreased in the sediment of the flood plain site. Fe(OOH)≈P decreased in all sediments at the end of each dry season and was not related to changes in the sediment redox potential. The differences in sediment composition between flood plain and open water sites were probably due to the effect of plant growth and not to the direct effect of drying.

3.6   Bioavailability of sediment phosphate

Golterman $et\ al.$ (1969) set up the first bioassays with sediment as follows: A culture solution was made containing all the elements needed for algal growth except phosphate. A few grams of wet, fresh sediment were added in suspension and $Scenedesmus\ obliquus$ was inoculated at rather low cell densities. Growth on gyttja sediment of Lake Vechten was monitored daily till it came to a standstill, while for humic rich sediment of Lake Loosdrecht the same duration was chosen, although some growth still occurred. Results are given in Figure 3.11, which shows that in Lake Vechten growth was exponential in the first week and then flattened to reach a plateau. In Lake Loosdrecht growth was slower and continued slowly even after 3 weeks. When growth came to an end, the addition of more o-P caused growth to restart and the same efficiency was obtained. Cell numbers obtained were directly related to the quantity of P_{sed} added; sterilization of the added sediment did not influence the amount of cells obtained, showing that the bioavailable fraction was mainly inorg-P. Growth efficiency, i.e., the number of cells per mg of P was measured in controls with KH_2PO_4 instead of P_{sed}.

Cells were counted in a simple blood cell counting chamber under a microscope with low magnification. Using the improved Neubacher cell counting chamber, cell counting is easy: it takes about 10 min and the mud particles do not present difficulties. Electronic cell counters take more time and difficulties are encountered with the mud particles, while the microscopic examination gives information about the health of the cells at the same time. An alternative method is to measure the concentration of chlorophyll a, but we found this less reliable as the chlorophyll a content per cell depends on the light intensity, and cultures tend to become yellow at the end of their growth, without P-loss. The resulting cell count was converted into P_{cell} by using the calibration curve made with cells growing on KH_2PO_4.

A few experiments were done with mud enclosed in dialysis tubes (Golterman, 1977), but

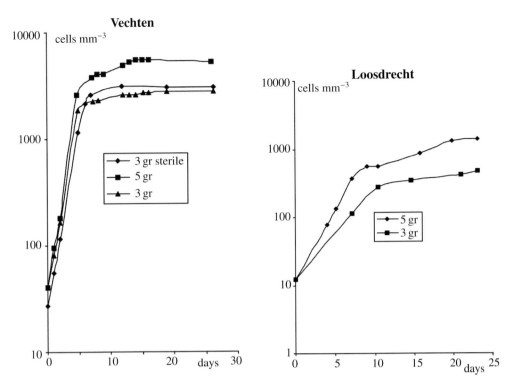

FIGURE 3.11: Growth of Scenedesmus on Lake Loosdrecht and Lake Vechten sediment. (Golterman *et al.*, 1969).

growth was very poor compared to growth on mud without these tubes. This is not due to the need for physical contact, but to the fact that the P is so strongly adsorbed onto the sediment that the diffusion through the dialysis tube becomes virtually zero. This was demonstrated by adding KH_2PO_4 to the mud suspensions in the dialysis tubes, after which poor growth was again obtained.

Bruning & Klapwijk (1984) modified the method for measuring the concentration of chlorophyll a by developing a derivative spectroscopic method to measure growth of algae in mud suspensions. It compensates for the loss of sensitivity due to turbidity caused by the sediment. In principle the chl a concentration is measured; the method therefore has the same disadvantage as the direct chl a determination. Klapwijk & Bruning (1984), using this method to estimate the percentage of P_{aa} in several Dutch lakes, found a high variability. Their results, however, cannot be compared with specific P-fractions as no fractionation was carried out. The results were used to predict in which lakes P-removal from the inlet would be most useful to avoid algal growth.

Grobler & Davies (1979, 1981) have shown that NTA-extractable P ($P_{\rightarrow NTA}$) in sediments of rivers and reservoirs in South Africa was strongly correlated ($R = 0.97$) with P_{aa} for *Selenastrum capricornutum*, but the P_{aa} was 5 times higher than the $P_{\rightarrow NTA}$. (It is possible that this difference was due to the interference of NTA with the P-determination.)

Grobbelaar (1983) has shown with sediment in the Amazon river (Brazil) that P_{part} was available to algae. As the Amazon river has low Ca^{2+} concentrations, most of the P_{part} was probably in the form of $Fe(OOH){\approx}P$.

Most work on bioavailability has been done on calcareous sediments. Except for one report (Golterman *et al.*, 1969) on Lake Loosdrecht sediment, very little work has been done on organic, or humic rich sediment. Figure 3.11 shows that in such cases slow growth may occur in the bioassay and may lead to wrong results. More work on different types of sediments may yield important results. The bioavailability of org-P_{sed} must be studied by experiments over longer periods combined with a careful fractionation of P_{sed}.

The same is true for the P-availability for macrophytes, about which hardly any information exists.

Carignan & Kalff (1979) labelled a 'mobile' fraction of P_{sed} with ^{32}P during 3 weeks (Lake Memphremagog, Quebec-Vermont). Three species of macrophytes were grown on this sediment; the specific activity of the macrophytes appeared to be identical to the mobile fraction. There is no chemical information on this sediment or on this mobile fraction, but it seems likely that this fraction is more or less identical to the sum of the inorganic phosphates; these experiments may indicate, therefore, that both $Fe(OOH){\approx}P$ and $CaCO_3{\approx}P$ are available.

Carr & Chambers (1998) demonstrated the influence of P_{sed} on the growth of a rooted macrophyte (*Potamogeton pectinatus*) in a Canadian prairie river (South Saskatchewan River). P_{sed} was the limiting factor for macrophyte growth, followed, after P addition, by N availability. Macrophyte growth increased with increasing concentrations of P_{extr} between 40 and $950 \, g \, g^{-1}$. P_{extr} was defined as extractable with 0.1 N $NaOH/NaCl$, which permits, however, no definition of its chemical composition.

Jensen *et al.* (1998), applying a mixed Psenner plus SEDEX method for P-fractionation (see page 79), came to the conclusion that the $CaCO_3{\approx}P$ pool was available for the seagrass *Thalassia testudinum* by acidification, either by acid root-exudates or by FeS oxidation. Acidification will certainly solubilize $CaCO_3{\approx}P$, but the pool size in this case was overestimated by the use of 0.1 M $NaOH$ prior to the extraction of $CaCO_3{\approx}P$ (see Section 3.4.1).

Smith *et al.* (1978) demonstrated that several naturally occurring apatites with varying crystalline structure were available to support several unialgal-mixed bacterial cultures, especially at lower pH values. 13 different algal species were used; incubation was 50 days. Apatite grain size was a dominant factor, smaller particles giving higher yields. The algal yield per mg of P indicated, however, a rather low efficiency of transfer, which might increase with increasing incubation periods. Mayer *et al.* (1999) stressed the importance of the interstitial water as a nutrient pool for macrophytes, but argued that a rapid turn-over is needed.

3.7 Comparison between bioavailable P and P-fractionation

When comparing the P_{aa} with the extractable P_{sed}, it must be understood in the first place that both procedures have been modified in the course of the years and that, even under the same name, different extraction methods have been used. In the literature there are even publications about "bioavailable-P" where this fraction is supposed to be equal to the quantity extracted by one extractant, e.g., strong acid. Furthermore, a 'good' correlation between the P_{aa} and the extractable-P is no indication that a certain chemical fraction is bioavailable as long as the slope does not approach 1 (one). We are looking for the identity of and not for the correlation between quantities.

There are several papers in which a chemical fraction is 'defined' to be equal to the biologically available one. E.g., House *et al.* (1995) used the iron-oxide stripping method to determine the P_{aa} without bioassays. Unless both methods are applied, such papers will not be considered further here.

Golterman *et al.* (1969) found no agreement between P_{aa} and $P_{\rightarrow alk}$ in Lake Vechten or Lake Loosdrecht, but the P_{aa} quantity approached the sum of the decrease in $P_{\rightarrow alk}$ plus $P_{\rightarrow ac}$. This is not identical to the sum of $Fe(OOH){\approx}P$ plus $CaCO_3{\approx}P$, as the NaOH extracts also contained hydrolysed org-P_{sed}. During the 3 weeks of the experiment the $CaCO_3{\approx}P$ may have re-equilibrated with the $Fe(OOH){\approx}P$, as both compounds are in equilibrium (Golterman, 1998). The total quantities of $P_{\rightarrow alk}$ plus $P_{\rightarrow ac}$ were not depleted during the incubation, $\approx 16\%$ had been used from Lake Vechten mud and $\approx 45\%$ from Lake Loosdrecht mud, which is due to the fact that these two pools also contained org-P, which is not bioavailable.

Using mud from 13 lakes in the Netherlands and the fractionation procedure with Ca-NTA and Na_2-EDTA, De Graaf Bierbrauwer-Würtz & Golterman (1989) found the following correlation between $X \ (= Fe(OOH){\approx}P + CaCO_3{\approx}P)$ and $Y \ (= P_{aa})$, both in $\mu g \ g^{-1}$:

$$Y = 0.97X + 5.9 \qquad (N = 14, \quad r = 0.9). \tag{3.14a}$$

This high correlation coefficient and the slope of 0.97 demonstrated the near identity of the quantities used by the algae and extracted. For Lake Vechten the extractable P was 50% of the Tot-P, a reasonable agreement with the quantity of P_{aa} found in 1969 (44%). (See Figure 3.12.)

Fabre *et al.* (1996), using sediment from the river Garonne (France), found that the sum $[Fe(OOH){\approx}P + CaCO_3{\approx}P]$ overestimated the P_{aa}. The difference with the Dutch lakes is probably due to the presence of detrital (rock) apatite, which is extracted by the EDTA method but is not available for the algae within three weeks. The possibility to distinguish between these two forms of apatite using the SEDEX method (Ruttenberg, 1992) must be further explored.

Williams *et al.* (1976, 1980) used the extraction method with 'CDB' followed by NaOH. Williams stated that $P_{\rightarrow alk}$ reflects the P_{aa}. However, in Williams *et al.* (1980) the amount of P used by the algae was not measured; it was only shown that a linear correlation was found between algal growth and $P_{\rightarrow alk}$, and the yield per mg of P of the algal culture used, *Scenedesmus*, is rather low: $\approx 20\%$ of what could be expected. These authors stated that they did not obtain a good agreement with Ca-NTA extractions, but they used a Ca-NTA extractant during a very short time only, and without the reducing agent. Golterman *et al.* (1969) and Fabre *et al.* (1996) have shown that $P_{\rightarrow alk}$ is not equal to the P_{aa}.

Using 5 different soils, Huettl *et al.* (1979) compared the amount of phosphate that can be adsorbed onto $Al(OH)_3$ affixed to a cation exchange resin in 24 hr with the amount of P_{aa}

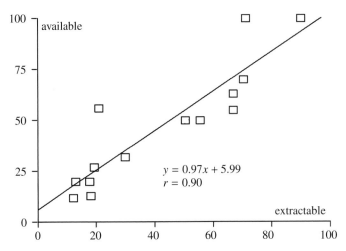

FIGURE 3.12: Phosphate availability against $Fe(OOH) \approx P + CaCO_3 \approx P$ in sediment from 13 lakes in the Netherlands and 1 marsh in the Camargue. Data in $P = \mu g\,g^{-1}$.

using a P-starved culture of *Selenastrum capricornutum*. The authors concluded that the P removed by the resin is, on average, 98% of P_{aa}. Several points remain, however, obscure in this paper. The quantity of exchangeable P depended on the duration of the extraction (it was still increasing after 48 hr and no constant level was obtained), the medium used and the ratio solution/soil. Neither is it clear whether the quantities of P_{aa} were taken from an earlier study or were new results. There is no information whether the cultures were aerated or not.

Dorich *et al.* (1984, 1985) studied P-bioavailability in 7 non-calcareous, suspended stream sediments, each taken at 4 different moments, and compared the quantity found with several sequential extraction methods. The procedures used were:

1) 0.1 M NaOH and 1 M HCl;
2) 0.5 M NH₄F, 0.1 M NaOH and 1 M HCl;
3) 0.01 M NTA;
4) hydroxyl-Al resin.

The percentages extracted from Tot-P were 52, 59, 45 and 17%, respectively. The extraction with NH₄F (22% of Tot-P) yielded a lower quantity of $P_{\rightarrow NaOH}$ and a higher quantity of $P_{\rightarrow HCl}$. *Selenastrum capricornutum* was used again and incubated for 2 or 14 days. The P_{aa} varied between 21 and 31% of Tot-P with an average of 25%. $P_{\rightarrow NaOH}$ correlated strongly with P_{aa} ($r = 0.95$), but overestimated the P_{aa} with a constant amount, the regression line being $P_{aa} = 0.85P_{\rightarrow alk} - 1.85$. It seems most likely that this overestimation is caused by hydrolysis of non-available org-P. The resin extractable-P was much lower than P_{aa}, while $P_{\rightarrow NTA}$ was higher. There remain a few problems: the culture still contained o-P after 14 d; it seems, therefore, that P was not the limiting factor for algal growth; it is not clear whether the cultures were aerated or not; pH values were not given. Lastly, the range of $P_{\rightarrow NaOH}$ and that of P_{aa} were very small – on a curve such as Figure 3.12 it would add only one point.

If, as in the new Williams method, NaOH is used as a P-extractant, the fractionation method is not different[2] from 'H&L', which is essentially the old Chang & Jackson method with a preliminary washing with NH_4Cl added at pH = 7, the function of which is rather unclear. The 'new' Williams method, therefore, need not be taken into account separately. The presence, in all sediments, of different forms of org-P which are partly hydrolysed by NaOH and HCl, is often not taken into account by geochemists, although their biological behaviour differs from that of the Fe- and Ca-bound phosphates.

Hosomi *et al.* (1981) studied P-release and P-availability (maximum growth of *Selenastrum capricornutum*) in 4 Japanese lake sediments. They used the Chang & Jackson extraction method, but with preliminary extractions with NH_4Cl and NH_4F, using a 2-chamber device without aeration. They found that anoxic P-release was strongly related to the concentration of $P_{\rightarrow alk}$ in the sediment and, furthermore, that about 30% of this phosphate was used by the algae. Tot-P decrease was identical to the decrease in $P_{\rightarrow alk}$. These results are similar to those from Lake Vechten, where $P_{\rightarrow alk}$ was not depleted either. Most likely this is the part of org-P hydrolysed by the NaOH (0.1 M, 17 hr shaking).

Premazzi & Zanon (1984) found that about 12% of Tot-P in sediment of Lake Lugano (Switzerland/Italy) was available for algae and that the sum of $P_{\rightarrow CDB}$ plus $P_{\rightarrow alk}$ was used by the algae, compared to only a minor portion of $P_{\rightarrow ac}$.

In a review, Sonzogni *et al.* (1982), summarizing studies on sediment from the Great Lakes (USA), considered P_{aa} to be equal to the sum of o-P plus $P_{\rightarrow alk}$ according to the Williams *et al.* (1976, 1980) extraction. They concluded 0.1 M NaOH to give a better agreement than 1.0 M NaOH, but they examined neither the influence of the extraction time nor that of the NaOH concentration, although they did mention the influence of the sediment/solution ratio; they recorded the degree of several extractions as 'incomplete', 'complete' or 'none', but did not give the criteria for these degrees. In the next review, Hegemann *et al.* (1983) concluded that bioassay techniques used so far do not measure all the P_{aa} and that the extraction techniques present no evidence that all the chemically extracted P in these fractions is bioavailable. They pointed out the difficulty of using membranes to separate algae from sediment, i.e., the slow diffusion from the mud to the algal compartment, and criticized the estimation of chlorophyll *a* as a measure of P-uptake. They suggested that protocols for long term experiments should be established, but did not give a solution for the problem of the growth of fungi under such circumstances.

Studying 27 different, but all non-calcareous, soils in the USA, Wolf *et al.* (1985) assumed that $P_{\rightarrow alk}$ equalled P_{aa}. They tested the so-called Equilibrium Phosphorus Concentration (i.e. the o-P concentration where there is no net gain or loss of P from the solution in a sediment suspension), which was then compared with different, classical, soil-P extraction techniques ($NaHCO_3$ and acid fluoride). They showed strong correlations between different fractions and bioavailability, which improved when soils were grouped together according to geographic location and taxonomic classification. Although the correlations were strong indeed, P_{aa} and P-extracted were not identical, the former fraction being 2–6 times higher than the latter. As these soils did not include calcareous soils, $Fe(OOH) \approx P$ and org-P_{sed} were the only phos-

---

[2]In the discussion comparing their method with that of Williams, Hieltjes & Lijklema changed the sequence of NaOH and CDB in the Williams method.

phates present, which will both be only partially extracted by the alkaline extractants. The classical P-soil extraction techniques do not offer an alternative for measuring P_{aa}.

Young and DePinto (1982), using the two-chamber device (i.e., a system in which mud and algae are in 2 different compartments, separated by a membrane), found that the P_{aa} ranged between 0 and 40% of P_{part} and correlated best with the $P_{\rightarrow alk}$. They found the following correlation between $P_{\rightarrow alk}$ and P_{aa}:

$$Y = 1.08X - 0.008 \qquad (r = 0.79), \qquad\qquad (3.14b)$$

but between $P_{\rightarrow alk}$ plus $P_{\rightarrow CDB}$:

$$Y = 0.692X - 0.07 \qquad (r = 0.84). \qquad\qquad (3.14c)$$

From the point of view of chemical fractionation this makes no sense: by setting X equal to the of sum $P_{\rightarrow alk}$ plus $P_{\rightarrow CDB}$, the slope becomes less steep, while the negative values at low P-concentrations are incomprehensible. The authors concluded, correctly, that

> "among the various chemical fractions of P_{part} which were compared to available fractions, no single fraction emerged which would quantify as a broadly-applicable surrogate and, thus, obviate the need to perform time-consuming bioassays".

This is certainly true for the period before chelating extractions were used. There are several objections against the two-chamber culture technique as used by Young and DePinto (loc. cit.); the membrane strongly decreases the diffusion of o-P, while the algal chamber was not aerated, causing a strong increase in pH.

Ekholm (1994), studying the availability of P_{part} in Finnish rivers, improved the 2-chamber method considerably: the algal medium was buffered at pH = 7 and the assay chamber was aerated (to avoid an increase in pH which would cause a change in P-fractions). Ekholm found that the P_{aa} was 5.1% (range 0–13.2%) of P_{part} in the river, 2.6% in the sedimenting material, but 0% in the bottom sediment. Ekholm attributed this low percentage to the high adsorbing capacity of Finnish acid mineral soils; this effect may well be enhanced by having mud and algae in separate chambers. The percentage bioavailable is, however, no lower than we found in humic rich Lake Loosdrecht sediment. Ekholm tested the diffusion rate through the membrane by using KH_2PO_4 instead of mud. It must be remarked, however, that this does not give a good idea of the diffusion barrier, because the o-P-concentration is much lower in the presence of adsorbing sediment than of KH_2PO_4, while the release rate from the particles may play a role as well.

The objections against the two-chamber culture device discussed above and against the anion-exchangeable P as a measure for the P_{aa} do not hold true for soluble P-compounds. Ekholm (1994, 2003) used the former with success to measure available P in diverse water samples and Hanna (1989) found a strong correlation between bioavailable and anion-exchangeable P (slope 1.05; $r^2 = 0.834$, but on log / log basis), although the dispersion was still large.

Boers et al. (1984), using the Hieltjes & Lijklema extraction procedure, found that the P_{aa} in humus rich sediment from Lake Loosdrecht was much less than the $P_{\rightarrow alk}$ and was nearer to the 'loosely bound' pool, i.e., a few per cents of Tot-P. Their results agreed with those of Golterman *et al.* (1969), but the authors did not investigate whether any further, slow growth still continued; neither growth curves nor a description of the methodology used were given.

Güde *et al.* (2001) addressed the question whether littoral sediment of a large, deep lake (Lake Constance, Germany/Switzerland) contributes to the P available for algae. They used the Psenner fractionation and found that littoral sediment had a much higher $P_{\rightarrow ac}$ (70%) than profundal sediment (22%). The sediment released up to $2\,mg\,m^{-2}\,d^{-1}$ of P during summer, when the redox gradient was within the first cm of sediment – this was explained by assuming that this P was bound onto reducible 'soluble' matter, such as $FeOOH$, of which P-compound some $3\,g\,m^{-2}$ was present – the release was therefore $< 0.1\%$ of the $P_{\rightarrow BD}$. P_{aa} was extremely low and was only in the order of magnitude of $P_{\rightarrow water}$, i.e., 1% of Tot-P. The paper gives no details about the algal growth experiments so that this discrepancy, which is probably the result of the methodology, cannot be explained.

The bioavailability of P-compounds in sediment for algal growth remains a research topic where much still has to be done. Chemical fractionation techniques are better advanced, but for both areas of research the function of org-P remains to be clarified. Bioassays are at the moment the only method to give some insight into the availability of P_{sed}, but must be replaced for routine work by a chemical fractionation. Only chelating extractants at the same pH as the sediment can give reliable results, but the presence of detrital apatite still presents a difficulty. Special attention should be directed at the presence, structure and function of humic-P and phytate.

Chapter 3

Sediment and the phosphate cycle

Part 2: Release processes

In this (sub)chapter only the P-release processes will be discussed, a subject on which as much literature has appeared since 1935, as on all those in Chapter 3, Part 1, together. There are several approaches to this problem, while many different chemical processes must be considered. The problem can be approached with lake balance studies, interstitial water analyses and experiments with isolated cores or lake enclosures. Often only the combination of two or more approaches will provide useful answers.

3.8.1   GENERAL ASPECTS AND STUDIES

3.8.1.1   Introduction

Three mechanisms may cause P-release under both oxic and anoxic conditions. When a P-containing sediment is brought into contact with a medium containing less or no P, part of the $Fe(OOH) \approx P$ will be released according to the equilibrium of Equation (3.2). Although this quantity may be small, it may approach the concentrations found in natural waters. Furthermore, $CaCO_3 \approx P$ is not entirely insoluble. Although the solubility product is low (see Section 3.2.3) at pH values as found in sediment (always slightly more acid than the overlying water), this quantity again is not negligible; Table 3.4 gives some maximal values. Lastly, part of the org-P_{sed} may also dissolve in the newly added water, especially when no Ca^{2+} is present. The dissolution of $Fe(OOH) \approx P$ and $CaCO_3 \approx P$ will happen in lake water experiments, but also in certain extractants such as NH_4Cl and dithionite/bicarbonate, causing some $CaCO_3 \approx P$ dissolution. (See Section 3.4.1.)

Release under anoxic conditions has always received more attention than release under oxic conditions, but the latter also takes place. In a literature review, French (1983) gave an overall range of release rate values between -10 and $50 \, mg \, m^{-2} \, d^{-1}$ for aerobic and 0–$150 \, mg \, m^{-2} \, d^{-1}$ for anaerobic conditions;[3] these figures represent minimal and maximal val-

---

[3] The negative values meaning P-uptake by the sediments.

TABLE 3.4: Maximal solubility of o-P in water with Ca^{2+} = 40 mg ℓ^{-1} as function of the pH.

pH	o-P_{max} (mg l^{-1})
6	480
7	3.4
8	0.071
9	0.0029
10	0.0001

ues and cover more recently obtained values as well.

Jensen & Andersen (1992) studied P-release from aerobic, P-rich sediments taken in 4 shallow, eutrophic Danish lakes. Influence of temperature, NO_3^- concentration and pH were measured in a continuous flowthrough system by varying one factor at a time. The thickness of the light-coloured, oxygenated surface layer varied between 3 and 15 mm and decreased when temperature increased. The interface between light and dark sediment was at the site of the steepest redox gradient, at a redox potential of $\approx$ 230 mV (calomel electrode). P-release from the oxygenated sediments was 15, 21, 33 and 100 mg m^{-2} d^{-1} (average summer values). Temperature and NO_3^- influenced P-release, temperature alone already explaining $\approx$ 70% of the seasonal variation. Q_{10} values varied between 4.1 and 6.8; these high values can probably only be explained by a combined influence on bacterial metabolism and diffusion. High NO_3^- concentrations increased the thickness of the oxygenated layer and thus reduced P-release. When during summer NO_3^- was depleted, NO_3^- additions increased P-release, supposedly by enhancing mineralization of organic matter. Increasing the pH had a limited effect. Only in one lake did high pH values (9.5–9.7) result in a significant P-release. In the sediment of this lake only traces of $CaCO_3 \approx P$ were present (which may well have been formed during P-extraction with NaOH), so this release was in agreement with the finding of the authors that $Fe(OOH) \approx P$ was the dominant source. In the other lakes precipitation of $CaCO_3 \approx P$ by the high pH will have masked the release from $Fe(OOH) \approx P$.

Jensen et al. (1995) followed P-sedimentation and release in a coastal marine sediment at 16 m water depth in Aarhus Bay (Denmark). Net sedimentation was $\approx$ 1.55–1.86 g m^{-2} y^{-1} and P-release 1.05 g m^{-2} y^{-1}, the resulting deficit corresponding to P-burial. $Fe(OOH) \approx P$ constituted up to one third of the P_{sed} and was the most dynamic P-pool. It decreased with depth and contributed only 3.5% to the total P-burial. In autumn P-release attained a maximum and was linked to sulphate reduction. Precipitation of FeS restricted upward flux of Fe^{2+}, but not of o-P, and resulted in a minimum ratio of $FeOOH/Fe(OOH) \approx P$ in the sediment layer in October. The authors considered $Fe(OOH) \approx P$ to be the most important factor controlling P-release from this sediment. The extraction scheme used was that of Psenner; i.e. NaCl, 'BD', 0.1 M NaOH and 0.5 M HCl); therefore their $Fe(OOH) \approx P$ pool may have contained some $CaCO_3$-P (see Section 3.4.1), and some org-P.

This P-release may counteract lake restoration measures by P-removal from influents, and sustain algal blooms. An example is presented by Granéli (1999) who noted the resilient high algal production in the eutrophic, shallow Lake Ringsjön (Sweden) after P-diversion, and measured P-release from the sediment in order to explain this. Tot-P_{sed} was high (2 mg g^{-1} of dw.) with concomitant high P-concentrations in the interstitial pore water and high concentrations of $Fe(OOH) \approx P$ and $CaCO_3 \approx P$ (estimated by the 'H&L' fractionation) in the sediment. The P-fractions were remarkably different in the three basins of this lake which, being shallow, is strongly exposed to wind causing an inhomogeneous distribution of the sediment and its grain size. P-concentrations during summer were highest, indicating release due to anoxic conditions inside the sediment. This study demonstrates once more the importance of P-release from sediment and the influence of its anoxia.

Mayer *et al.* (1999) showed that sediment can keep marshes (Point Pelee marsh on western Lake Erie) hypertrophic even when no external P-loading occurs. They measured different P-compounds and pH in pore water profiles. Tot-P profiles showed little diffference at the two sites with high P-concentrations of about 1200 mg kg^{-1} in the top 10–20 cm layer decreasing to 600 mg kg^{-1} at 40 cm depth. o-P-profiles showed a distinct maximum at about 10 cm and demonstrated upward diffusion. Ca^{2+} profiles showed a strong gradient between 1 and 10 cm with a maximum concentration of 150 mg ℓ^{-1}, while below 20 cm they decreased to 100 mg ℓ^{-1}. pH in the sediment decreased with 0.5–1 unit as depth increased to 10 cm, as a result of CO_2 production by mineralization, causing the dissolution of $CaCO_3$ plus $CaCO_3 \approx P$. Using model calculations, the authors stated that the pore water P-concentrations were strongly supersaturated with respect to $Ca_5(PO_4)_3.OH$. These authors used 10^{-44} for the $Ca_5(PO_4)_3.OH$ solubility constant, which seems to be rather high as 10^{-50} seems to be a more realistic value (see Section 3.2.2). However, using the data from their Figure 3, I arrive at a mean ionic product of $10^{-50.6}$, which is identical to the 'apparent' solubility of $Ca_5(PO_4)_3.OH$.

Release under anoxic conditions (see Section 3.8.4) is without doubt quantitatively the most important, especially in anoxic hypolimnia. Regardless of the chemical nature of the P-release process, we need to know how much is released. The process can be schematized as follows: The not easily biodegradable organic matter sinks through the hypolimnion, decomposing very slowly, and finally arrives at the water sediment interface. (See Figure 3.13.) Here the residence time increases, and slow processes now release measurable amounts of nutrients: nitrogen, phosphate and silicate (in the case of diatoms). The decomposing sedimenting material, having already lost most of its nutrients, no longer has a phytoplankton composition.

In the Orbetello Lagoon (Italy) Bonanni *et al.* (1997) found P-fluxes between 0.2 and 7 mg m^{-2} d^{-1} depending on season, temperature and sampling location. A high correlation was found between NH_4^+ and HCO_3^- in interstitial waters, indicating degradation as the source of these compounds, but the calculation of Redfield numbers[*] – which is often done – is of course not logical because of the many other reactions in which the elements are involved. The larger part of o-P will remain adsorbed onto the sediment, the CO_2 produced may react with the $CaCO_3$ of the sediment, while part of org-C and org-N will remain undecomposed. Nutrient flux studies should, but do not always, recognize this point.

Nutrient release, especially phosphate, may temporarily be inhibited by adsorption onto the sediment, depending on loading, depth of the water column, turbulence, pH, etc. In shallow waters the sinking period will be shorter, but is lengthened by resuspension resulting in

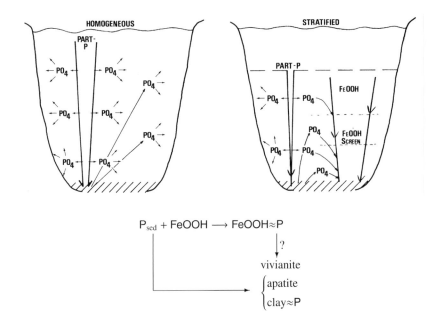

$$P_{sed} + FeOOH \longrightarrow FeOOH \approx P$$

FIGURE 3.13: Schematic sedimentation and P-release under homogeneous and stratified conditions. (From Golterman, 1984.)

the same effect.

The major problem remains how to measure P-release quantitatively. This problem is one of the most difficult ones, but answers are urgently needed.

There are three ways of attacking the problem:

a) Balance studies of the water/sediment fluxes, as a function of time, in a water/sediment column (e.g., of 1 m^2) or the whole lake. (See Section 3.8.1.2.)

b) Analysis of interstitial water by slicing cores or by using so called 'peepers'. (See Section 3.8.1.3.)

[A peeper is a device consisting of a series of small chambers mounted on a rack permitting the analysis of the chemical composition of interstitial waters at layers of increasing depth. For further details see Section 3.8.1.3.]

c) Quantitative studies on isolated samples with extrapolation to the lake situation, using either cores or benthic chambers. (See Section 3.8.1.4.)

[A benthic chamber is a device separating part of the sediment/water interface from the rest of the lake. Often bell-jars or square boxes are used which are pushed into the sediment with their open side down.]

In addition, ^{32}P can be used in these three methods.

3.8.1.2   Lake balance studies

P-flux has often been calculated from changes in the concentration in the overlying water with time. Lijklema & Hieltjes (1982), e.g., studied and modelled P-release in the 5.5 m deep eutrophic Lake Brielle (the Netherlands), with an average water residence time of 100 days. Loading was about $12 \, \mathrm{g \, m^{-2} \, y^{-1}}$ or $3 \, \mathrm{g \, m^{-2}}$ over the residence period. The authors described the phosphate sedimentation as follows:

$$\frac{d(\text{o-P})}{dt} = \frac{S}{L} - \frac{\Delta L}{L}(\text{o-P}), \tag{3.15}$$

where (o-P) = phosphate conc. $(\mathrm{g \, m^{-3}})$; S = sedimentation rate $(\mathrm{g \, m^{-2} \, y^{-1}}$ of P); L = mixing depth (m); ΔL = increase in sediment depth $(\mathrm{m \, y^{-1}})$.

Typical difficulties in such an approach are that the settling velocities have to be guessed from the literature (the authors took $0.3 \, \mathrm{m \, d^{-1}}$ from unquoted references) and that empirical values (which cannot be measured) for the diffusion constant must be used in combination with interstitial P-concentrations. Therefore, P-release was measured in column experiments and modelled in terms of Fick's law. This presents two problems: 1) eddy diffusion in a lake or its water column will be different from that in experiments; 2) it is automatically assumed that the diffusion rate controls the release rate but not the transfer from P_{part} to the interstitial water. (For Fick's law and eddy diffusion, see Section 5.3.) Besides these, unsolved, problems, the authors showed that the P-flux from the sediment depends much on actual loading levels. While annual loading estimates are already difficult to obtain, loading levels for short periods, e.g., a few months, can easily contain errors of 100% and cannot be calculated as $1/12^{th}$ of the annual loading. After increasing the loading by a factor 5, Lijklema found the P-flux into the water to decrease by a factor 15. Thus, any error in the loading estimate will automatically appear as a change in the flux into or from the water. Lijklema concluded that such an approach cannot simulate a short period (e.g., when sediment changes from oxic to anoxic) and that it is unlikely that a conceptual chemical equilibrium model can be developed that would be more representative of actual conditions than the 1982 model, which used empirical adsorption equations.

Balance studies based on calculations made per $\mathrm{m^2}$ may be imprecise when the lake bottom is not homogeneous, as is usually the case in shallow lakes like Lake Brielle. Schindler *et al.* (1977) measured P_{sed} in Lake 227 in the experimental lake area (Canada) and showed that the amount of P_{sed} in the sediment column varied largely with place and depth. Triplicates also showed a wide dispersion, which limits the usefulness of this approach. (See Figure 3.14.)

Ekholm *et al.* (1997) calculated P- and N-balances for Lake Pyhäjärvi (SW Finland), a lake loaded with agricultural run-off. The lake retained 80% of the external P-load; a remarkable export was fish catch, which removed 13% of the external P-load. Their balance model was:

$$\frac{dC}{dt} = \frac{L(t)}{V} - \frac{Q(t)}{V \, dt} \cdot C(t) - \frac{s(t)}{V} \cdot C(t) + \text{IL}, \tag{3.16}$$

where C = concentration, L = external loading, Q = outflow, s = settling velocity, IL = internal load and V = volume. For the IL, although high during certain periods, the annual value was set at 0, i.e., settling velocity refers to net sedimentation. Settling velocity was calibrated by minimizing the difference between the calculated and the observed mean annual concentrations. A value of $0.024 \, \mathrm{m \, d^{-1}}$ was found for Tot-P, about 20 times less than Lijklema's

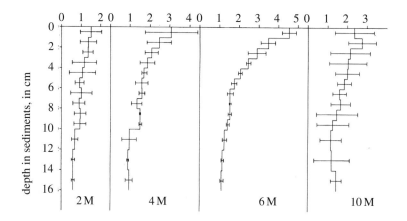

FIGURE 3.14: Tot-P_{sed} profiles from lake 227 sediment in May 1975. Horizontal bars indicate 95% confidence limits based on 3 cores at each depth. Cores were taken along three transects in different parts of the lake (modified after Schindler *et al.*, 1977).

value. Ekholm *et al.* (loc. cit.) showed a weak correlation between in-lake P-concentration and external load. *The weak correlation is mainly due to the internal processes not accounted for by the model. The idea of the modelling was to demonstrate that.* Although the work was carried out with great care, it shows the limited precision that can be obtained with whole lake studies. The average absolute error was estimated at 3.1 mg m^{-3}, with an observed value of 14 mg m^{-3}. Among other uncertainties, the external load remains difficult to estimate.

Knuuttila *et al.* (1994) established careful nutrient balances for two shallow, eutrophic Finnish lakes (Lake Villikkalanjärvi and Lake Kotojärvi) and discussed the influence of N- and P-input on phytoplankton dynamics. Mean Tot-P run-off from agricultural land was 1.2 kg ha^{-1} y^{-1} in both basins with total N loads of 19 kg ha^{-1} y^{-1}. Tot-P input into Lake Kotojärvi was 0.62 g m^{-2} y^{-1} and Tot-N input was 9.1 g m^{-2} y^{-1}. For L. Villikkalanjärvi these values were 3.1 g m^{-2} y^{-1} and 57 g m^{-2} y^{-1}, respectively. Annual variation was related to variation in run-off volume. In L. Kotojärvi $\approx$ 50% of Tot-P was retained (Tot-N: 33%) and in L. Villikkalanjärvi only 24% (Tot-N: 19%), the difference being due to the longer water retention time in L. Kotojärvi. The Tot-P concentrations were 120 μg ℓ^{-1} and 67 μg ℓ^{-1} in L. Villikkalanjärvi and L. Kotojärvi respectively, causing mean chlorophyll *a*, however, to be only a little higher in L. Kotojärvi (26 against 20 μg ℓ^{-1}); but the difference in blue–green algal composition was much larger. This was supposedly due to a higher P-release from the sediment in L. Kotojärvi, which was estimated to be twice the external P-load (in L. Villikkalanjärvi only 50%).

James & Barko (1997) compared P-retention with P-sedimentation, estimated from sediment cores and sediment traps in Eau Galle Reservoir (Wisconsin, USA). The authors distinguished between net and gross sedimentation. Net sedimentation was defined as the deposition of the outside P-load, and gross sedimentation as net sedimentation plus deposition of P recycled from the sediment via resuspension or focussing. In this study, cores were used to estimate the quantity of P accumulated above the preimpoundment soil, which for a reservoir is easier than for a lake. The mean annual retention was similar to the mean annual sedimentation as estimated from the cores, showing that sediment can indeed be used to reconstruct the P-loadings from the past. Sediment traps, although believed to be a promising tool, have the disadvantage that inside these traps mineralization of org-P (and org-N) is stronger than outside, while adsorption onto FeOOH may also occur. The use of these traps is often justified under reference to an extensive review of Bloesch & Burns (1980), but these authors stated that "this problem of artificial mineralization in traps is not yet solved" and later that "the question whether to use preservatives or not is not solved". They advise that exposure periods must be no longer than 3 weeks, but preferably shorter. The important point arising from their study is that simple cylinders are best, with an aspect ratio (= height/diameter) > 5, or in turbulent waters > 10, and that collars, lattices, baffles, lids or reference chambers must not be used.

James & Barko (loc. cit.) concluded that the P-budget for the Eau Galle Reservoir is far from complete. The sum of P retention plus internal loading was found to be about 65–75% of the sediment trap rate, indicating that in this case the traps overestimated sedimentation, if the other values are correct.

It is not surprising that a reasonable estimate of the P-flux from sediment cannot be obtained from field data. Although in principle it is a sound idea, in practice the accumulation of errors makes it rather uncertain, and errors might easily equal the quantities to be estimated.

In an older study the difficulties of a quantitative approach of a whole lake balance were shown. Burns & Ross (1972a, b, c) tried to estimate the P-flux from the sediment towards the hypolimnion and then to the epilimnion in Lake Erie and arrived at a release rate of $0.7\,\mathrm{mg\,m^{-2}\,d^{-1}}$ for oxic conditions during winter and $7.5\,\mathrm{mg\,m^{-2}\,d^{-1}}$ for anoxic conditions during summer, or about $300\,\mathrm{mg\,m^{-2}}$ for the stratification period, and assumed this quantity to re-enter into the life cycle of the lake. However, their balance showed a deficit of 31 tons, or $75\,\mathrm{mg\,m^{-2}}$, which may have been reprecipitated during the overturn, in which case there is no 'internal loading'. This study clearly shows the unavoidable errors that will occur with this approach. Diffusion estimates from field data are also rather cumbersome, although above the sediment concentration gradients will occur, of course. Data from Lake Vechten (the Netherlands), a stratified sandpit, are presented in Figure 3.15. This Figure shows that in the hypolimnion the o-P concentration increased from the bottom upwards with time, and that in August a concentration of $1\,\mathrm{mg}\,\ell^{-1}$ was reached near the bottom (10 m depth), while at 1 m above the bottom, $0.2\,\mathrm{mg}\,\ell^{-1}$ was reached. In this lake, well protected against wind and with a thermocline at 3–5 m, we found weak but distinct horizontal water movements such as probably occur in all hypolimnia. These movements, together with bioturbation, made it impossible to estimate eddy diffusion, which could easily be 10 times the molecular diffusion, the value roughly needed to explain this observed P-flux quantitatively.

The depth of the perturbation layer is often assumed to be no more than 10 cm, but there are no detailed studies on this subject in freshwater. For marine studies, see Section 5.3. Perturbation and bioturbation may be caused by wind and benthic animals, and are difficult to quantify. Wind influence is further treated in Section 3.8.2 and is most often only important in shallow lakes, where the wind alone may resuspend 2–3 cm of mud.

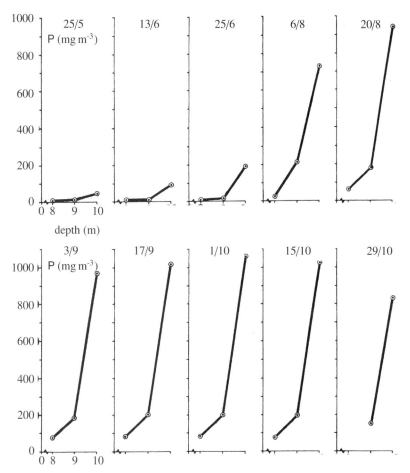

FIGURE 3.15: P-concentration above the sediment of the anoxic hypolimnion of Lake Vechten. (From Golterman, 1984.)

3.8.1.3 *Analysis of interstitial water*

Interstitial water can be analysed after slicing cores and squeezing out the interstitial water or by using 'Diffusion' or 'Dialysis' Chambers (also called 'Dialysers' but most often 'Peepers').

Peepers, introduced by Hesslein (1976), originally consisted of two sheets of clear plastic, one 0.3 cm and one about 1.5 cm thick, both 10–15 cm $\times 40$–50 cm, screwed together. (See Figure 3.16, from Ford *et al.*, 1998, Grigg *et al.*, 1999). In the thick sheet compartments or cells are milled (0.5–1 cm $\times 10$–12 cm) which can contain 10–15 ml of water. Between the two sheets a dialysis membrane is fixed; originally a 0.2μm filter (HT 200, polysulfone; Gelman Inc) was used, later other, incorrect, membranes (often Millipore membranes of 0.45μm, which have a too large pore size) were used as well. When the peepers are left 10–20 days in the sediment in an upright position, the water in the peeper equilibrates with the interstitial water; the volume of the chambers must be sufficiently large to permit analysis of pH and

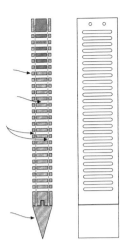

FIGURE 3.16: Diagram showing the construction of a 'peeper'. (From Grigg *et al.*, 1999.)

a few target variables, e.g., o-P, Tot-P and N-species. Different types of curves are depicted in Figure 3.17. The shapes of these curves depend on the composition of the sediment (such as the concentration of $CaCO_3$ and $FeOOH$), the input of o-P and org-P from the overlying water, up- and downward diffusion and characteristics such as the pH and the Ca^{2+} concentration in the interstitial water. Usually there are too many unknown factors to explain or predict the kind of curve that will occur. With the correct procedure peepers provide excellent profiles of interstitial components without too much variability. This is, however, not always the case. Villar *et al.* (1999) found wide variations in o-P, with values varying between $P = 0.1–1.1\,\mathrm{mg}\,\ell^{-1}$ in interstitial waters, without any discernible pattern, reflecting, among other processes, active in- and output by macrophytes.

Another way to obtain interstitial water is to cut a core sample into slices, under the necessary conditions, and to squeeze the water out of the sediment, e.g., by centrifugation or pressure. For anoxic sediment care must be taken not to introduce O_2; centrifuge tubes must be closed, etc. In our experimental reservoirs we found a good agreement between the two methods if Millipore membranes with $0.05\,\mu m$ pore size were used.

Wanatabe & Tsunogai (1984), analysing interstitial water from cores, found a strong seasonal variation for interstitial o-P. Maxima were found in spring just after a phytoplankton bloom sedimented, and during summer stratification. The P-maximum at about 10 cm depth corresponded with the layer having the steepest increase in alkalinity. The P-increase was accompanied by SO_4^{2-} reduction, indicating a strong mineralization of org-C. A likely explanation suggested by the authors was that adsorption/desorption coincided with the layer of active SO_4^{2-} reduction; the authors did not consider dissolution of $CaCO_3$ plus $CaCO_3 \approx P$ as a possible P-source.

Jacoby *et al.* (1982) investigated the possible effect of lake drawdown during summer on eutrophication and the role of the sediment in Long Lake (Washington, USA). They measured P-sedimentation and P-release in core experiments, and as differences in the mass balance in

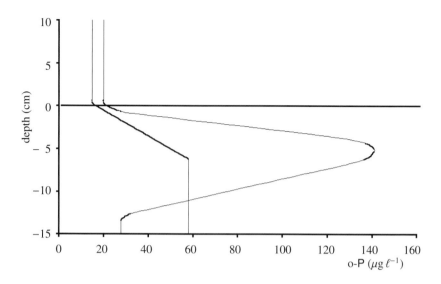

FIGURE 3.17: Two different schematized o-P-profiles, as often found in interstitial water.

a 2 m deep hypertrophic lake, taking into account all in- and outflows every 2 weeks. Their paper raised the important question as to whether the high pH during algal blooms does or does not favour P-release. The sediment in their core studies showed a high release at high pH values. It is not clear why the release following an increase in pH from 7 to 10 took as many as about 50 days; the chemical equilibria should be attained in a few days or a week. The authors estimated that the lake received as much phosphate from internal as from external sources during the summer period: positive and negative fluxes alternated, ranging from -7 up to $+7$ mg m^{-2} d^{-1} of P. The algal blooms probably decreased the concentration in the water so much that the effects of these blooms eclipsed the pH effect. The paper shows the need for calculations on the response time as a function of loading variations, perhaps by comparing peak concentrations in incoming and outgoing water; it moreover demonstrates how many assumptions must still be made in this kind of calculations. (The statement that recycling or internal loading is responsible for the present trophic status of a lake is not correct. It may be true for a short period, but the internal load is originally derived from the external load. Decreasing external loading will, in the long run, automatically decrease internal loading; internal loading or recycling should never be used as an argument against decreasing P-input, even though the effect may be delayed considerably.)

Masaaki Hosomi *et al.* (1982) showed that the amount of P-release was proportional to the decrease in Fe(OOH)$\approx$P under both aerobic and anaerobic conditions. Under anaerobic conditions, 90% of this fraction was released in 55 days. It seems likely that in their experimental methodology the released Fe(OOH)$\approx$P included important quantities of org-P and that the difference between aerobic and anaerobic release was caused by different bacterial activity.

A different device to sample pore water was used by Søndergaard (1990). He used an *in situ* incubation of ceramic cups (pore size: 2 μm, outer diameter 2.24 cm; inner diameter

1.76 cm), fitted with a gas inlet and a water outlet, and placed at intervals of 1 cm in the upper sediment layers, increasing to 10 cm in the deeper layers. The cups fill in about 10–20 min because of pressure from the surrounding sediment and can remain in position. In this way sampling at short intervals and resampling from the same position is possible. pH showed a sharp decrease with depth, while o-P increased from 0–2 mg ℓ^{-1} in the upper 5 cm to 3–6 mg ℓ^{-1} at 6–10 cm depth, suggesting the reduction of FeOOH$\approx$P as the main P-source, although Fe^{2+} was not found in the upper 20 cm. During summer the average P-gradient in the upper 6–8 cm was 1.0 mg ℓ^{-1} cm^{-1}, but in winter 0.2 mg ℓ^{-1} cm^{-1}. Net P-release was highly variable, with a summer average of 40 mg m^{-2} d^{-1}.

Gonsiorczyk *et al.* (1997) measured P-release rates in 20 mostly eutrophic lakes with an anoxic hypolimnion in North Eastern Germany in spring and in 12 lakes again in autumn. P-release was calculated using the o-P concentrations in the interstitial water and a diffusive flux model based on Fick's law. They used $D = 4.0810^{-5}$ m^2 d^{-1}, taken from Li & Gregory (1974). They found P-release rates of 0.0–4.7 (median: 0.5±1.3) mg m^{-2} d^{-1} during spring circulation and 0.2–5.8 (median: 2.3±1.7) mg m^{-2} d^{-1} at the end of the summer stagnation. These authors mentioned as possible P-sources the redox sensitive compounds, the pH sensitive compounds (i. e. CaCO$_3$$\approx$P) and mineralization of org-P, but maintained the lower redox potential as the cause of this process. Diffusion rates of o-P and Ca were positively correlated at the end of the summer period, showing the probability of CaCO$_3$$\approx$P dissolution.

In *theory*, diffusive flux can be calculated from the P-concentration gradient found in the interstitial water or between overlying and interstitial water. There are, however, several technical problems of interpretation. See Section 5.3.4, where a critical review of this technique by Urban *et al.* (1997) is quoted.

When using the interstitial water concentrations, e.g., to calculate P-release by diffusion, one finds, however, usually a strong gradient and a sharp decrease between the top compartment and the overlying water, and it is far from clear which concentration-gradient must be used for the calculations.

In a few studies P-release was calculated from the difference in concentration between overlying and interstitial water, and was supposed to be constant, sometimes for weeks or months. The major problem is that, as a result of diffusion, the interstitial concentration will decrease; it is then assumed that this concentration will remain constant because of dissolution of the particulate P-phase. We have, however, no information about the dissolution rate of Fe(OOH)$\approx$P or CaCO$_3$$\approx$P.

3.8.1.4 Studies on isolated samples

Experiments with isolated samples are usually carried out with cores, brought from the lake into the laboratory, or with benthic chambers, also called bell-jars, placed on the sediment, i.e. small enclosures (page 100). Core studies are useful to provide estimates about maximal release rates and about processes or conditions controlling this release. Bell-jar studies (with or without stirring) have the advantage of being carried out in the lake with the least possible disturbance of the sediment. Both core and bell-jar experiments can be done with or without the animal population present. In some studies P-release rates in bell-jars were compared with P-fluxes calculated from peepers.

The difficulties using bell-jars or core samples are, however, threefold:

1) The ratio water/sediment is changed and will cause an artifact, as the amount released enters a much smaller volume than in the lake itself. Among other problems, in the first place this will be the amount of CO_2, causing a significant change in pH. Furthermore, when substrates are added, the resulting concentrations will always exceed natural concentrations.

2) It is difficult to extrapolate from the experiment's duration to weeks or months.

3) The normal P-input by the sedimenting material is cut off. If part of the P-released comes from recent input, the release process is seriously underestimated.

Ad 1) Golterman (unpublished) filled a 60ℓ reservoir with about 30ℓ of mud and 25ℓ of overlying water. The reservoir was left for 6 months (only denitrification experiments and Cl^--diffusion studies were carried out in it during this period). After 6 months the pH was stable at 8.2 and the o-P concentration at $30\,\mu g\,\ell^{-1}$. Then a bell-jar with a height of one fifth of the water height was placed on top of the sediment. Within a week pH decreased to 7.2 and o-P increased to $150\,\mu g\,\ell^{-1}$ in the bell-jar, although in the reservoir itself no changes occurred. This was caused by the amount of CO_2 entering a much smaller volume, thus decreasing the pH and solubilizing the $Ca_5(PO_4)_3.OH$. The final o-P concentration was controlled by the concentration of FeOOH and the equilibrium with $Fe(OOH){\approx}P$, the concentration of which increased in agreement with Equation (3.7). (The decrease in $Ca_5(PO_4)_3.OH$ was too small to be measured.)

Core studies are useful to study processes in lakes and/or to obtain maximal values for these processes. E.g., Boström & Pettersson (1982) studied release from 8 different lakes and found that the results followed three distinct patterns:

   a) Some sediments did not release phosphate under any conditions; these lakes were characterized by a low adsorption capacity. It is not clear to what mineralogical properties this must be attributed; the results of the adsorption experiments might have been converted in terms of adsorption equations to improve insight in the underlying processes.

   b) Another category showed a high P-release when acetate was added, but the addition of nitrate prevented phosphate release. This pattern seems to follow Mortimer's theory, but FeOOH reduction was not measured, so the uncertainties about anoxic release are also present in these studies (see Section 3.8.4).

   c) The remaining sediments released o-P up to a constant level. This release was independent of acetate and /or nitrate additions; it may be suggested that it was controlled by Ca^{2+} and that a saturation value was obtained. Calculation of the ionic product may clarify this point.

Summarizing, no clear pattern regarding control by Ca^{2+} or FeOOH concentrations emerges from these studies; a comparison of the adsorption characteristics of these sediments with the differences in their mineralogical composition would have been preferable.

Another problem in this kind of approach is the fact that sudden addition of compounds like acetate or NO_3^- differs from the natural situation, where the supply is low and constant over prolonged periods. This causes a disequilibrium in bacterial population structure.

Ad 2) The difficulty of extrapolating in time is well demonstrated in a study by Grenz *et al.* (1991), who measured P-release in a benthic chamber (17ℓ) placed on the sediment of a Mediterranean lagoon (Etang du Prévost) and compared this release with the flux calculated from concentrations in the interstitial water ('peepers'). In the benthic chambers, active P-release was found to be about $4.65 \pm 1.5\,\mathrm{mg\,m^{-2}\,hr^{-1}}$; with the peepers it was $0.17 \pm 0.1\,\mathrm{mg\,m^{-2}\,hr^{-1}}$, a difference of a factor 27.[4] If such a high release had continued for two weeks, the lake concentration would have increased to $\approx 1.4\,\mathrm{mg}\,\ell^{-1}$ of P, an impossible value which indeed did not occur. This study shows the many pitfalls in this sort of work. The change in pH in the benthic chambers was not measured, apatite dissolution must have occurred, a diffusion constant ($7.9 \cdot 10^{-6}\,\mathrm{cm^2\,s^{-1}}$) was arbitrarily chosen from the literature, the flux was supposed to remain constant, etc.

A good design for a benthic chamber is given by Maran *et al.* (1995) who studied diffusion of metals from sediment into the overlying water. These authors pointed out the importance of knowing the rate of the chemical processes, which may strongly influence the simple diffusion process. The problem of the pH change induced by the chamber remained. A possible solution is to introduce a pH-stat, which is difficult in the field.

Nicholson & Longmore (1999) used benthic chambers to study CO_2, P, N and SiO_2 release from mud in three different places in Port Phillip Bay (Australia) under different regimes of external input. P-release varied between near 0 and $\approx 12.5\,\mathrm{mg\,m^{-2}\,d^{-1}}$ and was not different when clear or dark chambers were used, but varied largely over the seasons. Water temperature was supposed to influence P-release at the two sites near external inputs, but not at the centre of the bay. It is one of the few studies were SiO_2-release was also measured, which showed a strong similarity with P-release. No further analysis of this interesting phenomenon was given; it suggests P-release by algal material and not from $Fe(OOH){\approx}P$.

In a careful study on nutrient regeneration processes in 2 bottom sediments of the Sacca di Goro (Po delta, Italy), Barbanti *et al.* (1992) compared release rates in well stirred benthic chambers with those found by modelling concentration gradients in interstitial water, using 'dialysers' or 'peepers'. Concentrations of o-P and HCO_3^- increased in both sediments, in agreement with the decrease in pH (see Figure 3.18). Fe^{2+} showed a sharp maximum around 5 cm depth with much lower values below, indicating the importance of recent org-C as a reducing agent. The modelling approach accounted for bioturbation (a cumulative diffusion coefficient was used), but always showed lower release rates than the benthic chambers. Calculation of saturation degrees indicated the possible presence of apatite (with that of $CaCO_3$ and rhodochrosite, $MnCO_3$). Flux differences between stations could be attributed to the arrangement of the tubes of polychaetes, *Polydora ciliata*, in the sediment. The influence of a possible lower pH in the benthic chambers was, however, not taken into account and may have caused

---

[4]It is not clear whether the '±' referred to standard deviation or range in this paper.

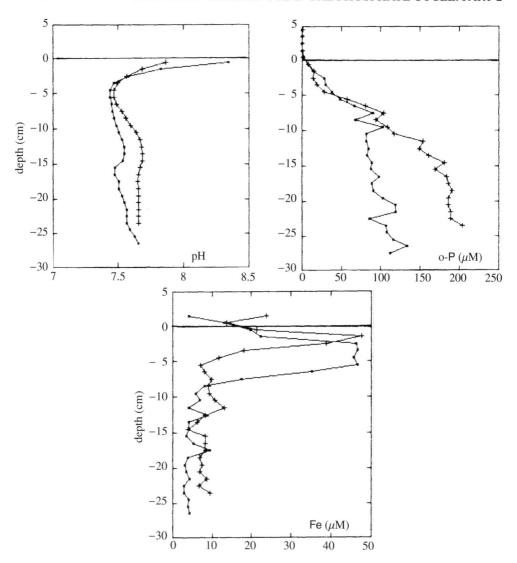

FIGURE 3.18: Concentration gradients of several chemical variables in pore water profiles, including bottom water, at two stations in the Sacca di Goro (Adriatic Sea, Italy). (From Barbanti *et al.*, 1992.)

the difference. Furthermore the authors studied the breakdown of organic matter by SO_4^{2-} and reported that SO_4^{2-} was the main oxidant in these sediments. By comparing $\delta(SO_4^{2-})$ with δN and δP they found an enrichment of N and P in the interstitial water as compared with the solid fraction. This enrichment decreased with depth as organic matter became more refractory there. However, the rapid adsorption of o-P released from organic matter remains an unknown factor in many of the calculations.

Cermelj *et al.* (1997) studied pore water N, P and SiO_2 concentrations in sediment of the Gulf of Trieste (Northern Adriatic) and compared the modelled fluxes with those

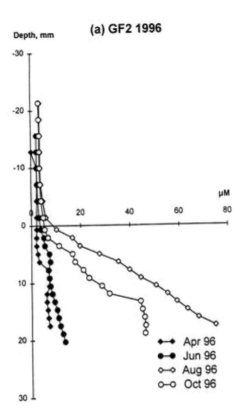

FIGURE 3.19: Pore water profiles of o-P in sediment of the Baltic Sea. (Mäkelä & Tuominen, 2003.)

obtained in benthic chambers (see Bertuzzi *et al.*, 1997). In winter they found an almost exponential increase in NH_4^+ and o-P with depth in the pore water profiles; in summer a maximum concentration between 3 and 6 cm and a minimum at about 8–10 cm was found. NO_3^- showed an exponential decrease in winter, and irregular high values in summer. The vertical variations were attributed to benthic infauna activity concentrated in the top 4–5 cm, which was lower in winter. The calculations from these profiles in order to obtain flux values are not clear; the concentration gradients were not curve-fitted and it is not clear which slope was taken. The agreement with the benthic chamber values is, therefore, only qualitative.

Feibicke (1997), studying P-availability in sediment after NO_3^- addition, is one of the few authors to show the wide standard deviations of pore water profile concentrations *in situ*. This variability is rarely taken into account in flux calculations and is not due to the peeper methodology, as we found small deviations with peepers placed in plastic tanks (unpublished results).

Mäkelä & Tuominen (2003) measured nutrient profiles in soft sediment of the northern Baltic Sea using cores and a squeezer technique permitting a mm-scale resolution. They showed low o-P concentrations, slightly increasing with depth in April and June, and high concentrations with a strong, nearly linear increase with depth in August and October; in this last series the concentration remained constant and below $\approx$ 15 cm (see Figure 3.19). In another station the concentrations were higher and the increase with depth greater. This spatial and temporal variability makes flux calculations over the summer or the whole year dubious, which is a clear disadvantage of this approach.

Nürnberg (1987) compared P-release as calculated from the o-P increase in 8 North-American lakes with release rates in cores. She found a correlation between these two rates ($r^2 = 0.85$; $n = 34$; several lakes were studied more than once), but only after a double log transformation. Real values differed by nearly an order of magnitude which illustrates the dubious value of either method. As the article gives little information about details of these lakes (many data are from unpublished sources), it is not possible to decide which method is best. She mentioned that some P sedimenting into the hypolimnion causes an overestimate of the hypolimnetic increase which renders the correlation even less.

All estimates in isolated cores or chambers are difficult to extrapolate to lake situations. Hydrological (physical) transport is different, while the constant input of recent sedimentation is cut off. Preferable is a device in which the sample of the lake bottom is isolated from the lake by a plexiglass cylinder or rectangular box. The cover of this box can automatically be opened, e.g., every 6 or 24 hr, and remains open for some hours in order to re-establish lake conditions in the box. The lid then closes again automatically, and the changing concentrations are monitored. As sampling now covers short periods, precise methods are needed, either by replicate sampling or by improving the precision of the (colorimetric) o-P-determination.

Fukuhara & Sakamoto (1987) examined the influence of two benthic animals on P- and N-release (see Section 4.3.1) in sediment from the eutrophic lake Suwa (Japan), placed in glass pots (height 17 cm, inner ϕ = 7.5 cm). Different densities of *Chironomus plumosus* and the tubificid *Limnodrilus sp.* were introduced and the overlying water was kept aerated. Tubificid density was set at 4500–72500 m^{-2} and chironomid at 1100–6800 m^{-2} corresponding to densities found in Lake Suwa. P-release was followed in the chironomid experiment for 117 hr and in the tubificid experiment for 132 hr. In the chironomid experiment P-release, up to 6–7 mg m^{-2} was higher in all pots with animals than in the controls; in the latter, however, it was still increasing after 132 hr. P-release did not depend on animal density – at the highest densities it decreased again. In the tubificid experiment, P-release increased with animal density. With the excretion measured separately, the ratio of excretion/release could be calculated and showed the greater importance of excretion for P-release.

Clavero *et al.* (1992) compared the influence of *Nereis sp.* on P-flux, using benthic chambers with P-flux calculated from pore water P-concentration gradients in intact cores. These gradients approached straight lines. The abundant presence of *Nereis* in the upper cm influenced the O_2 profiles in the overlying water, but not the redox potential in the sediment. The P-gradient ranged between 1.2 mg ℓ^{-1} cm^{-1} with 340 individuals m^{-2} and 0.6 mg ℓ^{-1} cm^{-1} with 900 individuals m^{-2}, while the P-flux varied between 6 mg m^{-2} d^{-1} in summer and 12 mg m^{-2} d^{-1} in October. P-flux increased nearly linearly with increasing O_2-flux, suggesting breakdown of org-C as the cause of this effect. The ratio of the *in situ* flux to the gradient-calculated flux increased from 1.7 at low to 5.8 at high abundance. P-flux was highest in winter and lowest in summer. For D_0 a value of $7.10^{-6} \, cm^2 \, s^{-1}$ was used. (For difficulties in this kind of calculation, see Section 5.3.) The quantification of bioturbulence on a lake scale, caused by local differences and patchiness, remains difficult – if not impossible. MacIntyre *et al.* (1999) studied the influence of weather conditions on nutrient exchange in Mono Lake (USA; $160 \, km^2$) and found that the coefficient of eddy diffusion was 2 to 4 *orders* of magnitude higher within 4 m from the bottom at an inshore station than at the offshore stations. Both horizontal and vertical mixing may enhance boundary mixing. The paper gives a good insight into the difficulties of studying temperature gradient structure and nutrient mixing.

The influence of benthic animals is limited to a water layer of a few cm's above the sediment and will be more important in core studies than in in-lake studies. Eddy diffusion may explain concentration gradients as shown in Figure 3.15, which could never be explained by molecular diffusion only. The steepness of the gradient in the hypolimnion (also found by Nürnberg & Peters, 1984) makes flux calculations extremely tricky, as the precision of the estimation of the quantity in the water column will be small.

Equilibria between sediment particles and interstitial water are established relatively rapidly (a matter of days or weeks probably) while exchange between interstitial water and overlying water is slow and therefore normally controls the flux (see Section 5.3).

Ad 3) The importance of fresh material input and even of contact with the sediment was demonstrated by, e.g., Andersen & Jensen (1992) with sediment from the shallow eutrophic Lake Arreskov (Denmark). In experiments with consolidated sediment or with seston suspensions they found that 100% ($\pm 10\%$) of P_{sest} was released when seston decomposed on top of the sediment, while after 30 days only 47% of the seston was decomposed when it remained in suspension. P_{sest} and N_{sest} were mineralized in the same proportion as in which they initially occurred in the seston and their release was strongly correlated with the O_2-consumption by the sediment. Chironomids were shown (Andersen & Jensen, 1991) to change the o-P release into uptake by the sediments initially, but after 15 days o-P was released again. The o-P release was higher in summer while the flux rate decreased with time. The experiments and the role of chironomids in mineralization of org-C and org-N are further discussed in Chapter 4. Hansen *et al.* (1998) evaluated the influence of *Chironomus plumosus* populations on nutrient exchange across the sediment-water interface in the same lake further. In sediment microcosm experiments 0 (controls) or 2825 larvae m^{-2} were added to sediment and a sedimentation pulse of organic matter was simulated by adding $36 \, g \, m^{-2}$ (dw.) of fresh algal material. In the controls the chironomids increased the O_2 consumption roughly 3 times more than expected from their own respiration, indicating an increased mineralization of org-C. Release of CO_2, NH_4^+ and o-P increased by the addition of the algae to the controls and corresponded to 65%, 31% and 58% of the algal nutrients respectively after 36 days. In the cores with chironomids these values increased to 147%, 45% and 73%, showing the enhanced mineralization of the org-C and org-P.

The increase in P-release was due to the metabolism of the larvae plus enhanced microbial and chemical breakdown of the algal material. O_2 input into the sediment was increased via ventilated burrows which added greatly to the observed effects.

It is clear that adding substrates once, but in high concentrations and therefore disturbing equilibria between bacterial populations, may give entirely different results from what happens in nature where a constant small flux is arriving on the sediment/water interface.

3.8.2 RESUSPENSION BY WIND

In shallow lakes resuspension of sediment by wind may cause important P-release. Søndergaard et al. (1992) studied the phenomenon in experiments with sediment samples of Lake Arresø (Denmark; surface = 41 km^2, depth = 2–4 m). Resuspension was simulated. Undisturbed samples, taken in May or August, released 12 or 4 mg m^{-2} d^{-1}, respectively, but a typical resuspension event in the lake (e.g., caused by a wind increase from 4 to 10 m s^{-1}) would increase the o-P release to 20–80 μg ℓ^{-1} or 150 mg m^{-2}. The P-release for the whole lake over the whole year was estimated at 60–70 mg m^{-2} d^{-1}, 20–30 times greater than the P-release without disturbance. Their result was partly based on a calculation in which a minimum algal sinking rate was taken into account, but not the P-release from the algae due to autolysis or bacterial mineralization during the sinking. The high release rate was supposed to be caused by the low Tot-Fe/Tot-P ratio and the high concentration of interstitial o-P. Re-equilibration of o-P$_w$ with the interstitial o-P did, however, not account for all P released. The quantity of actively adsorbing FeOOH was not measured. There was not only o-P release, but also P-uptake: when in August the o-P$_w$ was high (90–140 μg ℓ^{-1}), there was a negative release, or uptake, by the sediment. The authors noted that the o-P release depended largely on the o-P$_w$ concentration. This means a re-equilibration of phosphate between sediment and water; knowledge of the solubility diagram would have facilitated the interpretation. Kristensen et al. (1992) followed the resuspension in the lake itself. They found that during a storm event 2 cm of the upper sediment layer were resuspended; the concentration of suspended matter increased by 140 mg ℓ^{-1} and Tot-P by 0.48 mg ℓ^{-1}, while a flux from the sediment was calculated to be 560 g m^{-2} of suspended solids and 1.9 g m^{-2} of Tot-P. Although the residence time of suspended solids was short (7 hr, with an apparent settling velocity of 0.29 m h^{-1}), Tot-P in the water increased sharply with peak values between 0.8 and 1.4 mg ℓ^{-1}. An empirical model giving the relation between resuspended solids and Tot-P was developed:

$$\frac{dS}{dt} = \frac{K_{S_1}(W^{K_{S_2}})}{D} - \frac{K_d(S - S_b)}{D} \tag{3.17a}$$

and

$$\frac{dP}{dt} = \frac{K_{P_1}(W^{K_{P_2}})}{D} - \frac{K_d(P - P_b)}{D}, \tag{3.17b}$$

where S = mean concentration suspended solids (mg ℓ^{-1}); P = mean concentration Tot-P (mg ℓ^{-1}); W = wind velocity; D = water depth; K_d = settling velocity; S_b = background concentration of non resuspended solids (mg ℓ^{-1}); P_b = background concentration of non-resuspended Tot-P (mg ℓ^{-1}); K_{S_1}, K_{S_2}, K_{P_1} and K_{P_2} are constants, and were found by least sum of squares optimizing criteria to be 1.2, 1.45, 0.07 and 1.3, respectively. The authors quoted a value of 0.4 for K_{s_2} in the Dutch Lake Veluwemeer and a value of 1.0 for Lake Balaton (Hungary). The model showed that resuspension occurred during at least 50% of the time.

Resuspension is discussed in a series of papers edited by Bloesch (1994a). In his editorial, Bloesch warns that resuspension is more important than is often assumed and may even influence the measured settling flux in hypolimnetic traps, deployed at a distance from the bottom and considered to be sufficiently high to prevent disturbance. Bloesch (1994b) reviewed current methods to measure resuspension, but gives no solution for the question how to prevent wrong results for sedimentation rates of nutrients. In none of the papers influence on N- or P-release is discussed. Resuspension obviously accelerates the establishment of the chemical equilibrium between sediment and overlying water, and is therefore from a chemical point of view only an accelerator of the chemical processes. Like bioturbation it has a strong influence on the upward diffusion of compounds, which effect is rarely if ever taken into account.

3.8.3 OTHER ASPECTS: DRYING OF SEDIMENT AND RELEASE BY MACROPHYTES

3.8.3.1 Drying of sediment

Apart from the chemical processes controlling P-release many other factors intervene. Several, such as diffusion, bioturbation, resuspension, etc. have already been discussed. In temporal marshes in (semi-)arid zones and in hydro-electric reservoirs drying of sediment also plays an important role. This effect may be chemical, e.g., $FeOOH$ losing its adsorbent capacity because of water release and becoming Fe_2O_3 or, in hard waters, o-P precipitating with $CaCO_3$ as the Ca^{2+} concentration increases. Several studies have addressed the importance of drying on sediment in reservoirs, which has a strong influence on water quality, viz., eutrophication.

Fabre (1988) pointed out that the refilling rate has a considerable influence on resuspension and, therefore, on the P-distribution between mud and water. In experimental studies on the sediment of the Puyvalador reservoir (Pyrenees, France; 102 ha, volume $= 10^7 \, m^3$) he showed that the quantity of P re-solubilized varied according to the rate at which the water-level rose and depended on the initial P-concentration and pH.

De Groot & Van Wijck (1993) studied the drying of a marsh (1.5 ha) in the Camargue (Rhône delta, Sth. France), both in situ and in a laboratory experiment. They found considerable mineralization of org-P_{sed}, and complete oxidation of FeS, causing an increase in $FeOOH$ and $Fe(OOH) \approx P$. An increase in Tot-P in the top layer due to upward capillary flux occurred. This means that deeper layers down to 40 cm play an important role in temporary marshes. In contrast to Tot-P, C and N showed losses of up to 40–50% during desiccation. The authors discussed consequences for the management of temporary marshes and rice cultivation.

Qiu & McComb (1994) air-dried intact sediment cores at room temperature for 40 days and reflooded and incubated them at 20 °C. In previously air-dried sediment the P-release rate was higher than in wet sediment under aerobic conditions, but decreased again when aeration was stopped, suggesting a depletion of the source of this labile P-fraction. Under anaerobic conditions P-release from dried sediment was also higher than from wet controls. Accumulation of o-P during drying, due to the breakdown of org-P, and a decrease in the P-sorption capacity, due to this drying, were suggested as mechanisms. This phenomenon has, of course, a considerable impact in reservoir management; in the short term it may lead to increased algal productivity, but it could be used for long term improvement if the o-P or algal P could be removed by flushing. Besides the breakdown of org-P another chemical mechanism is decreased sorption capacity of $FeOOH$. The authors noted a pH increase of 0.5–1.5 in the dried sediment under aeration. Without aeration pH decreased, 0.5 pH unit more in the wet controls, but when aeration was resumed the pH in both experiments increased again, the wet controls showing the greatest change. The authors argued that this did not play a role in the P-dynamics. The greater pH shift in wet sediment ($8.2 \rightarrow 7.5$) compared with that in dried sediment ($8.2 \rightarrow 7.5$) is, however, more than sufficient to explain a considerable difference in P-adsorption or release. Qiu & McComb (1995) took the issue further and showed a 5-fold increase in o-P after air-drying of sediment cores from North Lake (Southwest Australia). Freshly killed phytoplankton released P before drying, but less when the sediment was air-dried. The way in which the plankton was killed must have induced autolysis followed by phosphatase activity, releasing most of the algal-P, while air-drying may have destroyed the algal phosphatases. When moisture was present bacteria incorporated o-P; the amount of P accumulated correlated with bacterial respiration of added glucose, not a natural (ideal) substrate. Upon drying, part of P_{bact} was returned to the water, this release increasing with the amount the bacteria had taken up. Air-drying killed about 76% of the bacterial biomass in sediment.

Turner & Haygarth (2001) showed bacterial immobilization to be a dominant process during soil drying, so that P accumulated in this biomass may be released into the water upon reflooding. These experiments show that the influence of bacteria on P-release is mostly through their influence on the breakdown of org-P and perhaps to a small extent by direct release or excretion of P_{bact}. In sediment from a eutrophic reservoir in New South Wales (Australia), Baldwin (1996) found a transition in P-affinity along a transect covering heavily desiccated, wet-littoral and submerged sediments, from both above and below the oxycline (the layer where the O_2 concentration changes strongly). Desiccation of sediment considerably reduced its o-P binding, which was explained partly by oxidation of org-C and partly by the conversion of $FeOOH$ to Fe_2O_3. 'Labile' Fe (i.e., extractable by 0.1 M HCl) increased from wet to dry sites, which suggests an increase in $FeOOH$ by oxidation of org-Fe or by hydrolysis of clay material. It was not indicated which substance was oxidized, but the oxidation of Fe^{2+}, and perhaps that of an organic-Fe complex, must be considered. Baldwin et al. (2000) studied P-release from ca. 6000 years old sediment and from more recent surface sediment (800–2000 y B.P.) of an Australian wetland (Ryans Billabon). The recent sediment had been either desiccated for 3 months, or constantly inundated, or was wet (from 25–30 cm below the surface). All sediments released small amounts of o-P on becoming anoxic. SO_4^{2-} alone stimulated P-release irrespective of previous hydrological conditions, while P-release from the deeper sediment was limited first by SO_4^{2-}, but after SO_4^{2-} enrichment by org-C. The SO_4^{2-} was supposed to act via the formation of FeS, which may cover the $FeOOH$ particles and in this way inhibit P-adsorption. (See Section 3.8.4.5.) P-adsorption could reasonably well be fitted on Freundlich equations. The A constant of Equation (3.2) varied between 1.5 and 8.9 under oxic and between 0.4 and 91 under anoxic conditions. In the deeper sediment the values under anoxia were much higher than those under oxic conditions, unlike in the recent sediment, where the latter ranged between 1.5 and 8.9 and therefore agree very well with the values found by Olsen and Golterman. Comparison can only be made with great caution, as it was not stated whether $CaCO_3$ was present, nor were pH values given. S-reducing bacteria were not adversely affected by desiccation or oxidation, although they are usually regarded as strictly anaerobic, but differences in species or adaptation may explain this controversy.

Mitchell & Baldwin (1998) showed that P-release from sediment exposed to air-drying was considerably less than that from non-exposed sediment. The same difference occurred when the sediment was treated with formaldehyde, suggesting a bacterial mediation of this release. Substantial P was released when S^{2-} was added to the non-exposed sediment, which release was reduced to one-fourth when the sediment was air-dried. Enhanced P-release from air-dried sediment was observed when glucose or acetate were added, but not when SO_4^{2-} was added. The effects were, however, not equally strong in the different sediments. These results suggest that the decrease in potential P-release was caused by shifts in bacterial communities, C-limitation as a result of the exposure and ageing of the minerals adsorbing P. A further P-fractionation might well have confirmed these results.

Watts (2000a) took the issue further and experimented with P-release from sediment under oxic conditions, with and without amendment with macrophyte material or acetate, either sterilized or not. She showed that under oxic conditions P-release from both macrophyte-amended and non-amended samples was maximal when the material was sterilized. Macrophyte addition to both fresh and dried sediment enhanced P-release more under anaerobic than under aerobic conditions, in agreement with what is said in Section 3.8.3. Watts suggested that Fe^{3+}-reducing bacteria transferred electrons from org-C to FeOOH, causing release of Fe^{2+} and o-P. This mechanism must still be proved to be quantitatively important. Re-flooding of sediment was shown to be a mechanism for aerobic P-release. Watts (2000b) compared two Australian reservoirs (Carcoar Dam & Lake Rowlands) with different degrees of *in situ* desiccation. Samples taken at 4 points along a transect, in both winter and summer, were compared. Along the transect o-P-release increased from the station below the water line to the station at the top water line; abiotic P-release was again greater from sterilized than from unsterilized sediment. Tot-P followed much the same pattern, although some back resorption occurred. Biotic P-release and uptake were greater in Lake Rowlands, where dense macrophyte beds are a source of much org-C. P-release appeared to be independent of season, showing the buffering capacity of sediment.

A few remarks must, however, be made. Amendments may lead to wrong extrapolations. If an, at present, limiting factor is added in relatively high amounts, another factor will automatically become limiting, which it would never have been in the lake. Additions of substrates like acetate and specifically glucose may favour bacterial strains quite different from those active in the lake, by sheer competition. Lastly, the manipulations needed in order to start the experiments may cause disturbances – e.g., shaking and aeration will put some P in suspension which with the 0.45 μm filters will not be removed quantitatively. This will certainly happen with $FeOOH/Fe(OOH) \approx P$ suspensions.

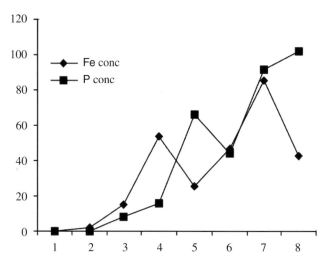

FIGURE 3.20: Quantities of Fe^{2+} and o-P released in Mortimer's tank experiments with Esthwaite Water sediment on 8 sampling dates. Relative standardized units.

3.8.3.2   Release by macrophytes

Macrophytes may take up phosphate from water or sediment, although no distinct picture has so far emerged. Furthermore, it is often suggested that some macrophytes 'pump' phosphate from the sediment in which they are growing into the overlying water. In order to be sure that no concomitant processes occur all other processes must be excluded. This is not often the case. Stephen *et al.* (1997) in a study on *Myriophyllum spicatum* and *Chara sp.* and on *Nuphar lutea* suggested that these macrophytes increased P-release from the sediments in two of the Norfolk Broads (UK). Release rates were affected by factors like the presence of plants, their species, and site and date. Measurements of the pH are, however, not even mentioned, and it is very likely that during these experiments pH changes were induced by treatment and photosynthesis. These pH changes, even if minor, might easily have caused release rates as found in these studies, which were, as the authors stated, not reproducible.

3.8.4   RELEASE UNDER ANOXIC CONDITIONS

3.8.4.1   Introduction

Release of phosphate from sediment becoming anoxic remains a much discussed problem in limnology, especially in discussions about lake restoration. Every year some 20–30 articles arrive on my desk quoting Mortimer's (1941, 1942) work about this subject as the source of the hypothesis that this release comes about by reduction of a $Fe(OOH)\approx P$ complex. Most of these 20–30 authors clearly have not read Mortimer's articles, as in the second publication "phosphate release" is not discussed. Only in his 1941 article do we find described an experiment in two tanks with sediment from Esthwaite Water, one of which remains oxic while the other one becomes anoxic. (It is surprising that his shortened version (Mortimer, 1971) is rarely quoted, although it is more accessible.) Actually, there is release of both Fe^{2+} and o-P

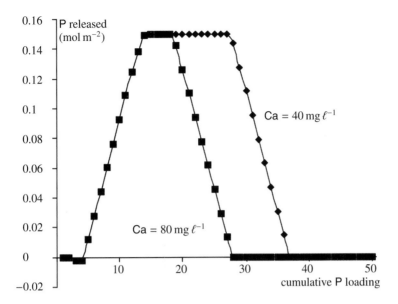

FIGURE 3.21: Calculated o-P release if the pH changes from 8 to 7 in an overlying water layer of 20 dm depth, respectively with $Ca = 40$ and $Ca = 80 \, mg \, \ell^{-1}$. (Golterman, 1998.)

in the anoxic tank. But when we look at the Fe/P ratio there is reason for doubt. Figure 3.20 shows the normalized quantities released. During the experiment, the Fe/P varied from 73 to 641, a difference of nearly an order of magnitude. In the hypolimnion (at 14 m depth, just above the sediment) this ratio varied between 11.5 and 316, covering the same range, but with an even larger variability. Mortimer himself wrote:

> "The sharp rise in phosphate concentration at the same time as disappearance of ferric iron [meant is colloidal Fe^{3+}] in the water (97–102 days) is in agreement with Einsele's (1938) observation that large amounts of phosphate are bound as insoluble ferric phosphate in the surface of oxidized muds, and liberated in soluble form on reduction".

Mortimer continued to point at

> "the possibility that an organic constituent, possibly of the humus type, forms a part of the adsorbing complex in oxidized muds".

This supposition is based on the increase in colour and organic matter occurring in the water after the mud surface had been reduced, but there is no experimental evidence for it. Mortimer also quoted values from Schleinsee (data from Einsele & Vetter, 1938), where the Fe/P ratio was much lower, while the variability was much less (4.4–6.1). Nowadays such extremely low ratios are, however, very difficult to understand, because the optimal ratio for adsorption is about 10.

Heaney *et al.* (1986) measured release of Fe^{2+} and o-P in the Esthwaite hypolimnion, again at 14 m, but in 1971 and 1972. These authors found much higher values for o-P release, but roughly the same for Fe^{2+}, and again no constant relation between these two elements.

The higher o-P release reflects a greater sedimentary accumulation of o-P adsorbed onto FeOOH and as org-P_{sed} between 1940 and 1971/72.

For a better insight into the Fe^{2+} and o-P release we have, therefore, to look at the articles by Einsele (1936, 1937, 1938) and Einsele & Vetter (1938). Einsele's article (1938) dealt with the *formation* of an $Fe(OOH) \approx P$ (iron phosphate complex). While he explicitly expressed the results of his *experiments* in terms of formation and precipitation of $FePO_4$, he proposed the Freundlich equation to describe the formation of the adsorption complex in the *hypolimnion*. He noted, however, that in his experiments some o-P always remained in solution, which is in agreement with an adsorption mechanism. His exponent in the Freundlich equation, 0.4, is not significantly different from the 0.33 we found in our experiments (Golterman, 1995a, b, 1998).

For the quantification of P-adsorption onto FeOOH we arrived in Section 3.2.2 at Equation (3.7):

$$\frac{P}{Fe} = 23600 \cdot 10^{(-0.416 \, pH)} \cdot (2.77 - 1.77 \cdot e^{-Ca}).$$

Using this equation it can be calculated that a change of 15% in P_{sed} is in equilibrium with a change of 50% in o-P. Unfortunately we do not yet have sufficient data to quantify the *rate* of release and adsorption processes, as these depend on (eddy) diffusion kinetics.

The reduction of Fe^{3+} by H_2S was studied by Einsele (1936), who found that the reaction was not stoicheiometric. In his experiments an 8-fold excess of H_2S was necessary to cause all the Fe^{3+} to appear in solution as Fe^{2+}. It is not clear why the first Fe^{2+} ions formed are not directly precipitated by the remaining H_2S. Einsele proposed as reaction

$$2Fe^{3+} + S^{2-} \longrightarrow 2Fe^{2+} + S,$$

which could partly explain the non-stoicheiometric reaction.

The Fe^{2+} appears in the water and is balanced by HCO_3^- ions. Qualitatively this mechanism seems probable, but the calculations must be repeated with the correct dissociation constant, as Einsele (1937) used a now out-of-date constant for H_2S.

3.8.4.2   *Arguments against* $Fe(OOH) \approx P$ *release by reduction of* FeOOH

Some doubt about this, now classical, picture was expressed by Golterman (1973b, c) who argued that, as Thomas (1965) showed, $Fe(OOH) \approx P$ does not go into solution even in activated sludge digestion. Lee *et al.* (1977) showed that the classical Einsele/Mortimer picture does not hold true for anoxic sediment from Lake Mendota. Prairie *et al.* (2002) analysed events in the hypolimnia of 10 lakes located in the Laurentian and Eastern Townships of Québec (Canada). When the hypolimnia of these lakes became anoxic, an increase in o-P was measured together with an incease in Fe^{2+}. Tot-P_{sed} was high in all these lakes (1.6–2.9 mg g^{-1}), and $P_{\rightarrow NaOH}$ and $P_{\rightarrow HCl}$ too. Only in two lakes a constant stoicheiometric release of Fe and o-P was found, with a ratio of $Fe/P = 9–12$ (ww.), consistent with the iron-reduction hypothesis. In a few other lakes P-release took place, but not at a constant ratio to Fe^{2+}. Release of o-P from org-P was suggested by the relation between bacterial activity and interstitial o-P concentration (measured as $P_{\rightarrow NH_4Cl}$). In one lake there was release of NH_4^+ and o-P at a rather constant stoicheiometric ratio, again suggesting mineralization as the source of the released o-P. Lastly, it may be supposed that there was release from the $CaCO_3 \approx P$ in these sediments, as an anoxic hypolimnion automatically becomes more acid.

3.8.4.3   Calculations

Exact calculations can help to understand the P-release problem better. In the first place we have to know how much reducing power reaches the sediment, and, secondly, whether this will indeed release P from the $Fe(OOH) \approx P$ complex or only the excess $FeOOH$.

3.8.4.4   Calculation of reducing capacity

Let us assume a shallow, stratifying lake with a hypolimnion of 5 or 10 m or an anoxic water layer above the sediment of this depth, and a primary production of about $100\,g\,m^{-2}\,y^{-1}$ of C, or $25–35\,eq\,m^{-2}\,y^{-1}$ of reducing capacity. After all other earlier reduction reactions have taken place, only about 2 to 3% of this capacity will be left for the reduction of $FeOOH$, i.e. about $1\,eq\,m^{-2}\,y^{-1}$. (For these calculations, see Table 2.7, taken from Golterman, 1984). This is sufficient to reduce $56\,g\,m^{-2}$ of Fe^{3+}, or about $5–10\,g\,m^{-3}$, which is very near to what can be observed in lakes with an anoxic hypolimnion. Assuming an 'active' sediment depth of 10 cm and a quantity of Tot-Fe in the sediment of about $5\,kg\,m^{-2}$ (i.e. 5% of 100 kg sediment per m^2), we see that only a small part of the Fe (1%) is reduced, insufficient to change the P/Fe ratio enough to cause a release. A deeper sediment layer will even diminish this percentage. A small change in $FeOOH$ concentration in the sediment will have no great influence, as P is adsorbed onto an excess of $FeOOH$. If the active, extractable, $FeOOH$ is about 10% of Tot-Fe, the reasoning does not change enough to invalidate these calculations, because the effect of the small amount of Fe reduced and lost from the sediment will be counteracted by the decreasing pH when the sediment becomes anoxic, which acidification enhances the adsorption onto $FeOOH$.

Goedkoop & Törnblom (1996), studying the constraints of bacterial activity by measuring ^{3}H-thymidine uptake in Lake Erken (Sweden), demonstrated that the production of bacteria was stimulated by deposition of diatom detritus. During autumn, even at decreasing temperature, they found an increase in C-incorporation per gram dw. from 1.1 to $3.1\,\mu g\,g^{-1}\,hr^{-1}$. They concluded that the deficiency of electron acceptors (i.e., the reducing capacity) for bacterial respiration was an important controlling factor, besides low winter temperatures. Goedkoop & Johnson (1996) showed that the increase in bacterial activity even had an influence on the benthic meiofauna (benthic invertebrates retained by a 50 micron mesh). Törnblom & Rydin (1998) continued these studies in an experimental design consisting of cores with a flow-through system. During 20 days bacterial mass and growth (heat production, measured in a microcalorimeter) were measured together with an ill-defined P-fraction. Microbial biomass and activity increased sharply after the addition and deposition of natural seston; a maximum was obtained after 7 days, followed by a linear decrease. Heat production suggested that about 11% of the added C_{sest} was oxidized in 20 days. The P-fractions studied were extracted with cold NaOH during 17 hr and separated in a reactive and a non-reactive fraction; it is not possible to say whether $Fe(OOH) \approx P$ or org-P was involved in the reactive fraction. Phosphate was trapped in the sediment when bacterial activity was high, but released when this activity was low. The identity of the non-reactive $P_{\rightarrow alk}$ remains obscure. The authors discussed the possibility of the presence of polyphosphates or phytate in this fraction, but there is no way of deciding this question from their results. The presence of bacterial polyphosphates in sediment must still be confirmed, that of phytate and humic-P is well established. Nevertheless, the article clearly shows the role of bacteria in P-release, but we have no argument that this release is caused by other factors than by their metabolism causing mineralization of organic matter.

Tuominen *et al.* (1996) followed the chemical and microbiological responses of sediment (selected from 10 m depth in Lake Vesijärvi, Finland) enriched with natural sedimenting material, taken at different moments, or with diatoms or cyanophytes. P_{sed} was labelled with ^{33}P and the added algal material with ^{32}P. They found that the seston collected in May enhanced o-P release, while material collected in July or August did not, nor did the additions of diatoms or cyanophyte material. The nature of the organic substrate may therefore have a strong influence on P-release from sediment. Bacterial activity was inhibited in half of the samples; in those samples no release was found at all. The authors suggested that inhibition of bacterial activity may decrease o-P release because decreased O_2 consumption may lead to increased adsorption onto the sediment. The double isotope technique looks very promising, but must be accompanied by a well-defined extraction technique. In this study, P-fractionation was carried out with 0.1 M NaOH (24 hr), so that no distinction is possible between org-P and $Fe(OOH) \approx P$, and it is quite likely that some org-P was converted via o-P into $Fe(OOH) \approx P$.

3.8.4.5   Calculation of energy needed

Golterman (1984) argued that the stability of the $Fe(OOH) \approx P$ complex protects it against reduction; it is a stabler complex than free $FeOOH$, which is available in excess, while moreover the reducing capacity is usually limiting. The reducing capacity due to sinking dead organic matter in a hypolimnion is usually sufficient for the reduction of only $\approx 1\%$ of the $FeOOH$ present. (See Section 3.8.4.4.) Thermodynamically, more energy is needed to reduce the $Fe(OOH) \approx P$ complex than $FeOOH$ alone. When phosphate is adsorbed onto $FeOOH$, energy is liberated, the amount depending on the Fe/P ratio. When the ratio was near 50 (mM / mM) the energy was about 32 J per mmol; when the Fe/P ratio was about 12 it decreased to 3.4 J per mmol (Miltenburg & Golterman, 1998).

Golterman (1995a) also studied the reaction between $Fe(OOH) \approx P$ and H_2S. In agreement with Einsele's experiments he found P release from $Fe(OOH) \approx P$ only if $\approx 75\%$ or more of the $FeOOH$ was converted into FeS. Even when 100% was converted, still not all P was released. Large quantities of H_2S are therefore needed; with lower quantities there was no P-release at all, although the $Fe(OOH) \approx P$ suspension turned an intensive black. The fact that a sediment is black is, however, no indication that large amounts of FeS are indeed present. A suspension of 20% FeS + 80% $FeOOH$ is as black as a suspension of pure FeS.

In theory FeS and $FeOOH$ cannot co-exist in suspension. Thermodynamically the following reaction must take place:

$$FeS + 2FeOOH + 2H_2O \longrightarrow S + 3Fe^{2+} + 6OH^-. \tag{3.18}$$

In natural sediment the two components may, however, co-exist. The reaction rate is slow: when in our experiments 20% FeS was present, it took generally several days before the suspension became red again, as S became visible. There is, furthermore, a constant new formation of FeS in sediment following the reduction of SO_4^{2-} to H_2S, if sufficient SO_4^{2-} is present. In a few lakes, the SO_4^{2-} concentration may be the limiting factor, but normally it provides – through H_2S formation – sufficient reducing capacity. In estuaries and lagoons under the influence of sea water the supply of SO_4^{2-} is even nearly unlimited. Furthermore, in Golterman's (1995a) experiments and in those of de Groot (1991), indications for the formation of vivianite, $Fe_3(PO_4)_2$, were found if much FeS was formed, which again will not lead to the release of o-P as it is a poorly soluble compound. Formation of vivianite in marsh sediment has been described by Postma (1982). De Groot (1991) reduced FeOOH in Camargue and Garonne sediment suspensions to FeS by adding different quantities of Na_2S and followed the changes in o-P and $Fe(OOH) \approx P$ with time. The o-P increase in the water was found to be smaller than predicted if all $Fe(OOH) \approx P$ was released and an iron-bound phosphate fraction was shown to remain present, even when Na_2S was added in excess. This is supposedly caused by the formation of an iron-phosphate salt: $FePO_4$ or $Fe_3(PO_4)_2$ (vivianite) during the reduction of the FeOOH, while the FeS formed on the outside of the particles may limit P-adsorption. Next, the changes in o-P were studied during the reduction of part of a FeOOH suspension onto which o-P was adsorbed and during the oxidation of part of the FeS in a FeS suspension. These experiments confirmed the results of the experiments with mud. In Section 3.8.4.5 it has been shown that, when $Fe(OOH) \approx P$ suspensions were treated with H_2S gas, only very small amounts of o-P were released, even when the H_2S converted about 90% of the FeOOH into FeS (Golterman, 1995a). Phosphate release under anoxic conditions does, therefore, normally not come from this $Fe(OOH) \approx P$ pool.

Roden & Edmonds (1997) studied P-adsorption and desorption in Fe-rich sediments of freshwater wetland, lake and coastal marine areas. They supposed that microbial FeOOH reduction solubilized 3–25% of P_{sed} in sulphate-free incubations and that precipitation of Fe^{2+}-P complexes occurred, e.g., $Fe_3(PO_4)_2$. Release of o-P into pore water occurred for 33–100% of P_{sed} during anaerobic incubations with abundant SO_4^{2-} and was correlated with SO_4^{2-} reduction and FeS formation. The paper leaves open important questions about the methodology: among others, about P- and Fe-determinations and extractions and P-fractionation. The subject merits attention, but with appropriate chemical methodology.

Clavero et al. (1997) measured the influence of SO_4^{2-} and chloramphenicol on CO_2 and o-P release from sediment under oxic and anoxic conditions, using sediment from the Palmones River Estuary (Spain). Using a mixed reactor in which 5 g of fresh sediment were mixed with 60 ml of overlying water, they showed that the addition of SO_4^{2-} increased the CO_2 and o-P release. P-release rates increased from 4.5 to 35.4 $\mu mol\, m^{-2}\, h^{-1}$ if SO_4^{2-} was added under anoxia and from 0.9 to 17.4 in the presence of O_2; maximal P-release rates were obtained when 3 mmol of SO_4^{2-} were added. When chloramphenicol was added there was no o-P release, either with or without SO_4^{2-} addition. It is clear from these experiments that release by bacterial metabolism is more important than P release from inorganic P-pools. The role of the SO_4^{2-} might be explained by supposing that adsorbing FeOOH particles became covered with FeS, loosing their adsorption capacity, but also that oxidation of org-C by SO_4^{2-} produces less energy than by O_2; the bacteria have thus to oxidise a larger quantity of org-C in order to obtain the same amount of energy. These studies demonstrated the occurrence of certain processes, but extrapolation from 5 g of sediment studied during 2 hr to the in-lake situation is impossible. Large enclosures are more suitable for *in situ* results. Clavero *et al.* (1999) studied the influence of bacterial activity on P-release further in aquariums with the same sediment as before. The aquariums, with mud, were sterilized or not and changes in o-P were followed in the interstitial water (as a function of depth) and in the overlying water. It was found that the o-P concentration decreased when bacteria were present, and that the flux from interstitial into overlying water decreased as expected. The authors assumed bacterial activity to have caused these results, but a noticeable decrease in the pH in the non-sterilized aquariums might easily have caused them. A comparison between measured flux and flux calculated by means of diffusion equations shows a large underestimate for the calculated one, being about 20-fold in the beginning of the experiment, and decreasing towards the end (when flux obviously became smaller). It shows the difficulty of calculating this flux. The authors used a diffusion coefficient of $7 \cdot 10^{-6} \, cm^2 \, s^{-1}$, a value which will be further discussed, with other errors, in Section 5.3.

Clavero *et al.* (2000) compared P-flux from pore water gradients in sediment of the shallow Palmones River estuary with P-release in benthic chambers in the field. In the benthic chambers O_2 flux varied between 62.5 and 87.4 mmol $m^{-2} \, d^{-1}$ and P-flux between 7 and 50 mg $m^{-2} \, d^{-1}$. Pore water o-P and NH_4^+ concentrations were high in summer and low in winter, but the calculated flux was 7 to 24 times lower (for NH_4^+: 21–55 times). This demonstrates several of the problems of the flux calculations, such as the true gradient to be used, the correct apparent diffusion constant and the uncertainty as to porosity (see Section 5.3), while, if the sediment contained $CaCO_3$, the pH shift may have caused overestimation of the P-flux in the benthic chambers – this can easily account for a factor 10. All these problems must be properly addressed when this sort of comparison is made.

3.8.4.6   *Apatite dissolution*

In sediment of eutrophic hard water lakes the presence of $Ca_5(PO_4)_3.OH$ is now well known, but it took till 1976 for this mechanism to become generally accepted (Golterman *et al.*, 1977). The formation of $Ca_5(PO_4)_3.OH$ may be caused by direct precipitation, by co-precipitation with $CaCO_3$ (House, 1990; House *et al.*, 1986) or by adsorption onto $CaCO_3$ particles (House & Donaldson, 1986; De Kanel & Morse, 1978). When sediment becomes anoxic, there is automatically a decrease in pH as a result of the CO_2 produced. This decrease will dissolve the $Ca_5(PO_4)_3.OH$, as the pH strongly controls its solubility. This was first shown by Stumm & Morgan (1970) and later recalculated by Golterman (1998). Part of the released phosphate will be re-adsorbed onto FeOOH which is also present in sediment. It can be demonstrated mathematically that the remaining release depends on the cumulative loading (Golterman, 1998). An example is given in Figure 3.21, where the release is depicted as a function of accumulative loading and Ca^{2+} concentration. Nürnberg *et al.* (1986) and

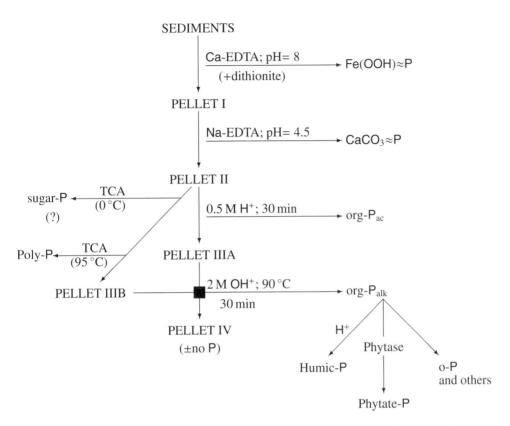

FIGURE 3.22: Fractionation scheme including fractionation of some org-P compounds.

Nürnberg (1988) found a significant correlation ($r^2 = 0.83$) between P-release and Tot-P in sediment, which might well be explained by the phenomenon described above. In the second paper, Nürnberg (1988) demonstrated that $P_{\rightarrow DB}$ correlated with Tot-P release, but contributed only 10% to the P-loss from the sediment. Evidence for P-release from apatite only was presented by Alaoui & Aleya (1995) who demonstrated in benthic chambers in a Moroccan reservoir that P-release was controlled by the pH and the solubility product of $Ca_5(PO_4)_3.OH$. Release from $CaCO_3 \approx P$ is probably more often important than is generally believed. Eckert *et al.* (1997), e.g., suggested P-release from $Fe(OOH) \approx P$ under anoxic conditions in cores of sediment from Lake Kinneret, but their fractionation method (NaOH and HCl) did not distinguish between $Fe(OOH) \approx P$ and org-P, while, as in Mortimers' experiments, the anoxic cores, not being flushed with air, had a more acid pH. This could easily have caused release of $CaCO_3 \approx P$, of which much was present.

It is clear that this acidity-related release did not occur in Mortimer's studies, as the Lake District waters are very soft, and in Mortimer's days P-loading was still low and therefore $Ca_5(PO_4)_3.OH$ was absent.

Brooks & Edgington (1994) suggested that the chemical equilibrium between o-P and

$Ca_5(PO_4)_3.OH$ controlled o-P concentration and algal growth in Lake Michigan (USA).

3.8.4.7   Mineralization of organic matter

When sediment turns anoxic, the aerobic metabolism of the bacterial populations is replaced by anaerobic fermentation. Not all organic matter will be oxidized to CO_2; a large part will remain as fatty acids and comparable compounds. This means that the bacteria, in order to obtain the same amount of energy, have to ferment a larger part of the organic matter, with a resulting P-release. Now of course a thin layer of FeS around the FeOOH particles will prevent P-adsorption onto FeOOH, increasing P-release. There is, however, very little experimental evidence, as the decrease in org-P can only be detected with a good fractionation method, which so far is lacking.

One of the compounds which may be involved is phytate. Its presence is shown in agricultural soil, sediment of lakes and near-shore coastal sediment. Suzumura & Kamatani (1995) showed mineralization of phytate under anoxic conditions, in contrast to De Groot & Golterman (1993), who showed a great stability of phytate due to its strong adsorption onto FeOOH. The quantity of FeOOH in the two different sediments may well be the explanation for this difference. There are few reports on the existence of phytate in freshwater sediment. Goedkoop & Petterson (2000) argued that phytate was never found in deeper lakes, but this is due to the fact that nobody looked for it in deeper lakes. In a few samples from Lake Geneva, I found about $50–60 \mu g\, g^{-1}$ of sediment (as dw.).

There are, often quoted, reports on P-release from organic matter in oxic sediment, e.g., by Boers & Van Hese (1988), Boers & De Bles (1991) and Sinke et al. (1990). The first reference (1988) develops a model for P-release supposedly coming from org-P by mineralization, but is based on three earlier papers, of which the first one (Boers et al., 1984) presents only a supposition of the role of org-P, whereas the reference to the two others is wrong. According to Sinke et al. (loc. cit.) the high correlation between calculated fluxes of NH_4^+, o-P and CH_4 is evidence that mineralization of organic matter is the driving force for P-release from sediment. All these studies used an inappropriate methodology (no proper P-fractionation) and were based on indirect arguments. While in the Boers & Van Hese paper (loc. cit.) temperature was found to be the most important factor controlling P-release, Sinke et al. (loc. cit.), working in the same lake, mentioned that mineralization and P-release were only weakly related to temperature.

Slomp et al. (1993) studied the effect of deposition of organic matter on P-dynamics in marine sediment from the North Sea, with and without addition of algal cells (*Phaeocystis sp.*). Pore water P-concentrations increased upon repeated additions and P-release rates concomitantly increased with a factor 3–5; when the algae were added once, the effect was limited. The increase was attributed to mineralization of the algal cells or to direct release from the algal cells. Macrofauna had no influence on P-exchange, although it did rework the sediment; $P_{\rightarrow NaOH}$, either org-P or $Fe(OOH) \approx P$, increased when the algal cells were added once.

Gächter & Meyer (1993), reviewing bacterial activity in sediment P-cycles, emphasized the role of bacteria in the release processes. Their calculations are not clear, and are based on isolated events or observations and inappropriate P-fractionations from other studies, e.g., in the discussion of the results of Boström *et al.* (1985). The latter authors used NaOH as P-extractant and, therefore, could not distinguish between $Fe(OOH) \approx P$ and org-P. From results in a series of Canadian lakes, Prairie *et al.* (2001) suggested too cautiously that mineralization of org-P may well be a source of P-release under anoxic conditions (see Section 3.8.4.2).

The difficulty of assessing the contribution of organisms to P-release is demonstrated by Brunberg & Boström (1992). These authors followed variations in microbial biomass and activity in sediment in Lake Vallentunasjön during 5 years. They suggested a strong co-variation between biomass of *Microcystis* and P-release, and therefore an important role of microbial processes (although in the last two years P-release increased again, while *Microcystis* did not). Low capacity of the Fe-system in the sediment to bind P was their explanation. In May 1981, $5 \, g \, m^{-2}$ of $P_{\rightarrow alk}$ was, however, present and in August 1990 still $3.2 \, g \, m^{-2}$, i.e., sufficient for an important P-release. Even more $P_{\rightarrow ac}$ was present; part of this may originally have been present as apatite and another part may have been formed by the previous extraction of $Fe(OOH) \approx P$ with NaOH. As there are no data on pH in this paper we cannot calculate the P-release that is possible from this source, but it will certainly have been important.

Brunberg (1995) followed these studies up by enriching sediment from the same lake with *Microcystis* colonies. Decomposition of the *Microcystis* colonies depended much on the physiological status of these colonies. Colonies having survived one year within the sediment and pelagic colonies from late August were rather resistant to decomposition, while those from a declining pelagic population were not. X-ray microanalysis showed that P_{cell} decreased strongly because of a decrease in both org-P per cell and in biomass, viz., mineralization. P-release from *Microcystis* and bacteria was equal to 75% of the decrease in org-P and to 65% of the decrease in $Tot-P_{sed}$ from the upper 1 cm. The fractionation scheme of Psenner *et al.* (1988) was used, which leaves doubt about the true values of org-P. During the extractions with NaOH part of the org-P will be hydrolysed to o-P, and the extraction with dithionite/bicarbonate was not repeated till depletion. The contribution of both bacteria and *Microcystis* was, although only qualitatively, well established.

Jensen *et al.* (1992), basing themselves on the analysis of a large Danish database, suggested that the Fe/P ratio explained 58% of the variation in the rates of aerobic P-release in 15 of these lakes. Lake sediments with Fe/P > 15 (w / w) released decreasing amounts of o-P, while lakes with Fe/P < 10 seemed unable to retain o-P. In about 100 lakes the Fe/P ratio was more strongly correlated with Tot-P than with any other variable. This approach can be much improved by not only regarding Tot-Fe, but also measuring the extractable FeOOH, the main compound for the P-adsorption.

Release from P_{sed} under oxic conditions is a process causing a delayed lake recovery after decreasing the P-load, just as often as P-release under anoxic conditions is mentioned. An example is given by Gächter & Müller (2003), trying to improve the trophic state of Lake Sernpach (Switzerland), who concluded after 15 years of experience that decreasing the external P-load decreased the P-concentration in the lake, but that increased hypolimnetic O_2 concentrations neither reduced the P-release from the sediments nor resulted in an increased P-retention. The authors formulated the hypothesis that oxygenation resulted in less FeS formation, but in more $Fe_3(PO_4)_2$ (vivianite) production. They suggested that the molar ratios of available $Fe^{2+}/S^{2-}/P$ in the anoxic sediment must be more important than the oxic sediment surface. Increased hypolimnetic O_2 concentrations automatically mean different shifts in pH before and after stratification, while decreased primary production due to the decreased P-loading will also affect the pH shifts. The article does, however, not give the relevant data to obtain a better insight into all different possibilities, but is important in promoting the idea that the old Mortimer theory may not be the only one applicable for lake management.

Olila & Reddy (1997) compared P-uptake as a function of redox potential by two sediments, from Lake Apopka and Lake Okeechobee (Sth. Florida, USA). Experimental redox levels were fixed between −235 and 555 mV; in Lake Apopka low levels increased o-P and 'loosely bound' P, but had a minimal effect on $P_{\rightarrow NaOH}$, which the authors correctly considered to be $Fe(OOH)\approx P$ and labile org-P. Calculations showed that $CaCO_3 \approx P$ had to be considered as an uptake mechanism at concentrations $\leq 3.4\, mg\, l^{-1}$, which is unnaturally high. No pH data are given, so it seems likely that some release came from this source, but this can no further be elaborated. On the other hand, in Lake Okeechobee o-P concentration increased exponentially with decreasing redox potential, while $P_{\rightarrow NaOH}$ decreased; it can, however, not be decided whether this was due to a decrease in org-P or in $Fe(OOH)\approx P$. The Okeechobee sediment had a strong affinity for P, both under aerobic and anaerobic conditions, in contrast to that of Lake Apopka. The uptake rate constant was found to be $0.51 \pm 0.05\, h^{-1}$, which is rather high for fixation onto $CaCO_3$. The authors suggested that Tot-Fe, when present in high amounts, is involved in P-uptake and geochemistry even in calcareous systems, but data for extractable FeOOH might have improved our understanding.

3.8.4.8   *Release of bacterial polyphosphates*

Hupfer *et al.* (1995) suggested that bacterial polyphosphates might be responsible for P-release under anoxic conditions. These phosphates are supposedly extracted with NaOH, in which they are present as so-called non-reactive phosphate (Baldwin, 1996; Hupfer *et al.*, 1995). NaOH (cold, 0.1 or 1 M) is very often used as it is supposed to extract $Fe(OOH)\approx P$, which it does, but it also extracts a considerable quantity of org-P.

The chemical evidence for the presence of polyphosphates is, however, weak. The blue colour obtained with toluidine blue is not specific for the phosphoric acid anhydride. "It identifies the mere presence of phosphate, but not phosphoryl groups linked by anhydride bonds" (Kulaev, 1979; Kulaev & Vagabov, 1983). Further methods were ^{31}P-NMR (Nuclear Magnetic Resonance) and scanning electron microscopy, but these methods are only correct if other interfering compounds can be excluded, e.g., by a preceding purification extraction. Phytate also has an NMR spectrum with 4 peaks (see annex I in Golterman, 2001), the shift frequencies of which cannot be compared with those of Hupfer, as different procedures and equipment were used. The round granules shown by Scanning Transmission Electron Micrography in Hupfer *et al.* (1995) might just as well consist of poly-hydroxy-butyrate or, less likely, elemental S in S-bacteria. (Both these compounds may accumulate to more than 30% of the dry weight which might explain the volume of the granules shown.) E. Vicente (pers. commun.) thinks these granules might well be staining artifacts. (A rough estimate based on bacterial numbers in sediment (see Section 3.10.2) shows that bacterial phosphate (including polyphosphate) is not likely to be present in quantities sufficient for an important release.) On the other hand phytate-P has been shown to be present in considerable quantities in all lake sediments studied so far. The concentrations range from about 30 to $150\,\mu g\,g^{-1}$ of dry sediment (De Groot & Golterman, 1993), but the source and fate of the compound remain unknown. The presence of active phytase-containing bacteria in sediment suggests active participation by bacteria. A fractionation scheme separating polyphosphate from phytate is presented in Figure 3.22; this could be applied before and after P-release in *in vitro* experiments. Further possibilities for analysis are gel filtration and chromatography, both however after pre-purification of the extracts.

Waara *et al.* (1993) experimented with ^{32}P labelled *Pseudomonas* bacteria, which were submitted to aerobic and anaerobic conditions in dialysis bags suspended in a water/sediment mixture in a 25 ml vial. P-fractionation of bacteria and sediment was carried out with the H&L method. The major part of the P_{bact} was recovered from the nr-$P_{\rightarrow NaOH}$ fraction, which was supposed to consist of polyphosphate. This is obviously not true, as many biochemical compounds from the bacterial cells will have been extracted by the 18 h extraction with 1.0 M NaOH. (In the old days shaking algae with 1.0 M NaOH was a way of measuring Tot-P_{cell}.) The study clearly showed that under anaerobic conditions only a rather small part of this P was released, not supporting the hypothesis that anaerobic release might come from bacterial polyphosphate. Waara *et al.* (loc. cit.) estimated P_{bact} in the sediment to be about 1.3% of P_{sed} or $9.6\,\mu g\,g^{-1}$, but they assumed a C/P ratio of 18, a rather high value.

3.8.5 KINETICS

Matisoff *et al.* (1981) tried to present a kinetic model for the release process in studies of the anaerobic decomposition of Lake Erie sediment, and showed that a kinetic model which can accurately describe the vertical profiles of Ca, Mg, Fe and Mn, cannot do so for C, N and P. Depositional flux significantly changed the concentration of these nutrients in the top layers of the sediment (Boström *et al.*, 1982).

Gunatilaka (1982) evaluated P-adsorption capacity using ^{32}P and ^{3}H-H_2O labelling. He used the Langmuir equation and showed two different linear adsorption regions with a separation at $P = 50\,\mu g\,\ell^{-1}$. The first region corresponded to a 1 min reaction and the second to a 1–10 min reaction, but the time-related curves are not shown. The isotope technique has a great potential, but the calculations are more complicated than often assumed, as besides adsorption isotope exchange also occurs.

Froelich (1988) described the adsorption mechanism as a reversible two-step sorption process, the first step, being adsorption onto the surface, with fast kinetics and the second diffusion of the adsorbed phosphate from the surface into the interior of the particles. The curves he presented resemble very much a curve like Figure 3.8. Again it looks as if there are two reactions, but this is not sure. The curves can be fitted equally well on a logarithmic and on a power curve:

$$P_{ads} = 2.37\ln(t) + 81.3 \qquad \text{or} \qquad P_{ads} = 81.4 \cdot t^{0.027}, \tag{3.19}$$

both with P_{ads} in % of Tot-P, t = hr, and both with $R^2 = 0.98$. (The number 81 reflects the percentage of P adsorbed already before measurements can start, a matter of a few seconds probably.) A smooth curve without any inclination point is then obtained. The problem is that it seems rather unlikely that an adsorption reaction like (3.8), to wit:

$$FeOOH + H_2PO_4^- \Longleftrightarrow Fe(OOH){\approx}P + OH^- + 2Fe_{inactive}$$

can be treated as a zero-, first- or second-order reaction, as if it were an ionic reaction. Froelich found different rates for the different substrata studies, much the same as we found in our own work with laboratory made FeOOH suspensions; particle size and age of the FeOOH suspensions changed the kinetics much, but after a week of ageing no further changes occurred. House & Denison (1997) described the kinetics of uptake or release of P by bed sediment by the following equation:

$$\frac{dn}{dt} = k \cdot A \left[\frac{(\text{o-P} - \text{EPC}_0)}{\text{EPC}_0} \right]^m , \tag{3.20}$$

where n = the amount of o-P, A = the area of bed sediment, m = an integer and k = the rate constant. House & Warwick (1999), in a fluviarium study, found the P-uptake to last about 1 d, and the P-release about 2 d. With $m = 2$, the constant k in one river varied between 0.02 and 0.28 and in other rivers increased to 12.7 nmol m^{-2} s^{-1}. The authors concluded that the relationship between adsorption affinities/rate constants and sediment properties is not yet clear. It seems likely that taking the solubility product of apatite into account may clarify this problem.

De Groot en Golterman (unpubl. results) found that the adsorption rate of o-P onto CaCO$_3$ suspensions was rather low, but increased considerably if traces of FeOOH were added. It was, however, impossible to obtain reproducible results, because of the problem to obtain reproducible sediment and FeOOH grain size. With FeOOH suspensions 95% of the uptake was finished in 24 hr, after which tailing occurred.

Davis *et al.* (1975), using ^{32}P in core studies, demonstrated a slight influence of burrowing tubificid worms on P-adsorption onto sediment, but no influence on release rate. As processes in lakes usually have to be considered in terms of days if not weeks, it seems that P-uptake processes will be more limited by the kinetics of the diffusion from overlying water into the interstitial water than by the kinetics of the chemical processes. Finally, in lakes, where events are usually measured in the time scale of weeks, it seems rather unnecessary to analyse events in minutes or hours.

3.9   Conclusions

It is possible to define some criteria which must be met in order to ensure that bioassay results are meaningful. The most important ones are:

1) Sufficient cell increase should be obtained. In the control culture (without mud) the final cell number must be at least 300 times the inoculum (this means at least 8 cell divisions, which in practical terms means an experiment lasting at least 8 days).

2) In order to prevent apatite formation, no important pH increase must occur during growth. This means that CO_2 must not be the growth limiting factor but should be supplied in excess. The cultures must, therefore, be intensively aerated.

3) A constant cell level at the end of the experiment is desirable. This may not be obtainable in organic (= humic) rich sediment; continuing increase in cell numbers is then an indication of the slow mineralization of org-P, a problem for interpretation.

4) During at least a short period an exponential growth rate should be obtained.

If these 4 criteria are met, such algal test assays can give important information about the availabilty of P_{sed} and, therefore, for lake management.

It is as yet not possible to quantify the bioavailable P-fraction chemically. Difficulties arise, e.g., fom the detrital apatite and the org-P compounds in the EDTA extracts. Chelating agents are better extractants for P-fractionation than NaOH, HCl and even dithionite/bicarbonate. All extractions must be checked for complete extraction or tailing.

Only experiments such as Mortimer's can lead out of the impasse of P-release under anoxic conditions, but they must be accompanied by a careful chemical P-fractionation. Two schemes based on specific extractions are at present available, the SEDEX method (Ruttenberg, 1992) and the EDTA method (Golterman, 1996a). Care must be taken to add small quantities of reducing agents in such experiments, e.g., once a week. More attention must be given to metabolic P-release; the correlations between NH_4^+ release from sediments and O_2 uptake indicate that breakdown of organic matter in sediment is a quantitatively important process, and when the bacteria have to adapt to anoxic conditions this will not stop, but more organic matter must be mineralized in order to obtain the same amount of energy. Furthermore, guided by the laboratory results, experiments in a lake with an anoxic hypolimnion might be carried out, or events followed, again accompanied by P-fractionation. Release from org-P and/or polyphosphate must be analysed with the appropriate techniques. A scheme to distinguish polyphosphate from phytate (see Figure 3.22) has been proposed by Golterman *et al.* (1998). A specific extraction of polyphosphates can be carried out with first cold and then hot trichloro-acetic acid, but only after removal of Fe and Ca bound phosphates. ^{31}P-NMR and gel filtration or gel permeation chromatography also seem useful methods, but only after pre-purification of the sediment. Extractions of org-P from the EDTA extracts with, e.g., butanol may be used to identify humic or fulvic bound phosphate. The often used extractions with NaOH and HCl are the worst possible ones to solve the problem of P-release from sediment. Laboratory and field studies must always go together, as Baas Becking (1935) suggested already (see Preface). If the necessary techniques are not available, they must be developed.

3.10   Appendices

3.10.1   APPENDIX I

Equation (3.1), $P_{in} = P_{sed} + P_{out} + \Delta(\text{o-P})_w$ has no predictive value for o-P as usually P_{sed} and $\Delta(\text{o-P})_w$ are unknown. Most studies on the relation between o-P and L_P, the P-loading

(P_{in} = $g\,m^{-2}\,y^{-1}$), were based on measurements of o-P_{in} only. If the lake is in a steady state, i.e. if $\Delta(o\text{-}P)_w = 0$, and homogeneous, Equation (3.1) can be rewritten, after algebraic calculation, as:

$$P_{lake} = \frac{L_P}{z(\rho + \sigma)},\qquad(3.21)$$

where P_{lake} = mean o-P concentration in the lake; L_P = annual P-loading; ρ = the water renewal time (τ^{-1}) and σ = the sedimentation rate (t^{-1}). Again, this equation has no predictive value, as σ remains unknown because it is difficult to measure directly. In the past it has been tried to relate it directly to the L_P, but this did not work. Trials were made to derive it from the retention coefficient, but this was a tautology (see Golterman, 1980). Vollenweider (1976) showed a vague statistical relation between σ and ρ and arrived at the well-known equation

$$P_{lake} = \frac{P_{in}}{1 + \sqrt{t_w}},\qquad(3.22)$$

which was, as a result of the OECD (1982) programme, changed into

$$P_{lake} = \frac{P_{in}}{(1 + \sqrt{t_w})^{0.8}}.\qquad(3.23)$$

Statistically this is not amazing: the confidence limits of the OECD model are rather wide, while t_w fell in nearly all of these lakes between 2 and 10 y, so that $\sqrt{t_w}$ fell between 2.4 and 4.15, equal to the range of the OECD data. The dimensions of this equation are physically non-sense, it was a purely statistical approach; its usefulness can be improved by replacing it by

$$\frac{P_{lake}}{P_{in}} = \frac{\sqrt{t_0}}{\sqrt{t_0} + \sqrt{t_w}},\qquad(3.24)$$

with t_0 reflecting a Constant, valid for (a group of) lake(s), just as the numerical value of K_s, the Monod halfsaturation constant, reflects specific bacteria. With $t_0 = 1$ (by chance?), Equation (3.24) is identical to Equation (3.22). The constant t_0 may depend on the chemistry of the lake, i.e., the Ca^{2+} concentration, the pH and the total amount of sediment entering. Rather unfortunately, the OECD data bank was made unretrievable and was never disclosed to the scientific community.

3.10.2   APPENDIX II: TENTATIVE CALCULATIONS OF CONCENTRATION OF TOT-P_{BACT} AND POLY-P_{BACT} PER g SEDIMENT (DW.)

In a few publications I have found the following abundance of bacteria (V. Collins, Kuznetsov): 10^6–10^8 per g dry weight of sediment ($\approx$ 1 ml). Assuming the highest bacterial numbers found (10^8), one can calculate as follows:[5]

---

[5]N.B. $1\mu = 10^{-3}$ mm; $1\mu^3 = 10^{-9}$ mm³; 1 mm³ = 1 mg (10^{-3} cm³ = 10^{-3} ml = 10^{-3} g = 1 mg). A layer of $1\,m^2$ with a depth of 10 cm contains $100 \cdot 100 \cdot 10$ cm³ = 10^5 cm³. Sediment expressed in dw.

assumption	calculation
10^8 bact. per g	
1 bact. = $1\mu\mu^3$ $\longrightarrow$	$10^8 \mu^3$ bacteria (vol.) per g sed = $10^8 \cdot 10^{-9}$ mm^3 g^{-1} dw. = 0.1 mm^3 bact. g^{-1} dw.
dw. = $0.1 \cdot$ ww. $\longrightarrow$	1 mm^3 = 1 mg ww. = 10^{-1} mg dw. per g. Thus, 0.1 mm^3 = 10^{-2} mg bact. g^{-1} = $10\mu g$ bacterial dw. g^{-1} sediment
C = 50% dw. $\longrightarrow$	$5\mu g$ C per g sediment
N = 10% dw. $\longrightarrow$	$1\mu g$ N per g sediment
P = 1% dw. $\longrightarrow$	$0.1\mu g$ P per g sediment

Assuming a polyphosphate accumulation to be 10 times the minimal amount (1%), we arrive at a maximal poly-$P_{bact} = 1\mu g\,g^{-1}$.

The following concentrations are often found in sediment:

$$org\text{-}C = 10\text{--}100\,\text{mg per g};$$
$$Tot\text{-}N = 1\text{--}10\,\text{mg per g};$$
$$Tot\text{-}P = 500\text{--}1000\,\mu g\,\text{per g}.$$

This means that poly-P can be at the most 0.1–0.2% of P_{sed}, with 0.2% being a high estimate.

Kuznetsov gave for C_{live} = 5% of C_{dead} and with C/N $\approx$ 10 in sediment, N_{live} = 0.1% of C_{live}. Thus, with C_{sed} = 20 mg g^{-1}, we arrive at N_{live} = $0.1 \cdot 0.05 \cdot C_{sed}$ = $0.1 \cdot 0.05 \cdot 20$ = $0.1\mu g\,g^{-1}$, and with N/P = 10, we arrive for P_{live} at $0.01\mu g\,g^{-1}$. This is one order of magnitude lower than found above. If we assume again that poly-P may be 10 times the minimal P concentration in bacteria we arrive at poly-P = $0.01 \cdot 10 = 0.1\mu g\,g^{-1}$.

Guesstimate of poly-P release:

The P-release from sediment is often found to be 10–100 mg m^{-2} d^{-1}. If we assume an active depth of 10 cm, we have 10^5 cm^3 = 10^5 g of sediment per m^2.

If the poly-P is indeed $1\mu g\,g^{-1}$, we can have at the most a P-release of 10^{-3} (mg g^{-1}) $\times$ 10^5 (g m^{-2}) = 100 mg m^{-2}. The supposedly present pool of poly-P_{bact} is then sufficient for a P-release during at most 1 to 10 days.

Chapter 4

Sediment and the nitrogen cycle

4.1 Introduction

In contrast to P, N occurs in the aquatic system in 5 oxidation states and there are several processes by which the different compounds are converted into each other. Most of these processes are mediated by bacteria; many of them occur in the mud and many may take place at the same moment. The electron transfer processes are the following:

Oxidation state of N:	$NO_3^- \rightarrow NO_2^- \rightarrow N_2O \rightarrow N_2 \rightarrow NH_3 \rightarrow$ org-N					
Electric charge of N:	+5	+3	+1	0	−3	−3
Thermodynamic stability	←					
Plant preference						→
Ammonification					←	
Nitrification	←					
Denitrification			→			
Dissimilatory NO_3^- reduction				→		
N_2 fixation				→		

Nitrogen compounds enter aquatic systems from different sources. The N concentration of igneous rock is only $46\,g\,t^{-1}$ (46 p. p. m.), while that of some sedimentary rocks may be ten times higher. Larger quantities come from erosion of both natural and artificially fertilized soils (see Table 2.7). These, older, data represent smaller watersheds and the period before N over-fertilization. Present data vary widely because of the large excess of N being used. For the watershed of the river Rhine ($185{,}000\,km^2$) a value between 4000 and 6000 $kg\,km^{-2}\,y^{-1}$ can be estimated from its N concentration and flow rate. In rivers under heavy influence of agriculture, N in the form of NO_3^- may be as high as $10\,mg\,\ell^{-1}$ of N. It is often the main source of N for lakes. In most sediments nitrogen is in an organic form, often > 90%. Although the N concentration in sediment is not particularly low, the *relative* release rates may be much lower than those of phosphate. N-release happens mainly in the form of NH_4^+ and perhaps as some unidentified org-N compounds. The origin of the NH_4^+-release is the decomposition of organic matter. It is known from agricultural research (Russell, 1973) that mineralization of the organic matter becomes more and more difficult when the C/N ratio increases towards 20, as the bacteria are then often N-limited. Finally, part of the org-N does not mineralize at all. A certain quantity of the remineralized NH_4^+ may remain fixed in the sediment owing to the "Cation Exchange Capacity*".

NO_2^- and NO_3^- may temporarily occur in trace quantities in interstitial sediment water, but as neither compound can participate in any adsorption complex, they will most often be converted into N_2 by denitrification, especially in those layers where the O_2 concentration approaches 0, and perhaps into NH_4^+ by dissimilatory NO_3^- reduction. While most of the N entering lakes is in the form of NO_3^-, most of the N in the lake is in the reduced form of R-NH$_2$ (with R = org-C or -H).

Primary production is the process by which the incoming load of NO_3^- is converted to the reduced form in plant material (org-N), for which light energy is needed. This is the main N-source for sediment. Since the N-input into sediment is largely in the form of this org-N deposition, its concentration in sediment depends mainly on primary production and is, therefore, strongly related to the org-C concentration. This was already demonstrated by Mackereth (1966). Mackereth found values for N-concentration around 0.2% of dw., with an org-C concentration of 5–10% in the oligotrophic Ennerdale Water and up to 1.5% in the eutrophic Esthwaite Water (both in the English Lake District), with org-C values of about 15% of dw.; these values, however, are on the high side of the normal range for lake sediment. The C/N ratio for these lakes ranged between 6 (at C = 3% of dw.) and $\approx$ 15 (at C $\approx$ 17%) and was proportional to the org-C concentration, falling off sharply at low C-concentrations. This was explained by glacial clay containing adsorbed NH_4^+ instead of K^+. The same was found for the C/N ratio increasing from 3.2 (at C = 2%) nearly linearly up to > 7 (at C = 17%) which was explained by adsorbed H^+ onto the clay material at low C concentrations.

In deeper waters most of the org-C_{sed} comes from decaying phytoplankton, while other, minor, sources are macrophytes and run-off from soils in the watershed. This day by day sinking and decaying phytoplankton will, while passing through the water column, lose major parts of its org-N and org-C already in the epilimnion, probably about 60–80%, but some 10–20% will enter the sediment, where it provides a continuous input of org-C_{sed} and org-N_{sed}. Input of decaying macrophytes in shallow waters will take place rather more abruptly, so that less material will be mineralized in the water and most of it in the mud.

Usually the organic matter concentration in unpolluted sediment ranges between 0.1 and 5%, while in shallow waters, supporting macrophyte growth, higher values are found. Serrano (1992) and Jaúrequi & Toja (1993) reported values of 40–50% in some sediments of the marshes of the National Park Doñana (Spain) and Paludan & Jensen (1995) in a Danish wetland up to 42% (see Section 3.5). Kleeberg *et al.* (1999) published values of about 35% in the surface sediment of the humic rich, eutrophic Lake Petersdorf (NE-Germany). Kleeberg (pers. commun.) sampled 50 cm long sediment cores from Lake Kleiner Mila and from Lake Großer Mila. In Lake Kleiner Mila, the 'Loss-on-Ignition' (LoI) increased from down-core from 60 to around 70%, and in Lake Großer Mila from 70 to 90% dw.! C/N ratios up to 20 were recorded by Poon (1977) who gave rather constant values for a marine sediment with different degrees of pollution, but lower C-percentages than the ones Mackereth found, in agreement with a probably lower primary production (see Figure 4.1).

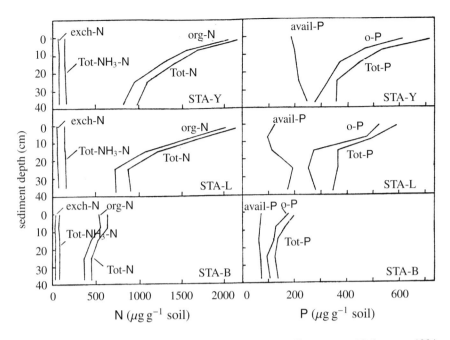

FIGURE 4.1: Nutrient distribution in more and less polluted sediment cores (Golterman, 1984, modified from Poon, 1977).

Mackereth (loc. sup.) also compared the org-C concentration (as measured from the CO_2 recovered after oxidation) with the, more often used, 'Loss-on-Ignition'. His data can be fitted onto the line:

$$\frac{LoI}{org\text{-}C} = 2.0 + 4.05 \cdot e^{-1.95 \cdot org\text{-}C} \qquad (R^2 = 0.85) \qquad (4.1)$$

except for two extremely low org-C values. The deviation from a constant value (theoretically ≈ 2) is caused by some H_2O adsorbed onto the clay minerals, which cannot be liberated by heating to 110 °C. Luczak *et al.* (1997), realizing how often the LoI is, nevertheless, still being used, proposed a standardized procedure, with an ignition temperature of 500 °C during 6 hr. His regression line: org-C $= 0.4321 \cdot$ LoI $+ 0.034$ ($N = 30$; $r^2 = 0.81$) also indicates a deviation at low org-C concentrations.

Sutherland (1998) compared org-C with the Loss-on-Ignition for 6 bed sediment grain size fractions (< 2 mm) from 113 sample sites (therefore $n = 678$) in a tropical stream (Hawaii). Using also the linear regression equation:

$$org\text{-}C = a \cdot LoI + b$$

(where a = slope and b = intercept) and reviewing 9 other publications including 479 data points, but neither Mackereth's nor Luczak's, Sutherland found slopes between 0.405 and 0.96 and intercepts between -9.36 and 0.015. The study showed different values for a and b for the different grain size fractions, a varying between 0.055 and 0.125 and b between -0.328 and 0.350. Negative values for b indicate statistical evidence of poor precision or a

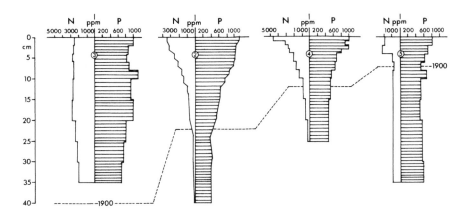

FIGURE 4.2: N- and P-concentrations in Lake Constance (Golterman, 1984, modified from Müller, 1977).

non-linear relationship. The low values for a (often around 0.1 or even smaller) found by Sutherland are difficult to explain, unless it means that much crystal water was still attached to the clays. The study, therefore, indicates that LoI is unsuitable for precise measurement and can only be used for rough estimations, and not at low org-C values or with certain clays, still holding water at 105 °C, present.

Three processes of the N-cycle take place mainly in the sediment: mineralization plus ammonification, nitrification and denitrification. In the first and third processes org-C plays an important role, either as a source of energy or as a reducing agent. Fractionation of org-N containing material has much less progressed than that of phosphate and the problem of its availability has been much less studied. This is nevertheless an important question regarding the growth of macrophytes, especially of crop plants such as rice.

N-concentrations in sediment usually range between about 0.05 and 0.5% (0.5 and 5 g kg^{-1}), the higher values reflecting increased phytoplankton growth. Assuming a sediment input into deep lakes of 0.1–0.5 mm y^{-1}, we arrive at an N-sedimentation rate of about 0.05–2.5 g m^{-2} y^{-1}. Müller (1977) measured N-concentrations in the sediment of Lake Constance (Germany/Switzerland) between 0.03% (pre-1900) and 0.35%, from which a present sedimentation rate of 2 g m^{-2} y^{-1} of N can be estimated (post-1970). During the 20th century, the N-concentrations of the sedimenting material in Lake Constance increased by a factor 5 to 10, roughly equivalent to the 10-fold increase in o-P in the water, causing a 5–10-fold increased phytoplankton production. (See Figure 4.2.) This compares well with an only two-fold P-increase in the sediment, as is indicated by the Freundlich adsorption equation, which shows the sediment P-concentration to increase with the cubic root of 10. (See Section 3.2.1.) org-N$_{sed}$ increases more linearly with primary production than P$_{sed}$, as its source is mainly dead algal material and much less adsorption of inorg-N.

There are only few studies that have measured the N-input into lakes. In the O.E.C.D. (1982) study on primary production and nutrient loading, the participants were so keen to

TABLE 4.1: Summary of data for the N-balance of Lake Baldegg and Lake Zug (North and South Basins). All data recalculated as $mg\,m^{-2}\,d^{-1}$ of N. From Mengis *et al.* (1997).

	Baldegg			Zug		
input	115			24		
output	24			2.8		
Δ inorg-N	−5.6			−2.8		
gross sedimentation	63			35		
net sedimentation	11.2			8.4		
denitrification (*)	85			15.4		
benthic flux	$-NO_3^-$ (f)		NH_4^+	$-NO_3^-$ (f)		NH_4^+
peepers	3.8		29.4–48	0.8–1.6		15.1–16.8
benthic chambers	59		49	21.8–27.3		24.5–37.4
C_2H_2 block (cores)	12.6–16.8					
	NO_3^- ammonification					
$^{15}N - NO_3^-$ (#) (cores)	50–55		2.4			
$^{15}N - NO_3^-$ (#) (chambers)	60		2.8			

(*) estimated from the N-balance; (f): all values negative; (#): isotope pairing.

demonstrate the importance of P that far fewer N-balance studies were done.

As sedimentation is small compared with input, it will be difficult to obtain reliable estimates of N-sedimentation from balance studies. See Section 4.2. In the past these balances have nevertheless been used at the same time, even for the same lake: a) to measure N_2 fixation; b) to measure denitrification, i.e., N-losses; and c) to measure sedimentation. Even if N_{in} and N_{out} are sufficiently known, we still have three unknown variables in the balance equation:

$$N_{in} = N_{out} + N_{sed} + N_{deni} + \Delta N, \tag{4.2}$$

and it is clear that these balances cannot be used to measure one component as long as only one of the other two variables is known.

4.2 N-balances

N-balances are even more difficult to establish than P-balances, because of the two processes involving N_2, viz., N_2 fixation and denitrification. Seitzinger (1988) gave data for N-burial and denitrification. It was suggested that each of them accounts for about $20 \pm 10\%$ of the N-input – with denitrification being a definite loss and N-burial being the source of future ammonification.

Mengis *et al.* (1997) presented N-balances of two eutrophic deep Swiss lakes, Baldegg and Zug. The relevant data are in Table 4.1. Denitrification seems to be a big loss, while the authors considered N-sedimentation (35–63%) also to be a loss. Denitrification was, however, measured from the mass balance and contains, therefore, a considerable error. Estimating the errors in the in- and output values to be 25%, a probable error of 50% can be calculated for the denitrification value. Net sedimentation was measured separately, but not many data are available and it is not clear why this was considered to be a loss, as NH_4^+ release from the sediment was mentioned to be 3–5 times higher. Dissimilatory NO_3^- reduction (see Section 4.5) appeared to be a minor factor in these lakes. Some processes were measured only a few times, even in different years, and at 1 or 2 sites only; thus it is dubious whether these values can be compared with the in- and output data, which are valid for the whole lake. Denitrification, including ammonification and coupled nitrification, are strongly season dependent processes, and values relevant for the whole lake are difficult to obtain.

Probably the most complete N-flux and N-balance model is that of Valiela & Teal (1979), who measured all important N processes in the wetland Great Sippewissett Marsh (Massachusetts, USA), the results of which were published in a number of papers summarized in 1979. They showed that quantitatively the most important processes were N-input, with both ground and surface water, N-output towards the sea, N_2 fixation and denitrification (see Table 4.2). Most processes were measured 10–12 times a year and mean values were taken. As the results of their experiments vary irregularly over time, one has to know the factors which control the variability in order to take a weighted mean. The variability and its controlling factors were, however, not taken into account. For example, Kaplan & Valiela (1979) measured denitrification, the variability of which depended largely on temperature, but it is clear from their results that other factors – viz., the reducing factors org-C or FeS, may be equally important. We have to know the effect of these controlling factors quantitatively to be able to apply them to other wetlands. The difficulty of quantifying the denitrification process is demonstrated in Section 4.4.

A careful study of a N-balance was made by Turner et al. (1983) in Lake Talquin (Florida, USA; $40 \, km^2$, $z = 4.1 \, m$, $z_{max} = 15 \, m$). Water, Cl^-, SiO_2, P and N balances were made over a three-year period. Over the whole period, Cl^- input and output balanced (49.3 and $49.4 \cdot 10^6$ kg, resp.), but within-year discrepancies of 11% were found, equal to the discrepancy in the water balance. The authors showed that the overall N-retention over this period was about 10% and concluded that balance studies can probably never be used to measure N-retention without measurements of sedimentation rate. In a summarizing table, they presented estimates for several lakes and current N-retention coefficients ranging between −0.1 and 0.4, with uncertainties of probably about 0.2.

Garnier et al. (1999) mentioned a deficit in the N-balance of three reservoirs in the Seine (France) watershed of 10–55% of N_{in} and supposed this to be equal to the denitrification. No standard deviations are given; it seems, therefore, that the lower values are not really reliable. Furthermore, the NH_4^+ balance showed great losses (100–350%), so that interconversions in the N-cycle must be taken into account as well.

Risgaard-Petersen et al. (1998) tried to establish an N-balance in sediment of Løgstør Bredning (Limfjorden, Denmark) of an eelgrass (Zostera marina) bed, but the quantities did in fact not balance. N-pools in vegetation and sediment were determined in April (low biomass, high production rate) and in August (high plant biomass). Vegetated sediment fixed

Table 4.2: Nitrogen budget for Great Sippewissett Marsh (USA). From Valiela & Teal, 1979 (rounded off values, in kg ha^{-1}; minor processes omitted).

Processes	Input	Output	Net exchange
Precipitation			
NO$_3$-N	110		
NH$_3$-N	70		
DON	190		
Part-N	15		
Total	380		380
Groundwater flow			
NO$_3$-N	2920		
NH$_4$-N	460		
DON	2710		
Total	6100		6100
N$_2$ fixation	3280		3280
Tidal water exchange			
NO$_3$-N	390	1210	
NH$_4$-N	2620	3540	
DON	16300	18500	
Part-N	6740	8200	
Total	26200	31600	5350
Denitrification		4120 (+2820)	−6940
Sedimentation		1300	−1300
Overall total	35990	39860	−3870

N in April and August; the N-pool in the vegetation doubled in this period. Furthermore, stirred benthic chambers with and without vegetation were used. The N-balance showed a large input of N$_{part}$. NH$_4^+$ and NO$_3^-$ fluxes were mainly determined by the vegetation and accounted for 60% of the plants' requirements. Coupled nitrification/denitrification, i.e., the denitrification of NO$_3^-$ produced by nitrification of NH$_4^+$ released by the org-N$_{sed}$, was studied with a ^{15}NH$_4^+$ perfusion technique and showed that the substrate for denitrification was mainly the NO$_3^-$ in the water and not that produced by nitrification. N-losses occurred mainly with leaf export. Denitrification was low in bare sediment, but increased in the rhizosphere in April, although it did not balance N$_2$-fixation; neither of these two processes was quantitatively important.

Mass N-balances in rivers are rare, and as they depend largely on the flow rate, which is often more imprecise than the chemical data, the low precision limits their utility. A careful balance was published by House & Warwick (1998) for the river Swale catchment area (Yorkshire, UK) during three campaigns in 1995/6. They showed an important loss of NH$_4^+$ and an increase in NO$_3^-$ which was, however, not sufficient to account for the NH$_4^+$ loss. In their N-balance a great uncertainty, therefore, remained.

Thus, the precision of N-balances depends, in the first place, largely on the precision of the waterbalance, as the concentration of the element involved can be measured more precisely. In the second place it depends on the knowledge of interconversions between different N-compounds.

4.3   Mineralization and ammonification

4.3.1   RELEASE OF NH_4^+ FROM SEDIMENT

At the sediment/water interface a considerable upward flux of NH_4^+ takes place, while upward flux of NO_3^- and NO_2^- hardly occurs; for these compounds the downward flux is more important. The release of N-compounds has been studied, like P-compounds, with benthic chambers, core experiments, including analysis of porewater profiles, and *in situ* studies. A difficulty not encountered in P-studies is that the NH_4^+ released is rapidly oxidized to NO_2^- and NO_3^- if O_2 is available, while other problems met in P-studies also occur in N-studies. The pH shift due to the release of CO_2 has, however, no importance for N-release.

The source of NH_4^+ released by sediment is the decomposition of organic matter, but which compounds are involved is not clear. We distinguish two kinds of organic matter in sediment: on the one hand 'humic compounds' such as humic acid, fulvic acid and humin compounds, and on the other hand the rest. Humic matter is yellow or brown; the original definition of humic matter was the brown material extractable with $NaOH$ from soils. There is a controversy between organic chemists and ecologists. The first consider humic material to be nearly N-free, while ecologists, thinking more of its function, will consider that some N, probably in the form of attached amino acids and such like, must be part of these molecules. (See Povoledo & Golterman, 1975.) The major problem in improving our knowledge of NH_4^+-release is that we know so little of the chemical nature of the material involved. Therefore nearly all studies on N-fluxes from the sediment focus on the released quantities and not on their origin, while studies on fluxes into the sediment focus on the quantities disappearing and not on their fate.

Dead aquatic plant material is remineralized by bacteria and zooplankton – several studies have addressed the processes. Dying phytoplankton, when not digested by animals, releases P and N, the first by autolytic action, the second mainly by bacterial breakdown (Golterman, 1975b). The autolytic processes are enhanced by the cell membrane becoming permeable causing a release of dissolved compounds. Phosphatases will then activate hydrolysis of org-P instead of their synthesis. Dead algal cells then lose their proteins rapidly, in a few days at the most. They are lost by the time the cells have sunk through epi- and metalimnion. During transport through the metalimnion and hypolimnion, other decomposing processes continue and what finally enters the sediment has a composition very different from phytoplankton; part of it will have been converted to humic material. The mineralization of org-N is due to the fact that phytoplankton contains a high concentration of proteins, with C/N ratios of ≈ 10, which will be used mainly as an energy supply for proteolytic bacteria and/or zooplankton. Its C- and N-compounds will be remineralized to NH_4^+ and CO_2 to deliver energy, and only a small part will be used for the new bacterial biomass (see Figure 4.3). For macro-

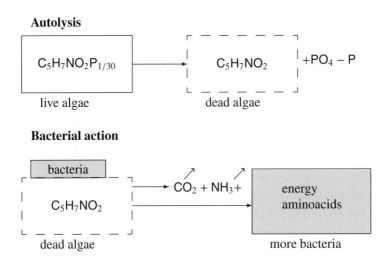

FIGURE 4.3: Schematic presentation of autolytic and bacterial decomposition processes on dead algal material.

phytes the situation is different: because of the high C/N ratio (> 20) in this plant material N is needed in order to make bacterial growth possible. The usual approach to the study of mineralization of this coarse material is to put dead material in so-called litter bags and to follow the decrease in dry weight in time, as is often done in terrestrial ecology. In this way Carpenter & Adams (1979) followed the influence of temperature and N-concentration on the decay rates of coarse particulate organic matter. They found decay rates of this material in the range of 0.01–$0.141\ d^{-1}$. The N-concentration in both water and tissue and the temperature had strong effects; for the temperature a $Q_{10} = 3$ was found, while extra added P had no influence. Furthermore, they showed wide differences between experiments in the field and in the laboratory, supposedly due to animals present in the field but not in the laboratory. Remineralization of the different constituting elements was, however, not studied.

A few examples of techniques used to analyse N-fluxes from sediment must be considered.

De Beer *et al.* (1991) used ion-selective micro-electrodes (tip size $7\ \mu$m) to measure NH_4^+ profiles in two different types of sediments (sandy and silty) of Lake Vechten (the Netherlands) and found a spatial resolution (10 points per mm!) superior to measurements of NH_4^+ after slicing the cores (2.5 or 5 mm slices). The values obtained from the slices closely approached the mean value of the micro-electrode data over the same distance, the difference being the same as between electrode measurements made a few m apart. The obtained flux data were supposedly confirmed by laboratory incubations in cores, but an interpretation problem in the calculation of diffusive flux is discussed in Section 5.3.4. In sandy sediment NH_4^+ was only consumed, while in a silty sediment both NH_4^+ production and consumption were observed depending on O_2 conditions. In the silty sediment, NH_4^+ production by anaerobic mineralization was not balanced by NH_4^+ consumption in the oxic surface layer.

Excellent laboratory apparatus to measure N-release was published by Schroeder *et al.* (1992). It consists of a thermostat in which 6 sediment cores can be placed, each with a diameter of 19 cm, and a total sediment area of 1700 cm^2 which can be flooded with $\approx 25\ell$ of water. It has the advantages of a large sediment surface, large water and gas volumes, a regulated laminar water flow above the sediment, and measurement and regulation of O_2 and other variables. The results are presented later (see Section 4.3.2).

NH_4^+-release from org-N and subsequent adsorption onto the sediment was studied by Kamiyama (1978) and Kamiyama *et al.* (1978), by estimating the rate of NH_4^+ generation. They showed that half the org-N arriving at the lake bottom was released as NH_4^+ and the other half accumulated as org-N_{sed}, but some more experimental data are needed to obtain a complete picture. They found that the ratio of NH_4^+ in the interstitial water over that in the whole sediment depended on the salinity of the water; comparisons between fresh and salt water should, therefore, be made only with great care; the equilibrium between interstitial and overlying water is probably influenced by other cations and pH as well.

Jones & Simon (1980, 1981) and Jones *et al.* (1982) examined the release of NH_4^+ in the sediment of Blelham Tarn (Lake District, UK), paying attention to the turn-over rate of org-N. Laboratory measurements showed that NH_4^+ accumulation, SO_4^{2-} reduction and methanogenesis were more active in the profundal than in the littoral sediment. Accounting for the relative surface areas of the profundal and littoral sediment, methanogenesis and SO_4^{2-} reduction were more important in the former ones, while NO_3^- reduction and aerobic respiration were greater in the littoral zone. They showed an increase in the C/N ratio when seston sinks, continuing when it enters the sediment, which leads to the low N-content of humic material. Redfield numbers have therefore no significance in sediment release studies. Further studies by Jones *et al.* (1982) showed a long turn-over time for added proteins, a shorter one for amino acids and a still shorter one for urea. Combining the turn-over times with pool size, they arrived at rates of NH_4^+ production of 3.5–7 mg kg^{-1} d^{-1} in both littoral and profundal sediment, only equalled by amino acids as N-source in the whole sediment. They took into account the role of urea in the N-cycle, which role was only established since the nineteen-eighties. They used, however, the analytical method of Newell *et al.* (1967), which does not seem to be specific for urea. Jones *et al.* (1982) argued that estimates of available protein are required before the decomposition rate can be truly estimated. Although the nature of the decomposing org-N compounds remains unknown, they are certainly not proteins (except for a small part in the bacteria). It seems unlikely that this problem will soon be solved. Jones & Simon (loc. cit.) used decomposition at high temperature to distinguish between chemical and biological processes, but the high temperature probably decomposed some organic matter and released the adsorbed NH_4^+. They demonstrated the presence of adsorbed NH_4^+ by exchange with KCl. This adsorption may cause a lag phase between production and release of NH_4^+. Reduced profundal sediment showed no adsorption at all.

Takahashi & Saijo (1982) modelled the mineralization of sinking particulate matter, using a one-dimensional diffusion-advection model for biological and chemical processes, and the seasonal changes in dissolved and particulate org-N. Eddy diffusion was calculated from the vertical temperature profiles. They estimated the decomposition rate of org-N from the NO_3^- produced by subsequent nitrification and arrived at a value of 0.8 μg ℓ^{-1} d^{-1} and a rate constant of $0.086\,d^{-1}$. Therefore, the amount of org-N_{part} decomposed per day was 8.6% in the metalimnion. They pointed out that the C/N ratio of organic matter decomposing in the hypolimnion was already much higher than the average value for phytoplankton. The study showed mainly that the inorg-N produced in the hypolimnion is not only derived from the sediment, but even from sinking material before it arrives in the sediment. The authors gave a careful discussion of all errors involved in these studies, one of which is the sinking velocity; identification or fractionation of N-compounds, during the sinking process, would greatly improve our insight. The measured breakdown rate in these studies is an average for different groups of compounds, each with a different rate which cannot as yet be identified.

Fukuhara & Sakamoto (1987), whose experiments on P- and N-release have been discussed in Section 3.8.1, measured N-release and the influence of tubificids and chironomids. N-release was mainly in the form of NH_4^+, although later some denitrification seemed to occur. N-release rate increased with animal density from ≈ 24 to $\approx 48\,mg\,m^{-2}\,d^{-1}$ (tubifex experiment) and from ≈ 20 to $\approx 80\,mg\,m^{-2}\,d^{-1}$ (chironomus experiment). It is not clear why, in the two controls without animals, N-release was significantly different and why denitrification occurred only in the chironomid experiment. N-release by bioturbation was more important than that by excretion by the animals.

Ullman & Aller (1989) measured NH_4^+ (and other elements') release rates of 6.3 up to $14.1\,mg\,m^{-2}\,d^{-1}$ of N in fine-grained sediment of the shallow Saginaw Bay (Lake Huron, USA). The N-release rate was highest in the 0–1 cm sediment layer and decreased with depth. Using the change in solute concentration at different depths in incubated samples during 17 days, net release rates were fitted as a zero kinetics function of depth with the equation:

$$R_{meas}(z) = \frac{R_0 \cdot e^{-\alpha z}}{1 + K}$$

with R_{meas} = the measured rate, R_0 and α are constants and K = the linear adsorption coefficient. The authors stressed the importance of benthic organisms (mainly *Chironomus*) for maintaining these high diffusion rates. These rates cover the values often found in lakes and wetlands and demonstrate, as for P-release, the importance of bioturbation. However, depletion of the substrate, org-N_{sed}, must finally decrease these rates after the incubation period, and extrapolation to longer periods is therefore not possible.

Andersen & Jensen (1991) showed the influence of fourth instar larvae of *Chironomus plumosus* on nutrient release in sediments of Lake Arreskov (Denmark, see also Section 3.8.1.4). They emphasized the point that the animals may increase O_2 uptake and CO_2 release, not only by their own metabolism (which may account for 8–15%), but much more by their pumping activity, causing an extra O_2 and sometimes NO_3^- supply to the sediment. In their study, initial O_2 uptake and flux rates of NO_3^- and o-P were directly related with the number of chironomids up to a density of $2380\,m^{-2}$. Addition of ≈ 1000 animals m^{-2} doubled the O_2 uptake rate. Opposite effects were seen for o-P: uptake of o-P was stimulated in winter, while release was seen in summer.

Svensson (1997) studied the influence of 4th instar larvae of *Chironomus plumosus* on O_2 uptake, NH_4^+ flux and denitrification in 7–9 cm cores of eutrophic sediment of Lake Sövdesjön (Sweden). About 2000 ind. m^{-2} were added, reflecting the density *in situ*. Denitrification was measured with both the acetylene blockage and the isotope pairing-technique. O_2 consumption increased 2-fold when larvae were present; the NH_4^+ efflux then became important as well. Larval excretion accounted for only 11–45% of the N-efflux, showing the importance of bioturbation. With increasing NO_3^- concentrations, denitrification of the NO_3^- coming from the water increased to a greater extent in the turbated than in the non-turbated sediment. Coupled nitrification/denitrification was unaffected. The acetylene blockage technique underestimated denitrification by $75 \pm 12\%$ compared with the isotope technique.

Svensson *et al.* (2001) quantified the influence of the oligochaetes *Limnodrilus sp.* and *Tubifex tubifex* on nitrification, NO_3^- consumption and denitrification rate in sediment of Lake Ringsjön (Sth. Sweden). Rates were measured in undisturbed sediment and in sieved cores to which different quantities of animals were added at high and low NO_3^- concentrations. NO_3^- consumption rate was positively correlated with oligochaete biomass with a slightly higher slope at high NO_3^- concentration, but was significantly higher in the undisturbed cores than that in the sieved cores with corresponding biomass. Nitrification followed a second-order polynomial regression with biomass at both NO_3^- concentrations, but was higher at the higher NO_3^- concentration, with a higher maximum. Denitrification rate, depending on NO_3^- from the overlying water, was higher at the higher NO_3^- concentration and was linearly correlated with biomass. In the undisturbed core it was similar to the rate in the sieved core at both NO_3^- concentrations. Coupled denitrification was significantly correlated to the biomass at low, but not at high, NO_3^- concentration. Sieving and homogenizing sediment had no effect on denitrification and nitrification rates compared to undisturbed sediment. In an extensive overview of the relevant literature the authors concluded that oligochaetes are less effective in mobilizing NO_3^- to deeper sediment layers than tube-dwelling chironomids, owing to the different ways in which the animals bioturbate the sediment layer. Quantification of bioturbation can therefore only be done after careful measurements in the lake in question.

Chironomids and other benthic animals have a strong impact on the N-cycle, stimulating NH_4^+ release and denitrification. In all these experiments extrapolation from laboratory experiments to field situations remains, however, impossible, because of the wide differences caused by the water movement, the change in the sediment/water ratio and the duration of the experiments, which is always short compared with the duration of the anoxic periods in hypolimnia. Depletion of substrate will certainly be an important factor there, modifying all the process rates.

The complexity of the org-N compounds and their reactions contrast sharply with that of P-compounds and must, therefore, be unravelled by using ^{15}N-labelled compounds. The advantage is in the first place that ^{15}N is a stable isotope and that there is much less physical exchange than is the case for phosphate isotopes, but the problem remains the preparation of natural source material. Using ^{15}N techniques, Blackburn & Henriksen (1983) studied N-cycling in different sediments from Danish marine waters. They measured an NH_4^+ production of around $0.7 \, mg \, dm^{-2} \, d^{-1}$ and showed that the active sediment layer is no more than $10 \, cm$ deep. The difference with Jones' & Simon's values illustrates the range of values that can be expected. Blackburn & Henriksen (loc. sup.) furthermore demonstrated that the rate of incorporation into N_{sed} was related to the decomposition rate, and assumed incorporation to be caused by bacterial uptake. Sterile experiments may prove this hypothesis to be true; their

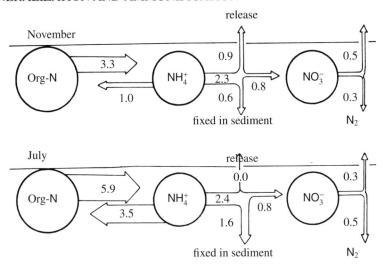

FIGURE 4.4: Model of N-cycling in sediment. Rates in mmol m^{-2} d^{-1} (Golterman, 1984, modified from Blackburn & Henriksen, 1983).

technique excluded adsorption (which was separately measured) as a possible mechanism for the incorporation, but not other chemical processes. The distinction between chemical and biological incorporation is not clear from this paper. The authors showed increasing C/N values with increasing depth, both in the water and in the sediment column. As they measured nitrification and denitrification, a fairly complete picture of the N-cycle was provided. (See Figure 4.4.) In their paper the importance of the adsorbed NH_4^+ pool is also stressed: it delays the NH_4^+ flux. In oxygenated waters, where the equilibrium NH_4^+ in the water is constantly being removed by nitrification, the adsorbed pool is less important.

Little is known about the nature of this adsorption mechanism; clays such as illite seem to be the only likely adsorbent, NH_4^+ being exchanged for K^+ (Logan, 1977). In a Camargue soil Golterman *et al.* (1998) found that NH_4^+ adsorption could, like phosphate adsorption, be described by a Freundlich equation:

$$(NH_3\text{-}N)_{sed} = 0.023 \cdot (NH_3\text{-}N)_w^{0.86} \qquad (r^2 = 0.9; N = 15). \qquad (4.3)$$

The high exponent (0.86) demonstrates the large capacity for NH_4^+ adsorption, although these soils contain only $\approx$ 30% of clay, most of which is illite, not a strong adsorbent.

Several studies analyzed the N-release from decaying macrophytes.

Hill (1979) followed uptake and release of nutrients by dead aquatic macrophytes and published decay rate constants of 0.03–0.04 d^{-1}, being the same for dry weight, N, P and C; why these rate constants were identical was not explained.

Forés *et al.* (1988) showed that the decomposition rate of rice-straw is much lower ($k \approx$ 0.008 d^{-1}), because of the high mineral content (mainly SiO_2) of the rice-straw. After 139 days of decomposition 75% of the biomass, 70% of the C, 50% of the N and 30% of the P remained in the rice detritus. Furthermore, they showed that the loss rate(N) < loss rate(C) < loss rate(P). The P-loss rate was high in the first 10 days, after which a small uptake took place, probably due to bacterial growth.

Menéndez *et al.* (1989) studied the decomposition of *Ruppia cirrhosa* in litter bags of 0.1 and 1 mm mesh. They showed that the nutrient loss rate could be described by a negative exponential function and was highest in the 1 mm mesh bags, where macroinvertebrates (*Gammarus aequicauda* and *Sphaeroma hookeri*) could enter. The rate constants were high during the first three days, and reported not to be different for C-, N- or P-decomposition, which was true for the 1 mm mesh bags, but not for the 0.1 mm ones.

In a critical discussion, Wieder & Lang (1982) pointed out that studies to measure decay rates in the field should be set up differently from studies trying to analyse the factors controlling these decay rates. In the latter case the set-up should allow a test of variance, which asks for a different approach.

The mineralization process of macrophytes in nature takes a long time, especially as after macrophyte death the temperature is often low. During this time burial of sediment may withdraw the org-N from the available layer, making quantification at this moment impossible.

One may wonder, however, how C and N can have the same disappearance rate, as is sometimes found, and whether the org-N will be remineralized to NH_4^+ or will partially enter other compounds of the org-N pool in the sediment. The bioavailability of org-N compounds in sediment is not equal, but fractionation of N-compounds is much less advanced than that of phosphate. Usually a separation is made by hydrolysis either with $NaOH$ or H_2SO_4, yielding two or three fractions, but the duration of the hydrolysis has not been standardized. Bonetto *et al.* (1988) showed that when a fractionation of the org-N_{sed} was carried out by a sequential extraction with $NaOH$ and H_2SO_4, three fractions with a different bioavailability for rooted plants (mainly rice) could be obtained. In this study the necessity of reaching a plateau, just as for P-fractionation, was formulated for N-fractionation. With cold $NaOH$, a plateau was obtained after 7 d, and only after this first extraction a second plateau could be obtained with H_2SO_4 at 97 °C after 30 hr. ($NaOH$ is supposed to extract humic-N, and H_2SO_4 is supposed to extract the so-called amino acid-N.) The cold $NaOH$ extracted a plateau value of $\approx 18\%$ of Tot-N, but the H_2SO_4 did not reach a plateau. From a Camargue sediment, 50% of Tot-N was hydrolysed in 13 hr if no $NaOH$ extraction had been carried out, but in 300 hr if it followed $NaOH$ extraction. 22% of the Tot-N was then found in this fraction, in contrast to the 49% without $NaOH$ pre-extraction. Bonetto *et al.* (loc. sup.) showed that the $NaOH$ and H_2SO_4 fractions do overlap to some extent. The one most biologically available was the residual fraction, at least when using rice as the test organism. These authors have, however, so far been unable (because of analytical difficulties) to find out into which pool the mineralizing org-N enters before it becomes NH_4^+, which may take a considerable time. Another unexpected finding of these studies was that the addition of inorg-N did not stimulate mineralization of rice-straw, contrary to a generally accepted idea in soil science (Russell, 1973). Moreover, Golterman (1995c) could not demonstrate any importance of the C/N ratio for the mineralization process. Leucine and lysine, each with $C/N = 6$, showed a quite different mineralization rate. The amino acid composition may be a factor influencing the bacterial growth on material with high C/N.

The NH_4^+ released by sediment may be an important direct N-source for plant growth. For example, Reddy *et al.* (1988) labelled a consolidated peat sediment (Lake Apopka, Florida) with $^{15}NH_4$-N, causing a slight enrichment, and noted that during the first 30 days water hyacinth metabolized 48% of its N-demand from NH_4^+ in the sediment, which decreased to 14% after 183 days, although absolute quantities were still increasing. About 25% of the sediment NH_4^+ was released into the overlying water, of which 17% was assimilated by the water hyacinth. Pore water NH_4^+ concentration was about 40 mg ℓ^{-1} below 25 cm sediment depth, decreasing to 0 at the top; these profiles did not change with time. The overall NH_4^+ flux was estimated at 48 mg m^{-2} d^{-1}, the soluble org-N flux at 58 mg m^{-2} d^{-1} and therefore the Tot-N flux at 106 mg m^{-2} d^{-1}, which fluxes are among the highest recorded.

4.3.2 PROCESSES INVOLVED

In this part I shall try to analyse the N-release processes further by comparing them with processes concerning other relevant variables (elements).

NH_4^+ produced in sediment has 4 different fates:
1) accumulation in pore waters and/or adsorption onto C.E.C.;
2) loss by upward diffusion, followed by
3) oxidation;
4) assimilation by vegetation.

Pore water analysis is a powerful tool for the study of N-dynamics – the advantage being that the analysis reflects values occurring naturally. In principle, N-fluxes can be calculated from pore water profiles, but with all the difficulties already discussed for P. Pore water analysis gives a better insight into the production by mineralization than N-release studies in cores or aquaria.

In a detailed study Val Klump & Martens (1989) analysed pore water in sediment of the Cape Lookout Bight (N. C., USA). Profiles of NH_4^+, o-P, ΣCO_2 and SO_4^{2-} showed clear seasonal cycles depending on temperature and org-C supply. Production and consumption rates and their depth and temperature dependence were combined with adsorption, diffusion coefficients and porosity to produce a diagenetic model predicting profiles from the initial conditions in spring. The model was based on the principle of the following equation:

$$C = (C_0 - C_\infty)e^{-\alpha \cdot z} + C_\infty,$$

where C_0 = the concentration at $z = 0$ and C_∞ = the asymptote for greater depth and α is a constant. Overall agreement was good, although sometimes deviations were found. Finer structures than 2 cm could not be analysed, a minimum NH_4^+ production rate could not be predicted, while for ΣCO_2 and SO_4^{2-} only 3, respectively 4, points were used. A combination with micro-electrode data seems a further possibility for studies, but commercial electrodes are often not sufficiently sensitive, while home-made ones are not free from interferences (Mäkelä & Tuominen, 2003). A few critical notes are in order here: it was incorrectly assumed that adsorption of NH_4^+ and o-P can be described by linear adsorption equations (this is only true for concentrations that are low compared with the K_s values of the adsorption equations); porosity data were not measured, but obtained by curve fitting. Checking by measuring Cl$^-$ diffusion might have revealed important information about the real value of the porosity.

Sumi & Koike (1990) studied ammonification and NH_4^+ assimilation in slurries of surface sediment of Japanese coastal and estuarine areas using $^{15}NH_4^+$ tracers. By the addition of known amounts of $^{15}NH_4^+$ its dilution can be measured (Blackburn & Henriksen, 1983) and used for the calculation of its production under aerobic conditions. Under anoxic conditions and in a closed system the pool size of NH_4^+ is determined by production and assimilation by benthic organisms. Under semi-aerobic sediment, where no denitrification occurred, a mathematical model was used to estimate ammonification rate. Incorporation of $^{15}NH_4^+$ was used to estimate NH_4^+ assimilation. Ammonification ranged from 80 to $\approx$ 3000 ng g^{-1} h^{-1} (dw.) and assimilation from $\approx$ 85 to $\approx$ 1600 ng g^{-1} h^{-1}. Rate values were found to depend on the supply of easily oxidizable org-C and O_2, while a correlation was found with the ATP concentration, as a measure of the metabolic activity of benthic organisms.

Schroeder et al. (1992) measured release rates of all inorg-N compounds from sediment of the river Elbe (Germany). Under aerobic conditions they found a NO_3^- flux between 56 and -49 mg m^{-2} d^{-1} of N, while under anaerobic conditions the flux increased to -10 mmol m^{-2} d^{-1} of N, the negative values indicating denitrification. Nitrification was limited to the zone of O_2 penetration, which was estimated to be only a few mm. O_2 flux was ≈ -14 mmol m^{-2} d^{-1}. After spiking with 0.25 mM NH_3, the O_2 flux was -6.5 mmol m^{-2} d^{-1} and the O_2 was consumed by nitrification, while NO_2^- was formed. The NH_4^+ flux into the sediment was nearly equivalent to the O_2 flux. NH_4^+ flux from the sediment was only found under anoxic conditions. N_2O was formed within the sediment under both aerobic and anaerobic conditions and its flux was highest when O_2 concentrations were low (0.1–0.6 mg ℓ^{-1}).

Vidal et al. (1992) studied NH_4^+ release and O_2 uptake at 7 stations in a Mediterranean estuarine bay (Ebro Delta, Spain), over a gradient imposed by freshwater input. NH_4^+ release fluctuated between 0.14 and 2.1 mg m^{-2} h^{-1} without any spatial pattern, while O_2 uptake decreased clearly over the transect. The authors noticed a correlation with the Fe/Mn ratio, but an equally strong correlation can be calculated between org-C and NH_4^+ release from their data ($R^2 = 0.8$), if the station closest to freshwater is omitted. Vidal et al. (1997) compared NH_4^+ flux and O_2 uptake in triplicate bell-jars at three stations with different org-C loadings in the same estuarine bay. Stirring inside the bell-jars strongly influenced the exchanges and was therefore adjusted to the minimal value at which stratification was prevented and resuspension avoided. O_2 uptake ranged from 0.3 to 2.5 mmol m^{-2} h^{-1} while NH_4^+ release ranged from 0.084 to 3.2 mg m^{-2} h^{-1}. Both variables were controlled by the distance from the freshwater input. The in site variability was weakly related, although statistically significantly, to water temperature and chl a derived pigments.

Svensson (1997) showed a strong correlation between NH_4^+ release and O_2 consumption (both in mmol m^{-2} h^{-1}) in sediment from the eutrophic Lake Sövdesjön (Sth. Sweden) ($NH_4^+ = .19 \cdot O_2 - 0.096$; $r^2 = 0.58$, $N = 16$) if Chironomus larvae were added (see below). The ratio of NH_4^+/O_2 was very low at low O_2 consumption, but strongly increased with O_2 consumption, up to 0.2, which means that in the org-C rich sediment the ratio C/N approached 5, a rather high value. It does mean that NH_4^+ release from sediment is a heterotrophic process, supporting evidence that the same must happen with org-P.

Mäkelä & Tuominen (2003) used a whole core squeezer technique to obtain nutrient profiles in sediment of the Baltic Sea with a 1 mm resolution. In the Gulf of Finland a chemical zonation in NO_3^- and NO_2^- concentrations was found with strong seasonal variations and with maximal concentrations of 400–700 $\mu g\,\ell^{-1}$ a few cm under the sediment/water interface, indicating nitrification potential just under the sediment surface and a deeper layer of denitrification. Replicates showed some variability, but the patterns were the same. Below 20 mm, NO_3^- was usually absent, whereas o-P, NH_4^+ and SiO_2 increased more or less linearly. The NO_3^- profile sometimes showed a minimum concentration just below the sediment/water interface, a higher concentration below that layer and a strong decrease in the deeper layers. The higher NO_3^- concentration below the layer of strong denitrification capacity can exist if the NO_3^- flux into the sediments is higher than the denitrification capacity, but only for short periods. (See Golterman, 2000.) Often, however, the interstitial water contains very small concentrations; examples are given by Barbanti *et al.* (1992). A few curves from Mäkelä & Tuominen (loc. cit.) and Barbanti *et al.* (loc. cit.) are given in Figure 4.5. In the Baltic Sea, NH_4^+ showed some sharp peaks. Because of the rapid changes in these N-profiles, N-fluxes into and out of the sediment layer cannot be estimated by calculation of the diffusion rate. The NO_3^- profiles agreed qualitatively with older data on NO_3^- fluxes into the sediment (2.8–14 $mg\,m^{-2}\,d^{-1}$ with a mean value of 7 $mg\,m^{-2}\,d^{-1}$) and with denitrification, which was also high in the Gulf of Finland (2.8–9.1 with a mean value of 4.2 $mg\,m^{-2}\,d^{-1}$), but low in the deep Baltic Sea, because of the low concentration of NO_3^- in these parts. The authors pointed out that denitrification is the most efficient N-removing process in moderately oxidized sediment, so that NH_4^+ released may be oxidized to NO_3^- first.

It is often automatically assumed that the released NH_4^+ comes from microbial decomposition of org-N compounds. However, benthic animals will also contribute, although their relative role is not easily estimated. Yamamuro & Koike (1998) analysed different N-fractions in pore water of sediment of a eutrophic estuarine lagoon (Lake Shinji, Japan). Compared were a shallow, sandy, aerobic site, with 4 species of macrobenthos, and a deep, anaerobic site devoid of animals. Org-N_{diss} varied between 2.4 and 21 $mg\,l^{-1}$ and was by far the major constituent, 10 times greater than NH_4^+ and 1000 times the NO_3^- concentration. The NH_4^+ concentration, 3–5 $mg\,\ell^{-1}$, was highest in the anaerobic mud and higher than in the aerobic sandy sediment. The org-N_{diss} concentration was higher in the aerobic than in the anaerobic sediment. Depth profiles of inorg-N_{diss} were nearly constant and org-N_{diss} was highest at the surface. The increase in org-N_{diss} and NH_4^+ was only observed where macrobenthos was found and suggested an influence of macrobenthos on the partitioning of the N-fractions through their motion and excretion. Obviously, N-excretion by animals must contribute to ammonification, but quantitatively it probably does so less than bacterial action, as by far the larger part of the primary production is re-mineralized in the sediment by bacteria, while the animals will mainly provide more O_2 for the bacteria through their pumping and turbating activity.

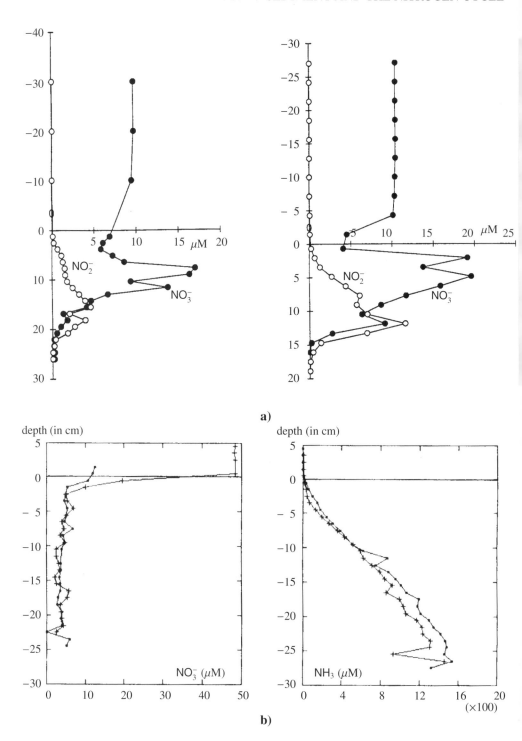

a)

b)

FIGURE 4.5: NO_2^- and NO_3^- profiles in sediment from a) the Baltic Sea (from Mäkela &Tuominen, 2003) and b)the Adriatic Sea (from Barbanti et al., 1994).

4.4 Denitrification[1]

4.4.1 GENERAL ASPECTS

Denitrification is the process by which NO_3^- is reduced, mainly to N_2; several species of heterotrophic bacteria are generally supposed to be responsible for this process, e.g., *Thiobacillus, Pseudomonas, Bacillus, Micrococcus, Achromobacter*. All of them are aerobes and use NO_3^- as electron acceptor only in absence of O_2 (Kuznetzov, 1970). *Escherichia coli* is an organism which is able to carry out only the reduction of NO_3^- to NO_2^-. Denitrification is a sequence of one-electron steps, as shown in Figure 4.6. Denitrification needs energy, which is derived from the oxidation of org-C or other reducing compounds, probably mainly FeS. O_2 inhibits the process because the same reductant will produce more energy with O_2 than with NO_3^-. At a given temperature denitrification rate is usually controlled by the concentration of the reducing agent or that of the NO_3^- present. Temperature does have an influence, but it depends on the concentration of the [H]-donor; if this is low, the temperature influence is restricted. (See further Section 4.4.3.)

Denitrification may sometimes produce N_2O as a side product. The quantities are usually small, and the factors that control the ratio N_2/N_2O are unknown – an imbalance between the availability of org-C and NO_3^- may be the cause. Minzoni *et al.* (1988) could not detect any N_2O in their extensive experiments in rice-fields in the Camargue. Blicher-Mathiesen & Hoffmann (1999) studied denitrification in a 15 m long transect from a hillslope towards a freshwater riparian fen (pH = 5.5–7). In the groundwater current, NO_3^- declined from 25 mg ℓ^{-1} to less than 0.14 mg ℓ^{-1}, while both O_2 and N_2O declined from 3.5 mg ℓ^{-1} and 56 μg ℓ^{-1} respectively to 0. Denitrifying activity was measured in chloramphenicol-treated sediment slurries and was about 100 mmol N (N_2-N and N_2O-N) $g^{-1} d^{-1}$ (ww.), of which 36% accumulated as N_2O. The N_2O formed uphill, however, clearly disappeared while the water flowed downwards. Such a 'geometric' influence may have occurred in other studies as well, but it may have been a 'time effect' or depletion of the reducing agent.

Either the actual denitrification rate can be measured, i.e., the rate in the field as controlled by one of its limiting factors (either the NO_3^- itself or the reducing agent), or, by adding NO_3^- at saturating concentrations, its potential or capacity. In this case the capacity is measured as limited by the reducing agent. If the factor limiting the rate in the field is added, this rate will increase as bacteria will proliferate. Tiedje *et al.* (1989) added chloramphenicol to stop bacterial growth in order to measure the actual rate, but adsorption of chloramphenicol onto sediment particles may decrease its activity. Denitrification can be measured in isolated samples, but then the extrapolation to the *in situ* conditions is difficult; it is much better measured in enclosures. Garcia (1975, 1978) determined denitrification activity by adding N_2O to samples in 250 ml serum flasks, measuring the disappearance rate of the N_2O by gas chromatography. Krypton was added as an internal standard. This method was extensively used *in situ* in rice-fields in Senegal. Because denitrification is then measured as mg $\ell^{-1} d^{-1}$, it is not clear how the conversion to m^{-2} must take place, but for comparison purposes the method is

---

[1]Two symposia had denitrification as their sole subject: "Denitrification in the nitrogen cycle" (Golterman, 1985) and "Denitrification in soil and sediment" (Revsbech & Sørensen, 1990).

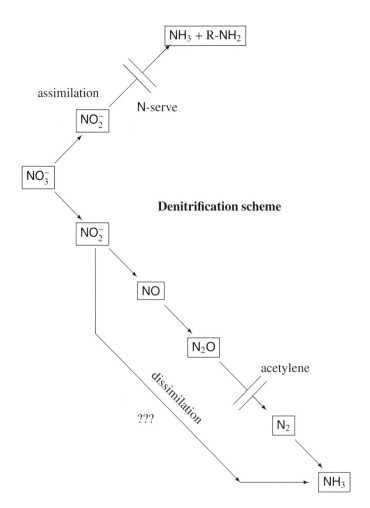

FIGURE 4.6: Different NO_3^- reduction pathways.

very useful. It was shown, e.g., that the rice rhizosphere had a positive influence (of a factor 2–4) on the denitrification rate. The coefficient of variation was high for the measurements *in situ*, but lower in pots filled with a homogenized sediment slurry, showing the inhomogeneity of the soils. Experiments in larger enclosures could have solved this problem. Garcia (1977) extended these experiments by adding samples treated with chloramphenicol and could distinguish between the initial activity of the enzyme N_2O-reductase and an activity potential, due to the addition of the substrate, after a lag phase of 4–6 hr. The existence of this lag phase is often not taken into account in this kind of studies.

Denitrification is essentially an anaerobic process, as O_2 will compete with the reducing agent and will produce more energy. Nevertheless, it will take place in well oxygenated waters, but in the sediment layer in or below which no O_2 is present. The rate of diffusion of NO_3^- through the sediment is therefore a controlling factor. In these cases nitrification and denitrification may occur in the same mud/water system. This is called coupled nitrification/denitrification. The nitrification is due to the oxidation of NH_4^+ released by mineralization of org-N in the top sediment layers still containing O_2. The physical distance between the two processes may be as little as a few mm's.

Denitrification can be quantified by measuring the disappearance rate of NO_3^-, if no other losses occur, or by measuring the compound(s) produced, usually N_2O. Logically, the increase in N_2 should be measured, but because of its ever-high concentration, this is often impossible, while, furthermore, the N_2 bubbles produced tend to stick in the mud. Most often the production of N_2O is, therefore, measured by inhibiting the reduction of this compound to N_2 by the addition of acetylene, C_2H_2. This inhibition is based on the resemblance of the chemical bindings in $N \equiv N = O$ and $HC \equiv CH$, which confuses the enzyme. This use of C_2H_2 should not be confused with the use of the same compound to measure N_2-fixation, where again the $HC \equiv CH$ confuses the nitrogenase, producing ethylene, CH_2CH_2.

Acetylene inhibits nitrification (Section 4.6) as well, and will therefore cause an NH_4^+ accumulation if the nitrification is quantitatively important. Using C_2H_2 and nitrapyrin (as nitrification inhibitors), Knowles (1979) demonstrated that in sediment from Lake St. George (Ont., Canada) coupled nitrification/denitrification (Section 4.3.1) was responsible for the N_2O production. Knowles showed that high rates of C_2H_2 reduction occurred under anaerobic conditions in the light, which is an important finding for field experiments with C_2H_2; it thus requires a sufficiently high concentration to be effective.

A third method to measure denitrification is to use ^{15}N, a non-radioactive isotope, which can be used to quantify the amount of N_2 produced by using the $^{14}N/^{15}N$ ratio. Nielsen (1992) introduced a modification in which $^{14}N^{14}N$, $^{14}N^{15}N$ and $^{15}N^{15}N$ are measured by mass spectrometry, permitting the distinction between denitrification of the added NO_3^- from the NO_3^- produced by NH_4^+ oxidation, the coupled nitrification/denitrification. Middelburg *et al.* (1996a) evaluated the isotope-pairing method by a simulation analysis with a set of fixed model parameters. The authors pointed out that the addition of $^{15}NO_3^-$ may increase denitrification rate and that therefore $^{15}NO_3^-$ should replace the $^{14}NO_3^-$ instead of being added. This is obviously true if NO_3^- is the limiting factor and the additions are large. Secondly, they questioned the need for the separation of denitrification from coupled nitrification/denitrification, as for the ecosystem's N-balance it makes no difference and this separation is supposed not to reflect a physically realizable situation. Nielsen *et al.* (1996) replied, correctly, that the two NO_3^- sources are regulated by different mechanisms, usually in an opposite manner. This differentiation may help to explain, e.g., the observation that a decreased O_2 penetration in the sediment may either stimulate denitrification, i.e., if the water-column NO_3^- is the major N-source, or inhibit it, if nitrification is O_2-limited. Middelburg *et al.* (1996b) maintained their objections; the question can only be solved experimentally, as the differentiation seems to be important for a better understanding of the ongoing processes. In the whole discussion, one point is missing, i.e. the percentage of NO_3^- incorporated into the denitrifying bacteria, which is the process by which sediment accumulates org-N despite denitrification. The percentage incorporated will most probably be different in the two cases. It is a problem usually overlooked in denitrification studies.

Other methods to measure denitrification include enumeration of denitrifiers and mea-

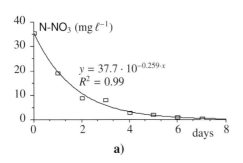 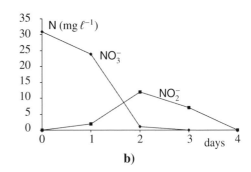

FIGURE 4.7: a) Disappearance of NO_3^- in an enclosure (ø= 4.5 dm) in a rice-field in the Camargue (from Golterman, 1995c); b) Disappearance of NO_3^- and production of NO_2^- in an enclosure (ø= 4.5 dm) in a rice-field in the Camargue (from Golterman, 1995c).

surement of the enzyme nitrate-reductase. Jones (1979) compared these methods and found that viable (most probable number) counts of denitrifiers could be used to show the difference between anoxic and surface waters, and between these and sediment, but did not correlate with differences between or within cores. This lack of correlation can be explained by the low precision of the MPN method and the fact that nitrate-reductase activity may be not only dissimilatory but also assimilatory. Jones used chlorate to separate the two activities and found that ≈ 60% of the activity was dissimilatory in sediment where the Eh was > +100 mV, and that this proportion increased to > 90% when the Eh fell < +50 mV. These Eh values are lower than those of Mortimer, but Jones used the Pt/calomel potential whereas Mortimer used the Pt/H_2 electrode. Nitrate-reductase was three to four orders of magnitude greater in freshwater sediment than in the overlying water; its activity was often at its maximum at a depth of 5–10 mm and coincided with a mean Eh of 210 mV in littoral (oxygenated) sediment. Under reducing conditions the Eh gradient moved upwards and nitrate-reductase activity was greatest at the water/sediment interface. Both MPN-enumeration and measurement of nitrate-reductase activity are rarely used in limnology because of the uncertainties involved, but seem to be promising if the technical difficulties can be solved.

Using small sediment cores from Vilhelmsborg Sø (Denmark), Seitzinger et al. (1993) compared three different methods to measure denitrification, viz., the C_2H_2 inhibition technique, which measures N_2O production, the ^{15}N tracer technique with and without C_2H_2 inhibition, and the N_2 flux method, by which the amount of N_2 produced is measured directly. The sediment was a silty clay, with 6% of org-C, a C/N ratio of 20 and a porosity of 0.86. With this method, pre-treatment during several days was needed to remove the atmospheric N_2 present, causing important changes in the sediment composition. The authors concluded that the three methods produced significantly different results. A combination of ^{15}N and C_2H_2 inhibition showed that the C_2H_2 inhibition of the N_2O reduction appeared to be incomplete, but this might well be due to the low quantities of C_2H_2 added; doubling the concentration might have given much information. Furthermore, it was shown that the coupled nitrification/denitrification was not included, because C_2H_2 inhibits nitrification of NH_4^+ as well. Denitrification rates of added NO_3^- as measured by the ^{15}N method were lower

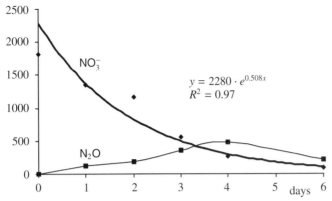

FIGURE 4.8: NO_3^- disappearance and N_2O production in an enclosure ($\emptyset$= 4.5 dm) in a rice-field in the Camargue.

(35% or less) than those measured by the C_2H_2 or N_2 flux methods. The different methods were, however, carried out on cores with different geometries; pre-treatment and the different durations of pre-incubations and incubations might well have changed the quantities of the available [H]-donor. No measurements of the NO_3^- concentrations were made; an N-balance might well have provided much information on what were the causes of the disagreements between the methods studied.

Kana *et al.* (1998) developed a high-precision membrane-inlet mass spectrometric method. They found that denitrification in a brackish river entering the Chesapeake Bay (USA) responded rapidly to changes in the NO_3^- concentration and that diffusion processes must be taken into account. They pointed out that there is not only a lag phase for bacterial response, but often also a delay caused by the geometry of the device used. They showed that there is often a linear relation between NO_3^- concentration in the water and denitrification, which once more emphasizes the fact that NO_3^- is often the limiting factor for denitrification. This points to the need for measuring denitrification not only as a function of the naturally occurring NO_3^- concentrations, because with high NO_3^- loadings the capacity will be high, but owing to this high capacity the NO_3^- concentration itself will be low.

If nitrification of NH_4^+ released from the sediment is a concomitantly occurring, quantitatively important process, that will also be inhibited by C_2H_2. This can, however, be taken into account by running parallel experiments.

Minzoni *et al.* (1988) showed that inorg-N, both NH_4^+ and NO_3^-, added to soils (mainly rice-fields) in the Camargue, was rapidly lost in 4–6 d (See Figure 4.7) – well before the rice could use it. In order to prove that the responsible process was indeed denitrification, the quantity of N_2O produced under inhibition by C_2H_2 was measured in closed plexiglass enclosures, fitted with the necessary in- and outlets. These authors could, however, only recover about 50% of the NO_3^- reduced as N_2O (see Figure 4.8). This discrepancy was caused by the set-up in the field, where enclosures were applied to which NO_3^- and C_2H_2 were added. It was, as usual, assumed that the N_2O formed would escape upwards as gas, like the N_2. However, as N_2O is much more soluble than N_2, downward diffusion, further stimulated by

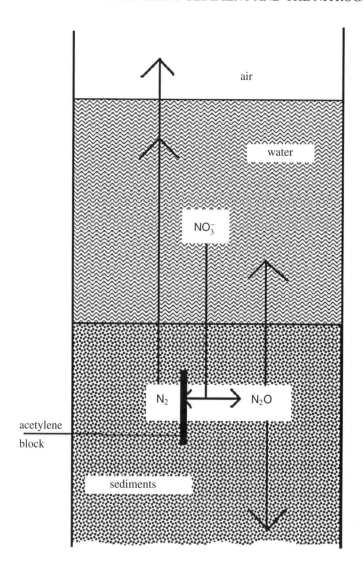

FIGURE 4.9: Upward and downward N_2O diffusion and upward N_2 diffusion in an enclosure in shallow water.

the irrigation water penetrating into the ground water, is as important as the upward diffusion (see Figure 4.9). Therefore N_2O and C_2H_2 were added to an identical enclosure; it was found that the N_2O decreased at a constant rate of $\approx 20\%$ per day. The real N_2O production rate could then be calculated from the quantity recovered in the overlying water plus the air space after correction for the loss of N_2O. The result is given in Figure 4.10; it can be concluded that about 70–80% of the NO_3^- added to these shallow waters is lost as N_2. Roughly 10% appeared as NO_2^-, as was found in all the experiments *in situ* and *in vitro*; the phenomenon will be discussed further below. (See Section 4.4.2.) This N balance leaves the possibility that

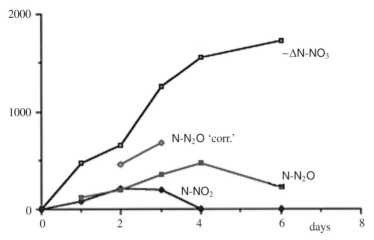

FIGURE 4.10: NO_3^- disappearance with NO_3^- and N_2O production in a rice-field in the Camargue. N_2O production: drawn line = measured values; after correction for the N_2O disappearance: $N-N_2O$ corrected (Golterman, 1995c).

about 10% of the fertilizers added to rice-fields will enter the underground water-table and will finally accumulate in the sediment. This was confirmed by using $^{15}NO_3^-$. Denitrification may, therefore, not be considered to be a complete protection against eutrophication in the cases where N is the limiting factor for primary production. The quantity of N taken up by the bacteria for their growth will accumulate in the sediment.

In the next step of these studies the rate controlling factors were quantified in order to allow extrapolation from one situation to another. Organic matter and NO_3^- concentrations, pH and temperature come to mind automatically. The pH has little effect in the range of natural freshwaters, i.e., 4–9. It may have some influence at extreme pH values (Chalamet, 1985). Each of the other factors has a strong influence. There is much discussion in the literature (see Chalamet, 1985) whether the NO_3^- concentration controls zero- or first-order kinetics. The solution for this problem is obvious: the rate of denitrification follows the Monod kinetics, therefore it depends on whether the NO_3^- concentration falls in the upward slope or in the plateau part of the Monod curve.

By adding NO_3^- to alkaline (pH = ±8) mud/water systems *in vitro*, El Habr & Golterman (1990) demonstrated that, if the sediment was relatively rich in organic matter (about 5%), the temperature had a strong influence on the denitrification rate, while if the organic matter was about 2%, this influence was feeble (see Figure 4.11). It is unexpected that the NO_3^- concentration decreased linearly with time: a bacterial process should cause an exponential decrease. This probably means that there is another limiting factor, a physical one, viz., diffusion. This point is further discussed in Section 4.4.5. When the NO_3^- concentration became low at the end of these experiments, a non-linear denitrification rate was found. As soon as in these experiments NO_3^- became the limiting factor for bacterial metabolism or growth, NO_3^- was added a second time to the same enclosures. Especially in the sediment poor in organic matter the results were different: the disappearance rate after the second addition was always

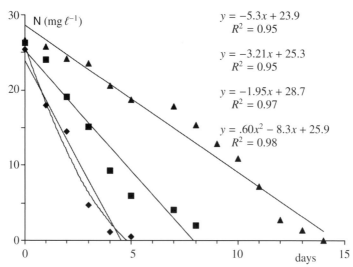

FIGURE 4.11: Influence of the temperature on denitrification rate in a mud/water system with Camargue sediment. Shown are linear rates at 0, 15 and 25 °C and the quadratic rate at 25 °C (Golterman, 1995c).

TABLE 4.3: Linear NO_3^- disappearance rate coefficients (in mg ℓ^{-1} d^{-1} of N) in two Camargue sediments with high ($\approx$ 5%) or low ($<$ 2%) org-C concentration. (From El Habr & Golterman, 1990.)

		Temperature				
		25 °C		15 °C		0 °C
NO_3^- addition		First	Second	First	Second	First
org-C conc.						
High ($>$ 5%)		5.28	6.49	3.77	3.68	1.45
Low ($<$ 2%)		1.91	3.59	1.58	2.59	1.07
High/low		2.76	1.81	2.39	1.42	1.35

higher than in the first experiment, but in the sediment poor in organic matter more than in the rich ones (Table 4.3). In the first instance it seemed likely that adaptation might cause this difference, but, if denitrification was indeed caused by a mixed flora of heterotrophs, such a sharp contrast should not occur. Therefore other possibilities had to be considered, and it was supposed that FeS might be a reductant for denitrification, besides the organic matter. In fact, when FeS was added to sediment, the denitrification rate increased considerably (Golterman, 1991b; see Figure 4.12). Furthermore it was shown that the amount of NO_3^- reduced roughly equalled the amount of SO_4^{2-} produced (see Table 4.4). It can be seen that in the first experiment the measured $\partial S/\partial N$ values were equal to the theoretical one. In the second experiment in which some denitrification with org-C occurred as well, this agreement was also found after correction for the amount of CO_2 produced. Denitrification with FeS will apparently

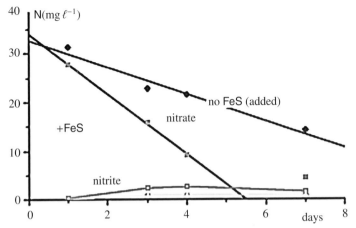

FIGURE 4.12: Denitrification rate with and without FeS added at ambient temperature (Golterman, 1991b, 1995).

TABLE 4.4: Denitrification experiments with FeS as the reducing agent in sediment of the Camargue. The sediment was enriched with FeS and NO_3^-; NO_3^-, sulphate and FeS were measured after 5, 9 or 14 days and are expressed in mmol ℓ^{-1}. (From Golterman, 1991b.)

First experiment						
FeS	∂NO_3		∂SO_4		$\partial S/\partial N$ (*)	$\partial S/\partial N$
	5 days	14 days	5 days	14 days		after correction for
--	2.36	4.43	0	0.37		"org-C denitrification"
--	1.71	4.71	0	0.38		
++	5.24	5.93	3.17	3.85	0.55; 0.64	
++	4.86	5.86	3.11	3.81	0.54; 0.64	
Second experiment						
	5 days	9 days	5 days	9 days		
--	2.5	2.5	0	0		
++	6.14	7	2.4	2.8	0.46; 0.4	0.66; 0.63
++	7.57	7.86	2.4	2.8	0.32; 0.4	0.58; 0.53

N.B. The $\Sigma(H_2S + H_2SO_4)$ was always recovered as $100 \pm 10\%$.
(*) The theoretical value of $\partial S/\partial N$ should be 0.56.

take place predominantly till FeS is depleted. The following reaction may be proposed:

$$5FeS + 9HNO_3 + 3H_2O \longrightarrow 5FeOOH + 5H_2SO_4 + 4.5N_2. \tag{4.4}$$

It seems likely that the reducing power of the Fe^{2+} will also be used, although this is not certain. For the N-balance of the ecosystem it makes much difference whether organic matter containing N or FeS is the substrate for the denitrification. In the first case the N for the growth of the bacteria may come either from the NO_3^-, or, more likely, from the organic matter:

$$\rightarrow\rightarrow \quad \rightarrow \qquad \rightarrow\rightarrow \quad \rightarrow\rightarrow\rightarrow \quad \rightarrow\rightarrow\rightarrow \rightarrow \rightarrow\rightarrow\rightarrow \rightarrow \rightarrow\rightarrow\rightarrow \rightarrow \rightarrow\rightarrow$$

$$\uparrow? \quad \text{or} \quad \uparrow? \qquad\qquad\qquad\qquad\qquad\qquad\qquad \downarrow$$

$$C_5H_7NO_2 \;+\; HNO_3 \;\;\Longrightarrow\;\; CO_2 \;+\; H_2O \;+\; N_2 \;+\; \Delta N_{bact}$$

In this case there is no increase in the N concentration of the sediment. However, if FeS is the substrate, the N_{bact} can be used only from the NO_3^-:

$$5FeS + 9HNO_3 \longrightarrow 5FeOOH + 5H_2SO_4 + (4.5 - 0.5x)N_2 + xN_{bact}. \tag{4.5}$$

It also makes a difference whether SO_4^{2-} or org-C is used as the reducing agent, i.e., a difference in pH shift following denitrification. Rust *et al.* (2000) noticed a pH increase of several units during denitrification according to the reaction:

$$4NO_3^- + 5CH_2O \longrightarrow 2N_2 + 5HCO_3^- + 2H_2O + H^+ \tag{4.6}$$

and proposed the use of encapsulated buffers. But if FeS is used instead of org-C as the reducing compound (Equation (4.4)), there is only a small increase in pH, while, moreover, in a $CaCO_3$ containing environment the pH will already be sufficiently buffered.

Kölle *et al.* (1983) found denitrification with FeS_2, pyrite, as the reducing agent in a reducing aquifer in northern Germany. The following reaction was proposed:

$$5FeS_2 + 14NO_3^- + 4H^+ \longrightarrow 7N_2 + 10SO_4^{2-} + 5Fe^{2+} + 2H_2O.$$

Evidence for this result was obtained by the increase in the SO_4^{2-} concentration supposedly originating from the FeS_2 oxidation and by laboratory experiments with *Thiobacillus denitrificans* cultures growing on slurries of FeS_2. In the laboratory experiments a strong competition with org-C occurred, and stoicheiometry was not observed.

Pyrite is often found in sediments of marine origin. It may be formed by reactions between FeS and H_2S or $S(0)$ (as S_8 or polysulfides). There are no reports about formation and oxidation of FeS_2 in recent lake sediments; it seems that more evidence is needed to establish the quantitative importance of this process.

4.4.2 NITRITE PRODUCTION

In denitrification experiments (Minzoni *et al.*, loc. cit.) in mud/water systems *in vitro*, large quantities of NO_2^- were produced and released into the medium (see Figure 4.10). In the above-mentioned experiments with FeS added to the Camargue sediment *in vitro*, high concentrations of NO_2^- were also found. In these latter experiments the disappearance rate of NO_2^- was also measured and it appeared that a rapid oxidation to NO_3^- occurred in the O_2 containing overlying water. When FeS was present, much less NO_3^- was produced, most likely because of the subsequent denitrification with FeS as the reductant. NO_2^- production depended on the presence of FeS and amounted in some experiments to concentrations of 30% of the NO_3^- added. García-Gil & Golterman (1993) showed that the reaction rate of this denitrification process followed Monod kinetics with the concentration of FeS as the controlling factor. It is certain that bacteria of the sulfur cycle are involved, but more research is

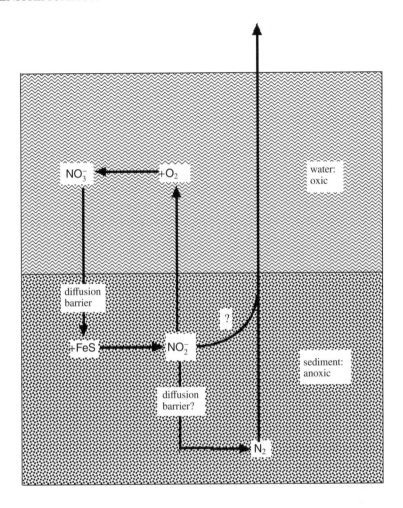

FIGURE 4.13: Scheme of NO_3^- reduction in a sediment layer and release of NO_2^- to the overlying water.

needed. From the mud used in these experiments, García-Gil (pers. commun.) later isolated a strain of *Desulfovibrio denitrificans*, which was more difficult to keep in culture than the usual strains.

The production of NO_2^- did not only happen in experiments *in vitro*. In the field, too, much NO_2^- was produced, which in marshes or wetlands is of course highly undesirable. The question must be raised why other bacteria than those of the S-cycle do not immediately use this energy source. An important variable that may influence the denitrification rate in the field is the rate of diffusion into the sediment; if the reduction of NO_2^- takes place in deeper layers than the reduction of NO_3^-, further downward diffusion will take time, while part of the formed NO_2^- will automatically 'escape' by upward diffusion (Figure 4.13). A theoretical study (Golterman, 1995d, 2000) has shown the great importance of the geometry of the experimental design, including the depth of the sediment layers (see Section 4.4.5).

Stief & Neumann (1998) incubated 17 sediments from running waters with 14 or 28 mg ℓ^{-1} of N-NO$_3^-$ in upward perfusion experiments, both long-term and short-term. The perfusing water was sampled from distinct horizons via horizontal tubes. In 11 sediments, regularities in the time courses of NO$_2^-$ and NH$_4^+$ were found, but in the remaining 6 only irregular patterns emerged. N-NO$_2^-$ concentrations up to 0.5–2.5 mg ℓ^{-1} were found in 5 cores depending on flow rate, O$_2$ and NH$_4^+$ concentrations, while in the other 6 (out of the 11) only N-NO$_2^-$ $\approx$ 0.3 mg ℓ^{-1} was found. In the first week, NO$_3^-$ was completely consumed in the first 4 cm and NO$_3^-$ and NO$_2^-$ were undetectable on the 9 and 16 cm perfusion route. The addition of NH$_4^+$ decreased the NO$_2^-$ production, demonstrating that the NO$_2^-$ was not derived from NH$_4^+$ oxidation. Denitrification rates in batch sediment samples gave N$_2$O production rates between 0.42 and 1.82 mg kg^{-1} d^{-1} (ww.). Temperature variation gave a Q_{10} of 2.8 (9–15 °C) and 4.7 (15–19 °C) for NO$_3^-$ consumption and a somewhat wider range for NO$_2^-$ production. A strong decline of the NO$_3^-$ concentration during the first 2–3 days was followed by a continuous, considerably weaker decrease, probably because of depletion of the reducing agent, much in agreement with El Habr's results mentioned above (page 159). In both El Habr's and Stief's & Neumann's experiments the concentrations of NO$_3^-$ added may seem high – as does the concentration of NO$_2^-$ produced. These NO$_3^-$ concentrations do, however, occur in European rivers, and Stief & Neumann report rivers with comparable NO$_2^-$ concentrations, causing considerable ecological harm.

Kelso *et al.* (1999) found high concentrations of NO$_2^-$ in sediment of rivers in North Ireland and measured denitrification and dissimilatory NO$_3^-$ reduction in these sediments after adding different org-C sources as the [H]-donor. Without additions the org-C in the sediment was already sufficient to reduce 45% of the added ^{15}NO$_3^-$ in 48 h and resulted in an NO$_2^-$ peak of N = 80 μg ℓ^{-1}. Glycine appeared to be an N-, but not an energy source for the bacteria involved. Glucose addition increased the NO$_2^-$ peak to 160 μg ℓ^{-1}. Formate and acetate stimulated a continuously linear NO$_2^-$ production rate; formate increased the NO$_2^-$ peak to 1300 μg ℓ^{-1}, but acetate only to < 100 μg ℓ^{-1}. The authors discussed physiological differences between various denitrifiers, but the use of the chosen substrates demands more physiological work to explain the results with the different [H]-donors. Whether FeS was present or not was not established. In the two above-mentioned studies it might well explain the high NO$_2$ concentrations.

In the first place, it must now be shown whether the FeS using denitrification exists in other waterbodies than those of the Camargue. In the second place, the influence of the FeS concentration and temperature on the denitrification rate must be measured before extrapolations from one waterbody to others can be made.

Measurement of denitrification in rivers is more difficult than in lakes because of the constantly changing water. Budget studies, i.e., measuring the flow of NO$_3^-$ at two stations, upstream and downstream, is imprecise as all errors will accumulate in the calculated loss between the two stations. Steinhart *et al.* (2001) worked with samples taken at different places (an organic debris site, and a sand and gravel site) in 5 different USA streams and measured both activity and actual rate after addition of chloramphenicol. They found low rates in unamended cores, not different for the two sites, but a higher capacity in the organic rich sites. At both sites the addition of glucose plus NO$_3^-$ increased N$_2$O production. In organic rich streams denitrification may be an important loss of NO_3^- even if the water is O_2 saturated.

Christensen *et al.* (1990) measured denitrification in sediment cores of a NO_3^- rich Danish lowland stream by C_2H_2 inhibition. NO_3^- from the overlying water was always the major source for denitrification, which under light conditions was reduced to 85% because of photosynthetic O_2 production extending the oxygenated zone through which the NO_3^- must diffuse. A simple diffusion-reaction model of denitrification, based on the O_2 respiration rate and the O_2 and NO_3^- concentrations in the water, predicted the measured rates and their seasonality reasonably well. In this case the ratio between denitrification and O_2 respiration was found to be relatively constant and a value of 0.8 was used. Denitrification rate varied strongly during the year and was different in light- and dark-incubated cores. O_2 production from benthic photosynthesis reduced denitrification by up to 85%, a factor to be taken into account in laboratory studies of denitrification.

4.4.3   RATE OF DENITRIFICATION

Denitrification rates depend on factors such as the concentration of NO_3^- or the reducing agent, and furthermore on temperature, O_2 concentration and the geometric characteristics of the system studied. The rates may vary widely; an upper limit is probably $\approx 1000\,\mathrm{mg\,m^{-2}\,d^{-1}}$, but normally the values fall between 15 and $100\,\mathrm{mg\,m^{-2}\,d^{-1}}$, with a few reports of up to $150\,\mathrm{mg\,m^{-2}\,d^{-1}}$. Seitzinger (1988) made a table involving about 30 lakes and coastal marine areas where rates fell in this range; more recent data generally fall in the same range. Comparing the different ecosystems, Seitzinger suggested that denitrification is higher in rivers, while in freshwater and marine coastal ecosystems it is roughly equal, with a tendency for the freshwater systems to fall at the low end of the range. Differences in methodology make these statements, however, dubious. Denitrification may also occur in hypolimnetic waters, but is there lower ($\approx 30\,\mu\mathrm{g}\,\ell^{-1}$) than in sediment. Obviously there is no lower limit. As some of his experiments lasted only 4 hr, it is not always permitted to extrapolate to 24 hr, as the concentration of the limiting factor will decrease – even if this is not obvious during these first 4 hr. There is no simple mathematical equation to describe the dependence of the quantity of NO_3^- disappearing, or of N_2 appearing, on time. J. M. Andersen (1985) found equal r values ($r = 0.96$–0.99) for the dependence on the NO_3^- concentration for a reciprocal and an exponential curve. He also demonstrated a nearly linear influence of the sediment O_2 uptake, a measure of the org-C concentration. This influence depended on the NO_3^- concentration. Reviewing older data he demonstrated an initial linear decrease in the NO_3^- concentration, flattening at the end, but it can be seen that the data points fit even better the model presented later (see Section 4.4.5).

Minzoni *et al.* (1988) found that in enclosures in the very shallow waters of rice-fields as much as $50\,\mathrm{mg}\,\ell^{-1}$ of N could disappear in about 4 days, when the season was well advanced, viz., in August, with water temperatures approaching $30\,°\mathrm{C}$. Earlier in the year it took about 8 days, but no single mathematical equation described all cases. Some examples were:

$$N = 2400 \cdot e^{-0.077\,t},$$

$$N = 2150 \cdot e^{-0.508\,t},$$

$$N = 775 \cdot e^{-0.98\,t},$$

with N = N-NO$_3$ (mg enclosure^{-1}) and t = time (days). The increase in the coefficient of time was probably due to the increase in both temperature and org-C in the sediment.

In some cases linear fits also gave R values > 0.8, but no proposal for a mathematical model could be made. In marshes of about 1 m depth, El Habr & Golterman (1990) found lower, and often linear, rates varying with temperature and the org-C concentration of the sediment (see Table 4.3). Svensson *et al.* (2000) measured denitrification rates in sediment cores from 4 sites representing different input situations, in the Lagoon of Venice (Italy), in order to identify the relative importance of different areas of this Lagoon, using the 'isotope pairing technique'. Ambient NO$_3^-$ concentrations varied between 60 and $18350\,\mu g\,\ell^{-1}$, the NO$_3^-$ arriving from a freshwater stream. The highest denitrification rate was found at the places with the highest NO$_3^-$ concentrations and was around 2.8–4.2 mg m^{-2} d^{-1} of N with st. dev. = 0.7–1.05 mg m^{-2} d^{-1}. The rate decreased towards the lagoon outlet to values as low as 0.15 mg m^{-2} d^{-1}. O$_2$ consumption rate varied between 0.7 and 2.4 mmol m^{-2} h^{-1}, while it was very low at a station near the outlet of the lagoon. Benthic fauna was shown to be important and was responsible for 30% of the denitrification in the intertidal area. About 50% of the denitrification was due to denitrification of NO$_3^-$ produced by nitrification. It seems likely that the wide differences were due to different org-C concentrations and temperatures.

Villar *et al.* (1998) analysed interactions between river and flood plain in the Lower Paraná River and measured denitrification rates ranging from 2.1 to 3.6 mg m^{-2} h^{-1} of N in enclosures in the flood plain marsh. The flood plain thus protects the river against high NO$_3^-$ concentrations originating from agriculture.

Several system constants control the denitrification rate. These include:

1) *physical factors*, such as molecular diffusion and its temperature dependence, and an acceleration factor caused by eddy diffusion.
2) *bacterial factors*:
    2a) the bacterial inoculum
    2b) their growth rate and the influence of the temperature
    2c) growth limiting factors, i.e., the concentration of either NO$_3^-$ or the reducing agent, viz., org-C or FeS.

The influence of these factors was analysed by Golterman (2000) using a numerical model to simulate denitrification from 10–200 cm deep water. These depths were chosen since denitrification is quantitatively important in marshes and other shallow wetlands, while 2 m is probably the water layer above the sediment in anoxic hypolimnia where denitrification may take place in deeper lakes. It appeared that denitrification can only be modelled if all these system constants are sufficiently well known. (See further Section 4.4.5.) In temporary marshes and systems like rice-fields, desiccation also influences the denitrification rate.

4.4.4   Desiccation of sediment and denitrification

Just as on P-metabolism, drying of sediment may have a strong influence on denitrification. When N-losses must be prevented, e.g., in rice-fields, dry periods can usefully be introduced to improve the rice yield. This may be attributed to the oxidation of FeS, which will be enhanced when O$_2$ can penetrate into the sediment layers. Farmers in South America use this practice (Bonetto, pers. commun.). Mitchell & Baldwin (1998) noticed a high bacterial mortality as a result of sediment desiccation and a C-limitation for microbial processes. Mitchell & Baldwin (1999), reviewing literature, reported that N mineralization and nitrification increased following rewetting of desiccated sediment and that desiccation reduced the potential for methanogenesis (e.g., Boon *et al.* 1997). Mitchell & Baldwin noted an increase in denitrification rate when NO_3^- was added to sediment of Lake Hume (an Australian reservoir), but a much stronger increase if org-C was also added, either as glucose or as acetate. Denitrification did, however, not significantly increase following desiccation (see Qiu & McComb, 1996). The sediment did not show N-release in the form of NH_4^+ or NO_3^- after rewetting, in contrast with the results of Qiu & McComb (1996). In this sediment there was negligible nitrification before as well as after rewetting, which may have been caused by killing of the nitrifying bacteria followed by a long lag phase. The difference between Qiu & McComb's and Mitchell & Baldwin's experiments seems to be time-related, depending on when the samples were taken. Nutrient release is a short-term and suddden event (hours/days). Microbial activity will be subsequently activated during coming flood events and re-immobilize the nutrients into cell biomass again. Qiu & McComb (loc. cit.) found a lag phase of 4 days, but this appeared not to be the case in the Lake Hume sediment. The authors feel that desiccation alone is not a mechanism to exclude nitrifiers. The main difficulty in this kind of dialogue is that neither the substrate for NH_4^+ production followed by nitrification nor the reducing agent for the denitrification were identified or measured. FeS, e.g., will be much more rapidly oxidized when exposed to air than org-C, which in sediment is obviously of a rather refractory nature. Glucose and acetate are definitely not the natural, and therefore not the best, reducing agents in this kind of studies. Mitchell & Baldwin (loc. cit.) discussed the controversy around the still problematic relationship between denitrification and methanogenesis. The more realistic way to resolve this problem is to study not only the products of these processes, but the source materials or the natural substrates as well.

4.4.5 MODELLING THE DENITRIFICATION RATE

Golterman (2000) developed a numerical model to describe the disappearance of NO_3^- caused by bacterial denitrification, which occurs mainly in the sediment, from 10–200 cm deep water layers, either shallow waterbodies or the deepest layer in a hypolimnion. The model employs the molecular diffusion constant, an acceleration factor describing eddy diffusion, and three bacterial growth constants, viz., the inoculum size, the maximum growth rate and the halfsaturation constant for the hyperbolic process. The values of these system-constants were varied over a wide range. The curves obtained were compared with two well-defined cases, i.e., two simple situations in which diffusion takes place:

a) without any reaction in the sediment (a 'conservative' compound, e.g., Cl^-);

b) with a complete, immediate reaction in one of the sediment layers (leading to a concentration = 0 directly upon the NO_3^- entering that layer).

These cases have analytical (i.e., correct mathematical) solutions, and were simulated by the numerical model 'DiffDeni' (see Golterman, 1995d). This model was based on a numerical iteration, with different assumptions, but nevertheless perfectly approached the analytical

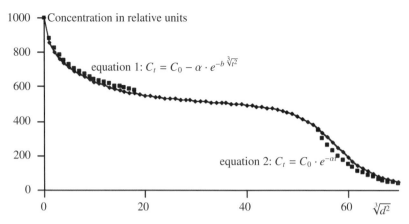

FIGURE 4.14: Denitrification curve showing the two different equations, i.e. the values calculated with the two equations (dots) (from Golterman, 2000).

solution for the cases a) and b). A negative exponential equation, (Figure 4.14, Equation 1: $C_t = C_0 - \alpha \cdot e^{-b \sqrt[3]{t^2}}$) was found to describe the first phase of this denitrification, i.e., when bacterial activity is low and NO_3^- behaves as a conservative compound. When the bacterial growth rate increased and attained a critical value, a negative exponential curve (Figure 4.14, Equation 2: $C_t = C_0 \cdot e^{-\alpha t}$) described the second phase of NO_3^- disappearance sufficiently well. The change-over moment from phase 1 (no NO_3^- reduction) to phase 2 (complete, immediate reduction), a moment that may vary between day 1 and day 50, cannot be described analytically (there is no mathematically correct solution). The influence of temperature on denitrification was assessed in this model; it was shown that this influence is equally strong as regards bacterial activity and diffusion. If only a few points are available (as often happens in field work) the two parts of the curve may be fitted on a straight line (see Figure 4.15), with still $R^2 > 0.9$, a satisfactory result for field work. Close inspection of Figure 4.11 suggests the presence of these two parts, but there are too few points for correct curve fitting. The same can be shown using the results of Messer & Brezonik (1983), who compared the annual denitrification rate for Lake Okeechobee (Florida, USA) as obtained by

a) incubation experiments with sectioned cores; or
b) a diagenetic flux model using first-order kinetics, with the value obtained from the N-mass balance.

Denitrification in the mass balance was estimated as ($N_{in} - N_{out} - N_{sed}$) and ranged from $N = 0.5–1.3\,\mathrm{g\,m^{-2}\,y^{-1}}$, representing 9–23% of the annual N-input. N_{sed} was estimated as P_{sed} times the ratio of $N_{sed}/(P_{in} - P_{out} - \Delta P)$, a remarkable approach. (In this way, denitrification values accumulate all errors in the other estimates.) If the sediments were often resuspended, incubation of sediment isolated from its NO_3^- and org-C sources produced low rate estimates, while the diagenetic model produced unrealistically high rates, but incubation techniques were generally in agreement with the mass balance value. The authors described denitrification with first-order kinetics, because $\log[NO_3^-]$ versus time gave a near-linear curve. If plotted linearly, however, a curve with two bends is obtained (see Figure 4.16), like Figure 4.15,

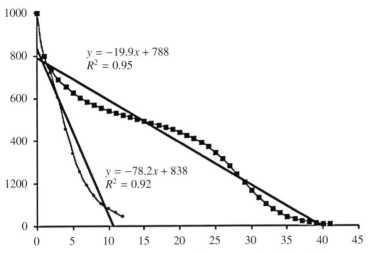

FIGURE 4.15: Two cases where linear fits gave satisfactory results, although the curves follow the two different equations of Figure 4.14 (Golterman, 2000).

showing that denitrification follows more complex kinetics than first order.

Nicholson & Longmore (1999), using stirred, benthic chambers (clear and dark) in a marine embayment (Port Phillip Bay, Australia), found that the variability in denitrification rate decreased with decreasing temperature, although other factors like org-C supply also had influence. On the whole, there was little influence of the type of chamber. These authors defined the concept of denitrification efficiency, i.e., the percentage of N originating from the mineralization of organic matter. The efficiency was usually > 80% at the central site and decreased to $\approx$ 30% at the two sites more under the influence of the external loadings. At one site, near the river inflow, denitrification was lowest when org-C input was highest, rather contrary to what could be expected. Because the inorg-N produced was not measured, it was calculated with the Redfield ratio (C/N = 6.6), which is too low for sediment and is not a constant. The concept of 'denitrification efficiency' seems useful for the comparison of different ecosystems, but the efficiency should be measured or at least compared with controls in which denitrification was inhibited in order to have a direct estimate of the inorg-N produced. Furthermore, denitrification efficiency includes the efficiency of the nitrification. For release of P (and SiO_2) in the same experiments, see Section 3.8.1.

4.4.6   DENITRIFICATION IN DIFFERENT ECOSYSTEMS

Saunders & Kalff (2001) reviewed denitrification activity in wetlands, lakes and rivers and showed that wetlands retained the highest proportion of Tot-N loading, the word 'retaining' being used by them as the difference between N-input and N-output. For the whole population of waterbodies analysed (52) denitrification was 63% and sedimentation 37% of the total N-loading. Both showed a weak relation with the N-loading, with water discharge as the controlling factor. Although the number of data for the increase in N_{sed} was much smaller

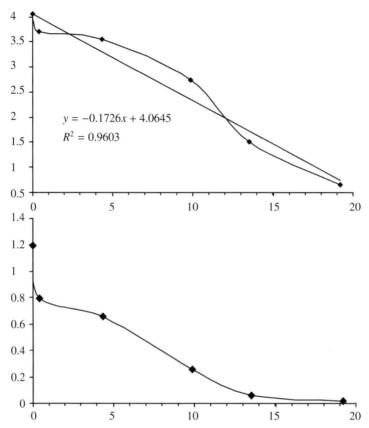

FIGURE 4.16: NO_3^- disappearance in incubation experiments with sediment from Lake Okeechobee, plotted on a log (original figure; top curve) and linear scale (recalculated from Messer & Brezonik, 1983).

(< 15) and the data not very precise, this shows that denitrification does not give a safe protection against eutrophication in water systems with N as the limiting factor of primary production, as are most wetlands.

In a series of 6 papers, Whitton and colleagues addressed the question whether denitrification can be important in river systems. Mass balances have been used in the past, but are subject to great uncertainties. Pattinson *et al.* (1998) studied denitrification rates (C_2H_2 blocking method) in the sediment of 9 intact cores at five sites along the river continuum of the Swale Ouse river system (Yorkshire, UK), together with one site on a highly eutrophic tributary, the river Wiske, once a month for 17 months. Time course experiments confirmed linearity over 4 hr, but results were not shown. Rates varied between $N = 280\,\mathrm{mg\,m^{-2}\,h^{-1}}$ (headwater site 2.5 km upstream of the start of the main river) and $9\,\mathrm{g\,m^{-2}\,h^{-1}}$ of N (at the tidal limit). Seasonal peaks were found and attributed to NO_3^- concentration and temperature. There was a strong correlation (see their Figure 6) between denitrification and temperature plus NO_3^- concentration ($r^2 = 0.91$, but no regression factors were calculated) for the whole database, but not for the particular sites, which was tentatively attributed to other environmental variables. Denitrifying bacteria were enumerated (Most Probable Number) and increased from 240 to 900,000 g^{-1} (dw.) in passing downstream.

García-Ruiz *et al.* (1998a) measured denitrification in the sediment and water of the freshwater tidal river Ouse, using both the C_2H_2 blocking method and $^{15}NO_3^-$ enrichment. Denitrification rates varied between n. d. and 8 mg m^{-2} h^{-1} of N, and increased with NO_3^-, but not with glucose additions. This may mean that sufficient org-C was available, or that glucose is the wrong substrate. In one, polluted, place denitrification was expected, but undetectable; as denitrification is sensitive to toxic compounds, pollution may well have been the cause. It is one of the problems in measuring denitrification rates in polluted areas. They also measured the kinetic parameters of denitrification in the Swale Ouse river system in intact sediment cores, using the C_2H_2 blocking method. Denitrification rate followed a Michaelis–Menten-type of curve with K_m values increasing from N = 0.18 to 1.25 mg ℓ^{-1} when going downstream and much higher values in a polluted tributary. Maximal rates increased from 0.5 to 1.5 mg m^{-2} h^{-1} over the same range. In the polluted tributary the maximal rate was N = 16 mg m^{-2} h^{-1}; N_2O production accounted for from about 25% at the headwater to 80% at the polluted site. These values are among the highest recorded and the high N_2O production indicates a serious imbalance in the ecology of these sites.

García-Ruiz *et al.* (1998b) examined N_2O production and potential for denitrification in sediment slurries and in cores of the River Wiske, a lowland, nutrient-rich tributary of the River Swale. Both ^{15}N and C_2H_2 methods were used, but the duration of the incubation was only 4 hr, which leaves open the possibility of a lag phase influence. Denitrification decreased with 40% from the 0–1 cm to the 1–2 cm zone, while maximal denitrification capacity (5.3 mg m^{-2} h^{-1}) on fine sand plus silt covered by plant debris was only 4 times the minimum, 1.4 mg m^{-2} h^{-1} in fine sand. Denitrification rate in cores measured monthly showed the lowest values in winter, a peak in late spring and a subsequent decrease through summer and autumn. N_2O production showed a similar trend and accounted for 0–100% of total denitrification, with an average of 40% of the N-gases produced by NO_3^- reduction. Temperature had a different effect, according to whether assays were carried out with cores or with slurries; in each case denitrification rates increased with temperature up to 30 °C. Adding org-C had no significant effect on denitrification rate, but decreased the proportion of N_2O formed as a result of the NO_3^- reduction significantly. It seems therefore that org-C was present above limiting concentrations, while the difference between slurries and cores is probably due to the influence of diffusion, which will also partly contribute to the temperature effect.

Finally, García-Ruiz *et al.* (1998c) investigated denitrification and N_2O production as functions of ambient NO_3^- concentrations in 50 sediments in a range of physical and chemical conditions during 3 hr incubation. In more than 90% of the samples N_2O concentration was below 28 ng g^{-1} h^{-1} (dw.) and production rates were equally low. In 90% of the cases denitrification rates were below 0.5 μg g^{-1} h^{-1}, with a few exceptions up to 3.5 μg g^{-1} h^{-1}. In some streams the rate increased downstream. Correlation analysis showed the importance of NO_3^- concentration and water content of the sediments, which explained together 64% of the variability, while the percentage of fine particles had some influence. Only Tot-C was determined and not org-C, which may have been an important factor, while prolongation of the assays could have shown a possible lag phase.

Laursen & Seitzinger (2002) developed a new method to estimate denitrification in rivers. The increase in N_2 concentration is measured, while the losses to the air are estimated by injecting non-reactive traces (i.e., propane and butane) and calculating a first-order transfer rate constant for N_2. The calculation model was validated with the naturally occurring Ar concentration – the agreement was good in several, but not in all cases. The authors arrived at values between 3.8 and 220 mg m^{-2} h^{-1} of N, but with great errors at the lower limit. The method seems well suited to analyse the influence of limiting factors such as temperature, O_2 and NO_3^- concentrations and flow rate. Estimates of the errors involved should, however, be extended.

Cornwell *et al.* (1999) reviewed denitrification in estuaries and coastal ecosystems, in which denitrification represents an important N-loss. (See also Seitzinger, 1988.) Denitrification was controlled by water residence time and O_2 and NO_3^- concentrations, the O_2 inhibiting the bacterial response and controlling the depth of the layer in which denitrification can occur. These authors attached no importance to org-C, probably because in the eutrophic systems in which nowadays denitrification is important, NO_3^- and not the org-C or FeS concentration is the limiting factor. In Section 4.4.2 more factors, biological and physical, which may also play a role, have been discussed.

Berelson *et al.* (1998) measured biogenic remineralization and solute exchange in benthic chamber experiments in Port Phillip Bay (Australia) which receives half of Melbourne's wastewater. Exchange of O_2, NH_4^+, NO_3^-, NO_2^-, o-P, SiO_2 and Tot-CO_2 was monitored in the chambers over, generally, less than 24 hr, as O_2 values decreased rapidly to near 0, depending on the site. Variability of flux values within a site was comparable to year to year variability, both ±50%. Highest flux values (6–9 mg m^{-2} d^{-1} of N) for NO_3^- plus NO_2^- were found in the northern part of the bay, receiving most of the wastewater; most of the flux values levelled off already after 10–15 hr. Measurements of [222]Radon concentrations showed that fluxes were 3–16 times the molecular diffusive flux caused by bio-irrigation (= bioturbation). Sediment denitrification efficiency decreased roughly linearly with increasing org-C mineralization from about 100% to near 0% at an org-C mineralization value of 1.2 g m^{-2} d^{-1} of C. Denitrification was, however, not measured directly, but calculated from an expected N-flux, calculated in turn from the C-flux and the Redfield C/N ratio – an unreliable method. Variability of the N-fluxes was caused by variations in bottom water O_2 concentration or penetration depth, differences in benthic primary production at the sediment-water interface and differences in bio-irrigation. Heggie *et al.* (1999) continued these studies and measured denitrification and N-fluxes in benthic chambers in sediment of the same bay, which received ≈ 3 g m^{-2} y^{-1} of N, causing serious problems concerning water quality. As the N_2 increase in these chambers was expected to be 1–5% of its ambient concentration, they used high precision mass spectrophotometry, compared with gas chromatography using N_2/Ar ratios. As earlier data had suggested that denitrification decreased at places with high C-loadings, attention was given to the C-loading as well. Overall agreement between mass spectrophotometry and gas chromatography was good ($r^2 = 0.9$; $n = 12$), although the slope was 0.8. Denitrification rates of up to 18 mg m^{-2} d^{-1} of N were found at places where org-C loading was 180–300 mg m^{-2} d^{-1}, with denitrification efficiencies of 75–85% indicating that most N processed through the sediment returned to the water as (unavailable) N_2. It was indeed found that, at org-C loadings $> 1.2\,g\,m^{-2}\,d^{-1}$, denitrification rate and efficiency were practically 0, and N was released only as NH_4^+. It is not clear whether this was due to lack of O_2 to oxidize NH_4^+ or whether a dissimilatory NO_3^- reduction took place.

Gran & Pitkänen (1999) measured denitrification in cores in order to analyse N-losses in the Gulf of Finland (Baltic Sea). N-loadings here amounted to $10\,g\,m^{-2}\,y^{-1}$, of which about $5\,g\,m^{-2}\,y^{-1}$ were shown to be lost from the mass balance. Denitrification was measured after the addition of $1.4\,mg\,\ell^{-1}$ of N, in cores taken from 8 stations, using the isotope pairing technique, at an ambient temperature of 3–6 °C during 3 hr. Rates varied between 0.1 and $17.6\,mg\,m^{-2}\,d^{-1}$ of N, and were highest in stations with highest redox potential. NH_4^+ fluxes calculated from pore water concentrations varied between -0.4 and $14.9\,mg\,m^{-2}\,d^{-1}$ and were highest at stations with low denitrification rates. Although the isotope pairing method was used, the authors did not report which part of denitrification was due to nitrification of NH_4^+. The short incubation time may also have a certain influence, because a possible lag phase may influence the results. Highest denitrification rates were found in areas where bioturbating macrofauna (*Monoporeia affinis* and *Pontoporeia femorata*) occurred in high densities. It was suggested that this is due to an increased oxygenation of this sediment – an extensive discussion on the role of benthic fauna is given.

Gilbert *et al.* (1998) examined the influence of macrofauna on the denitrification rate in Mediterranean coastal sediment in cores replaced *in situ*, after 1, 4 and 6 months. Natural and potential rates (with KNO_3 and glucose additions) increased with 160 and 280%, respectively, in comparison to the controls that were defaunated. Bioturbation was measured by placing cakes with luminophore particles (40–60 and 150–200 μm) on top of the cores and following the displacement of these particles. After 1 month no burial of luminophores occurred in the controls, after 6 months some occurred. In the inhabited cores, 17.7% of the luminophores were recovered from 2–10 cm depth after 1 month. This technique merits more attention than it has received up to the present.

Tuominen *et al.* (1999) measured N-fluxes in sediment of the Gulf of Finland (Baltic Sea) using cores and the isotope pairing method, permitting a separation between coupled nitrification/denitrification and denitrification with added $^{15}NO_3^-$. Coupled nitrification/denitrification was reduced from 2.8–4.2 mg m^{-2} d^{-1} to 140–280 $\mu g\,m^{-2}\,d^{-1}$ of N after the addition of dead algae ($340\,mg\,m^{-2}$ of C) to the cores, but the addition of the amphipod *Monoporeia affinis* (1500 ind. m^{-2}) restored denitrification, although the effects were somewhat different between day 4 and day 12. The effect of the algae was a decrease in the O_2 concentration, causing NH_4^+ release without further oxidation to NO_3^-; the effect of the fauna was to circulate O_2 rich water into the sediment. NH_4^+ concentrations in the pore water were highly increased by the addition of algae, especially after 4 days, causing NH_4^+-flux towards the sediment. The addition of the benthic fauna caused a decrease below control values. NH_4^+ excretion by the benthic fauna was responsible for 5–10% of the total NH_4^+ released. Denitrification measured by $^{15}NO_3^-$ addition followed first-order kinetics and test incubations showed the addition of $1.4\,mg\,\ell^{-1}$ of ^{15}N to be optimal. Denitrification coupled with nitrification showed a hyperbolic function with the concentration of the added $^{15}NO_3^-$, but did not reach its plateau value. The study showed the great importance of carefully separating coupled denitrification from that depending on added NO_3^-.

Risgaard-Petersen (2003) analysed the influence of benthic microalgae on coupled nitrification/denitrification using data from 18 European estuaries. Sediments were grouped in 4 classes (fully or net heterotrophic and highly or net autotrophic) according to the benthic trophic state index, i.e., whether O_2 fluxes in light and dark were smaller or greater than 0. Field data supported the hypothesis that in autotrophic sediments lower rates of coupled nitrification/denitrification are present. In fully heterotrophic sediments 90% of the data fell between 0 and 1.3 mg m^{-2} h^{-1} with a median of 0.3 mg m^{-2} h^{-1}, while in highly autotrophic sediments these values were 0–0.5 mg m^{-2} h^{-1} with a median of 0.06 mg m^{-2} h^{-1}. Experiments with ^{15}N and microsensor (NO_3^-) techniques supported the evidence from the field data. Coupled nitrification/denitrification in the autotrophic sediments was between 4 and 51% of the activity in fully heterotrophic sediments, depending on the N-load. Induction of N-limitation of the nitrifying bacteria is thus a major controlling mechanism of coupled nitrification/denitrification.

4.5   Dissimilatory nitrate reduction

Besides the reduction of NO_3^- to N_2 there are reports that the reduction may proceed to NH_4^+, a process called 'dissimilatory NO_3^- reduction', or sometimes 'NO_3^- ammonification'. For the ecosystem there is an important difference with denitrification: while N_2 is lost from the ecosystem, NH_4^+ is not. The energy aspect is also different; the process needs more energy. It may easily be confused with N-accumulation in bacteria owing to NO_3^- uptake, followed by death or autolysis of the bacterial cells releasing NH_4^+.

Caskey & Tiedje (1980) isolated a culture of *Clostridium sp.* from soil and showed an increase in molar growth yield of this bacterium on glucose of $\approx$ 16% as compared with fermentative growth. The bacterium showed no regulatory features in common with the features of assimilatory nitrate reductases, demonstrating the dissimilatory character of this process. No comparison was made with enzymes of dissimilatory denitrification. The culture was maintained on the usual rich nutrient medium, at a very low redox potential, with added amino acids, glucose (2 g ℓ^{-1}) and Na-thioglycollate (0.5 g ℓ^{-1}), which makes extrapolation to field conditions rather unreliable.

One of the earliest reports of an *in situ* experiment came from Sørensen (1978a, b) who measured NH_4^+ production in marine sediment from Limfjorden (Denmark). NO_3^- ammonification was usually less than, but sometimes equal to the denitrification rate in $^{15}NO_3^-$ amended sediment at rather low redox potentials. Incubations with C_2H_2 ($7 \cdot 10^{-3}$ atm) caused N_2O accumulation in the sediment equal to N_2 production without C_2H_2. Denitrification capacity decreased rapidly with depth, whereas the ammonification was significant also in the deeper layers. Buresch & Patrick (1981) showed that the conversion of NO_3^- into NH_4^+ in an estuarine environment increased with decreasing redox potentials. They argued that, when NO_3^- is diffusing into natural sediment, the redox conditions are not very low; under these conditions only 15% of NO_3^- was recovered as NH_4^+.

Jørgensen (1989) measured denitrification and dissimilatory NO_3^- reduction (called NO_3^-

ammonification) in 4 layers of 1 cm of sediment of the Norsminde Ford estuary (Denmark) using C_2H_2 and $^{15}NO_3^-$ during 4 hr incubations. Denitrification capacity showed two maxima, one from May to August (N $\approx$ 1.4 mg cm^{-3} h^{-1}), and a second, smaller peak from October to December (N $\approx$ 0.7 mg cm^{-3} h^{-1}), while NO_3^- ammonification showed a much lower maximum, usually at greater depth, in late summer ($\approx$ 0.5 mg cm^{-3} h^{-1} of N), when the sediment was more reduced. In intact sediment denitrification accounted for 11–27% and ammonification for 8–29% of total NO_3^- reduction. The curves of NO_3^- disappearance, N_2O production and total-N production showed rather different patterns, sometimes straight lines, sometimes hyperbole-like curves and for N_2O production even a sigma-shaped curve. Changes of denitrification rates were found especially in late summer, reflecting synthesis of denitrifying enzymes. This lag phase was eliminated in experiments with mixed sediment by pre-incubation with NO_3^-. In mixed sediment the recovery as Σ(N-gases) was 30–80%, whereas NO_3^- ammonification was 4–11%. Incorporation in org-N was never higher than 5% and the total NO_3^- recovery was 45 ± 2% in intact cores and 44–94% in mixed sediment. $^{15}NH_4^+$ incorporation in clay lattices, $^{15}NO_3^-$ production and incorporation of ^{15}N in dissolved organic compounds were discussed as the possible origin of these experimental losses. Why recovery in mixed sediment was better than in cores is not clear in this case.

Bonin *et al.* (1998) measured simultaneous occurrence of denitrification and dissimilatory $^{15}NO_3^-$ reduction in sediment near the French Mediterranean coast; the activities were N = 0 to 135 μg ℓ_{sed}^{-1} d^{-1} and 32 to 1150 μg ℓ_{sed}^{-1} d^{-1} respectively. $K^{15}NO_3^-$ was used, increasing the NO_3^- concentration by only 10%. Denitrification was highest in early spring, and dissimilatory NO_3^- reduction was highest in autumn, but values > 100 μg ℓ_{sed}^{-1} d^{-1} were only found on one occasion, and not at all stations. However, NO_3^- loss was recovered as N_2O plus NH_4^+ sometimes only for < 40%, and sometimes up to 100%, so it seems that technical difficulties still render these measurements difficult, while the rather short incubation time (< 5 hr), the addition of 1 mmol of NH_4^+ (to suppress assimilatory NO_3^--uptake) and other technical problems raise some questions concerning methodology.

There is still doubt about the occurrence of dissimilatory NO_3^- reduction in the literature. In some reports this process is quantified using ^{15}N techniques, while in other studies it is taken as the difference between total NO_3^- reduction and denitrification. Thus, Jørgensen (1989), Jørgensen & Sørensen (1985, 1988), studying a Danish estuary sediment, stated dissimilatory NO_3^- reduction to be quantitatively important, and sometimes even more important than denitrification, while Binnerup *et al.* (1992) and Rysgaard *et al.* (1994), using the same sediment, found mainly denitrification. This study was carried out in a steady state continuous flow-through system at different O_2 concentrations, while for the N-cycle studies ^{15}N was used, and $^{30}N_2$ and $^{29}N_2$ were separately measured. At O_2 concentrations below 100%, NO_3^- from the overlying water was the main N-source for denitrification, while above 100% of O_2, NO_3^- derived from nitrification was the main source. The authors suggested that at low NO_3^- concentrations (such as in coastal waters), higher O_2 concentrations will stimulate denitrification, while at high NO_3^- concentrations the opposite will occur. However, more experimental data are needed to confirm this hypothesis. Substrate and sediment characteristics like grain size or org-C concentrations will play a role as well.

Kelly-Gerreyn *et al.* (2000), working on a large database from the NO_3^--rich (mean N =

5.7 mg ℓ^{-1}) Gt. Ouse estuary (North Sea, UK), suggested that dissimilatory NO_3^- reduction (to NH_4^+) can account for a high proportion of the total NO_3^- reduction. In this system, N-flux into the sediment by total NO_3^- reduction was N = 4.3 mg m^{-2} hr^{-1}, while NH_4^+ release was N = 3.8 mg m^{-2} hr^{-1}. The authors developed a one-dimensional model of 150 compartments with a depth of 0.1 cm, which differentiates uncoupled NO_3^- reduction into a) denitrification and b) dissimilatory NO_3^- reduction. Temperature controlled this proportioning of the NO_3^- reduction. The result is a function showing that dissimilatory NO_3^- reduction and denitrification occur at all temperatures, but that dissimilatory NO_3^- reduction is the favoured pathway at the extremes of the observed temperatures (< 14 and > 17 °C), while denitrification was optimal only in a narrow temperature range (14–17 °C). The mechanism might be an adaptive response of different NO_3^--reducing bacteria to temperature. This temperature effect implies that during an extended warm summer in temperate estuaries receiving high NO_3^- inputs, NO_3^- reduction may contribute to eutrophication rather than counteract it. Models of the N-cycle in coastal areas should, therefore, include dissimilatory NO_3^- reduction. Because the temperature effect is in fact a seasonal effect, it might be suggested that the reducing agent for the NO_3^- reduction also changes with the seasons, and that the different energy contents of the reducing agent have to be considered as well. So far, there are no reports on dissimilatory NO_3^- reduction in freshwater. This difference may be explained by a lack of interest from limnologists, or a different bacterial flora or substrate. The high salinity in the estuaries will certainly cause different bacterial activities, although there are no studies on this subject. On the whole, there is a big gap in the knowledge of bacterial physiology between the laboratory and the field. One wonders whether different org-C compounds as reductants in the marine environment may cause a different bacterial metabolic pathway from that in freshwater.

4.6   Nitrification

Nitrification is the process of NH_4^+ oxidation via NO_2^- to NO_3^-. Nitrifcation is blocked by C_2H_2 in experiments where N_2O is measured to estimate denitrification. It can also be blocked by inhibitors such as nitrapyrin, allylthiourea and N-serve. Minzoni et al. (loc. cit.) found that N-serve was rapidly inactivated in field experiments, probably by adsorption onto the sediment. This was not the case with allylthiourea. Both inhibitors were used at concentrations of 10 mg ℓ^{-1}. De Bie et al. (2002) measured nitrification in sediment of the Westerschelde estuary (the Netherlands) with inhibitors and the $H^{14}CO_3^-$ incorporation technique. They showed that, in the field, C_2H_2 (and CH_3F) had different effects from those in cultures of Nitrosomonas europaea; they can, therefore, not be used as nitrification inhibitors in the field. This kind of difficulty, caused by shifts in bacterial metabolism or species, is probably more often important than realized when inhibitors are used that have been shown to be 'specific' in cultures. The authors used N-serve without problems, while thioallylthiourea was shown to be satisfactory as well. A scheme to measure NH_4^+ release, nitrification and denitrification together can be set up by measuring ΔNO_3^-, ΔNH_4^+ and ΔN_2O in control experiments, experiments with C_2H_2 and with N-serve or allylthiourea as follows:

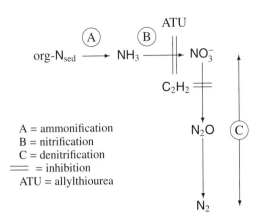

A = ammonification
B = nitrification
C = denitrification
══ = inhibition
ATU = allylthiourea

Pauer & Auer (2000) showed that in sediment of the hypertrophic Onondaga Lake (Mich., USA) nitrification also occurred and that it was more important in the sediment of the lake than in the water column. They attributed this phenomenon to the high density of nitrifiers in the top cm of the sediment. They suggested simple zero-order kinetics as a first approach to model this process, in which, however, the O_2 concentration must certainly also find a place.

Qiu & McComb (1996) studied the influence of drying and reflooding on NH_4^+-release followed by nitrification in sediment cores from a small shallow lake in Southwest Australia. Direct NH_4^+ release into the overlying water of the intact (control) core was high, ≈ 0.5 mg ℓ^{-1}, doubled in two days and was rapidly oxidized to NO_3^- on aeration. Air drying followed by reflooding caused a release of 6 mg ℓ^{-1} NH_4^+ after 4 days followed by an oxidation to NO_3^- again about 4 days later. O_2, NH_4^+, NO_2^- disappeared completely in non-aerated control cores, but NO_3^- levelled off from > 2 mg ℓ^{-1} to 0.8 mg ℓ^{-1}, where it remained stable. In the non-aerated, reflooded cores, NO_3^- decreased from > 5 to ≈ 1 mg l^{-1}, NH_4^+ and NO_2^- remained constant near 0 mg ℓ^{-1} and O_2 levelled off at 6 mg ℓ^{-1}, showing that the reflooding had caused considerable removal of reducing substances by oxidation. Argentine rice producers have used drying and reflooding of rice-fields in order to increase rice production (Bonetto, pers. commun.). The effect may be the oxidation of FeS more than that of org-C.

Nitrification may be prevented by live benthic algae (situated on top of the sediment layer or even mixed with the top few cm's) taking up the NH_4^+ released before it can be oxidized – and then denitrified again.

4.7 Combined studies

In nature, different processes always occur together, such as ammonification followed by nitrification, followed by denitrification. If only one process must be analysed, there are different options: one substrate may be added at such high concentrations that other processes are quantitatively 'overruled', or specific poisons are used to inhibit processes not under study. In the past, many mistakes have been made with the latter solution, e.g., the addition of $CHCl_3$ to study activity or metabolism of CH_4 bacteria, while it was not realized that $CHCl_3$ kills all other bacteria as well.

A few studies have found other solutions. Lorenzen et al. (1998) applied a micro-biosensor

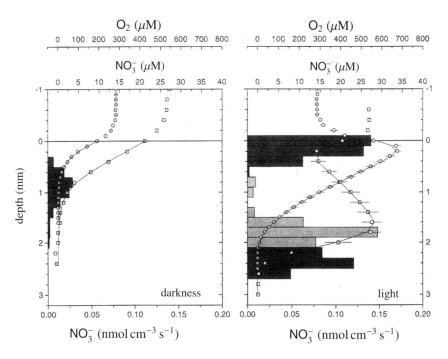

FIGURE 4.17: Mean O_2 and NO_3^- concentrations (with standard deviations), calculated profiles (lines) and NO_3^- assimilation (gray, top), nitrification (light grey) and denitrification (gray, bottom) rates. A: during darkness. B: during illumination. (From Lorenzen, 1998.)

technique to study several concomitant processes. The basic equipment consists in an O_2 sensor and a microscale NO_3^- biosensor made from a N_2O micro-sensor ($30 \mu m$), covered by an immobilized layer of N_2O-reductase-deficient denitrifying bacteria. Several N-cycle processes were studied at the same moment in sediment covered with benthic diatoms. (See Figure 4.17.) It was found that there was no nitrification in darkness, but a high rate in the light, coupled with a high rate of denitrification, plus an important NO_3^- assimilation. Nitrification during darkness could be induced by O_2 supply, showing that photosynthesis by the diatoms was responsible for the nitrification in light. NH_4^+ addition did not stimulate nitrification during darkness, O_2 being limited to the upper 1 mm of sediment and NO_3^- to a 1.5 mm layer. Denitrification maxima were at 1 ± 0.5 cm and at 2.5 ± 0.5 cm with an active layer of nitrification in between. Denitrification occurred at O_2 concentrations up to $20 \mu M \approx 0.6 \, \mathrm{mg \, l^{-1}}$), but below that concentration it may have been masked by overlap with nitrification. In the light, net denitrification was not detected when the O_2 concentration was higher than $1 \mu M$, while above that concentration nitrification occurred. Consumption and production rates were calculated using a diffusion model; for NO_3^- the molecular diffusion constant of $17 \cdot 10^{-6} \, \mathrm{cm^2 \, s^{-1}}$ was used and multiplied by the squared porosity, i.e., $0.8^2 \cdot 17 = 11 \cdot 10^{-6}$. The influence of the temperature on the diffusion coefficient was not taken into account. (For problems of diffusion calculation see also Section 5.3.) This will have introduced a serious error, so that comparisons with other methods or publications are not allowed. Their statement that the

C_2H_2 technique underestimated denitrification must be considered with caution.

Jensen *et al.* (1990) studied NH_4^+ and NO_3^- flux following the sedimentation of a diatom bloom (mainly *Skeletomena costatum*) in Aarhus Bight (North Sea, Denmark) early in April 1988. Release fluxes of NH_4^+ and NO_3^- were comparable (N = 2.8 to 5.6 $mg\,m^{-2}\,d^{-1}$) in winter. After mass sedimentation of the algae NH_4^+ release increased to 21 $mg\,m^{-2}\,d^{-1}$ and NO_3^- flux changed to an uptake of 11.2 $mg\,m^{-2}\,d^{-1}$. In summer, NH_4^+ fluxes decreased again to 5 $mg\,m^{-2}\,d^{-1}$ while NO_3^- uptake changed to 2.1 $mg\,m^{-2}\,d^{-1}$. The intense denitrification was attributed to the low O_2 concentration in the bottom water, which also inhibited nitrification. Later, in autumn, lower maxima were found. The annual N-budget indicated that 25% of the total inorg-N loss was caused by denitrification in the sediment layer; 50% of the NO_3^- consumed was provided by the bottom water.

NH_4^+ recycling, nitrification, denitrification and NO_3^- and NH_4^+ fluxes were studied by Kemp *et al.* (1990) in sediment of Chesapeake Bay (Maryland, USA). C_2H_2 and *N*-serve were used as the usual inhibitors. There was a midsummer flux of NH_4^+ to the overlying water and a May peak of NO_3^- removal. Nitrification and denitrification both showed high values in spring and autumn, and were virtually absent in summer. Nitrification was limited by the slight O_2 penetration into the sediment, less than 1 mm. The N_2 loss due to denitrification was equal to the recycled NH_4^+ flux to the overlying water in spring and autumn, but, as in summer denitrification is negligible, NH_4^+ recycling was enhanced. Denitrification rates using the C_2H_2 block method were supposed to underestimate actual rates based on calculations of the N-balance with a factor 2–7. These calculations, in which estimations of the errors involved are missing, imply, however, some rough estimates based on data obtained from other sites or studies.

Hopkinson *et al.* (1999) analysed nutrient recycling at 4 sites of a freshwater to marine transect in the Parker River-Plum Island Sound estuary (Massachusetts, USA). O_2, CO_2, N- and P-fluxes were measured in cores, while denitrification was quantified using stoicheiometry of C- and N-fluxes calculated as diffusive flux based on porewater gradient data – not a reliable method. At the freshwater site org-C in the sediment was 10.3%. In the sandy marine sediment it was 0.2%, the C/N ratio varying from 23:1 to 11:1. In the pore water, NH_4^+ was highest in the upper and mid-estuarine sediment, attaining values $> 14\,mg\,\ell^{-1}$. NO_3^- was frequently absent from porewater, except in the marine sandy sediment, where it could reach $\approx 100\,\mu g\,\ell^{-1}$. Denitrification rates were low in the upper and mid-estuarine sites, in agreement with the oxidation grade of this sediment, and were about 100% of the org-N_{sed} recycled at the marine site. Benthic remineralization could supply < 1–190% of the N-requirements of the phytoplankton, but only 0 to 21% of its P-demand, at least during the rather short incubation times of the cores (6–48 hr).

Stepanauskas *et al.* (1996) studied ammonification, nitrification and denitrification as a function of water infiltration rate in wetland soil cores ($\approx 10\,cm$) from an artificially flooded meadow near Vomb (Sth. Sweden), using ^{15}N dilution and N-pairing methods. Infiltration rates varied between 72 and 638 $mm\,d^{-1}$; these rates seem extraordinarily, if not unrealistically high, for a '*wetland*', be it natural or artificial, as it gives values between 14.4 and 128 m for a summer season of about 7 months. It probably means that the meadow is dry for the major part of the year. Infiltration rate controlled the amount of O_2 and NO_3^- supplied to these soils. At the highest infiltration rate the outflow equalled the inflow in chemical composition. The mean annual denitrification rate of $16.8 \, mmol \, m^{-2} \, d^{-1}$ (with a standard deviation of 1.8) remained constant at the low infiltration rates, but decreased twofold at $638 \, mm \, d^{-1}$. Assimilatory NO_3^- uptake was low as compared with denitrification at low infiltration rates, but equalled denitrification at $638 \, mm \, d$. Ammonification was nearly constant at $10.4–16.1 \, mmol \, m^{-2} \, d^{-1}$; at $72 \, mm \, d^{-1}$ only 0.5% was nitrified. Dissimilatory NO_3^- reduction to NH_4^+ was always $< 1\%$ of the total. The ratio of CO_2 production over O_2 consumption was low (0.5 at low infiltration rate, but 0.25 at $267 \, mm \, d^{-1}$) and was explained by chemolithotrophic bacterial metabolism. It might be suggested that oxidation of FeS by NO_3^- or O_2 could play an important role. Davidsson *et al.* (1997) made a follow up to these studies by measuring the vertical patterns of the N-transformations. In a sandy soil all O_2 was consumed in the top 14 cm and 70% of the NO_3^- was consumed in the longest (28 cm) core. In a peaty soil all O_2 was consumed in the top 7.5 cm and all NO_3^- in the top 20 cm. Denitrification was counteracted by org-N and NH_4^+ release. In the peaty soil, N-release was equal to N-removal in the 14 cm cores, but exceeded removal in the deeper layers, leading to a Tot-N increase of 100% in the effluent of the 38 cm core. Infiltration rate and depth seem to be two factors that may strongly influence N-removal in wetlands, but the influence of the infiltration rate and incubation time must be studied better. Ammonification dropped off within 1–2 days and for the other variables the article gives no information, as a steady state was assumed to have been reached. This seems unlikely, certainly as far as the reducing capacity is concerned, which is always present in limiting amounts.

4.8   Links with other elements

4.8.1   LINK WITH SALINITY

In estuaries, salinity may have a strong effect on several processes in the N-cycle. Rysgaard *et al.* (1999) showed NH_4^+ release when sediment from the Randers Fjord estuary (Denmark) was suspended in solutions with increasing salinity, and found decreasing amounts of adsorbed NH_4^+ extractable with KCl when salinity was increased from 0 to 1%. The effects were largest at 1% NaCl, and did not change when the salinity increased further to 3%. *In situ* nitrification and denitrification (measured as $^{15}N_2$ production) also decreased when the salinity was increased to 1%. The authors suggested that the latter two effects were a physiological effect of the salinity, but as most experiments lasted 4 hr, this may have been the result of an adaptation.

4.8.2   LINKS BETWEEN N, P AND S CYCLE

In connection with denitrification caused by FeS, Golterman (1991b) drew attention to an unexpected direct link between S, N and P cycles: denitrification with FeS as the reducing agent produces FeOOH, which will strongly enhance P binding onto sediment, which in turn will become more oxidized and lose its toxic character for some parts of the vegetation (Golterman, 1991b; De Groot, 1991). Figure 4.18 represents a schematic link between these

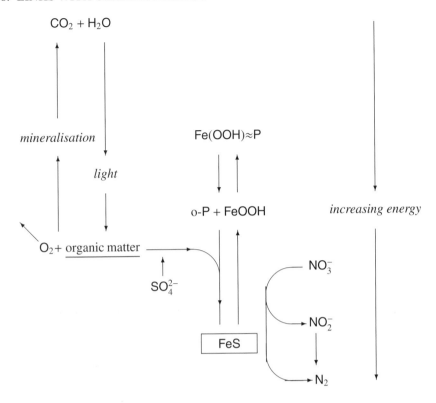

FIGURE 4.18: Schematic relations between primary production, denitrification and phosphate binding onto sediment (Golterman, 1995c).

processes, but it remains difficult to develop an overall mathematical model combining them. The overall result is a loss of N-compounds and a stronger P-fixation, which together will produce a strong pressure towards blue–green algal development. This will especially be the case in productive wetlands, where it is not desirable. The practical consequences of the relation between denitrification and P adsorption onto $FeOOH$ were realized and applied already long before this chemical mechanism became clear. It has several times been discussed or applied.

Ripl (1976) and Ripl *et al.* (1979) emphasized the possibility of adding NO_3^- to decrease the O_2 consumption of polluted lakes and to prevent P-release. Andersen (1982) evaluated this effect in 31 Danish lakes and found a positive correlation between high NO_3^- concentration and low P-release from sediment. The cause of this coupling was often said to be due to a high redox potential, without specifying the chemical processes. Tirén & Pettersson (1985) analysed this coupling further and related it to the prevention of P-release from $Fe(OOH){\approx}P$. Besides a smaller P-release they showed in some sediments an increased release of o-P and NH_4^+. This might be caused by increased bacterial metabolism (denitrifiers also need to oxidize organic matter for their energy) or by a shift from NO_3^- to Fe^{3+} as final electron acceptor for the denitrifiers, when the NO_3^- was depleted.

Andersen (1985) mentioned increased NO_3^- loadings as a cause of improved P-binding onto sediment and suggested that this may have a desirable effect on phytoplankton species composition. He emphasized that NO_3^- must not be used for lake restoration as an alternative to the reduction of P-loading. This is correct as the increased P-binding in the sediment will counteract long-term measures.

The after-effects of NO_3^- depletion are often overlooked and may, in fact, have results different from those expected. Strong denitrification will also have a strong effect on P-release from sediment dominated by $CaCO_3 \approx P$ by decreasing the pH. Reliable P-fractionation will increase the predictive value for restoration measures.

Feibicke (1997) treated the sediment in an enclosure in the brackish Ford Schlei in the Baltic Sea (Germany) by injection with an excess of commercial calcium-salpeter, probably 95% $Ca(NO_3)_2$. There was a significant release of gas, mainly N_2 and CO_2. Average N-consumption rates were 0.8 and $1.8 \, g \, m^{-2} \, d^{-1}$ in the two treatments, with maximal values somewhat higher. In the Schlei Ford, Tot-P concentrations increased from $0.14 \, mg \, \ell^{-1}$ in April to $0.74 \, mg \, \ell^{-1}$ in August, but in the enclosures Tot-P reached 0.11 and $0.15 \, mg \, \ell^{-1}$, respectively. As a response the phytoplankton biomass also decreased to about one third of that outside the enclosure. It seems clear that denitrification by FeS and adsorption of P onto the FeOOH formed may explain these results. P-fractionation of the sediment might have given the ultimate proof.

McAuliffe et al. (1998) demonstrated that NO_3^- addition of 5 or $10 \, mg \, \ell^{-1}$ of N considerably decreased P-release from sediment cores from the Harvey Estuary (Western Australia). Even at a station with relatively low O_2 demand, the o-P decreased from $160 \, \mu g \, \ell^{-1}$ to 10–$20 \, \mu g \, \ell^{-1}$. In other cores the addition of $75 \, mg \, \ell^{-1}$ of N as NO_3^- caused the o-P concentration to decrease from about $900 \, \mu g \, \ell^{-1}$ to $> 250 \, \mu g \, \ell^{-1}$ which is equal to the concentration at high O_2 concentrations. Other results of the NO_3^- additions were a lower O_2 demand and lower, stable redox potentials. Different mechanisms were discussed, but the possibility of FeS interactions was not yet taken into account.

In their experiments on denitrification (see Section 4.4.6), Tuominen et al. (loc. cit.) found, after the addition of algae, much higher o-P concentrations in the pore water as compared with the increase in the NH_4^+ concentration, than in the control experiments, after both 4 and 12 days. The addition of the benthic fauna decreased these concentrations even below the controls. Anaerobic P-release from the algae seems a reasonable explanation, especially because o-P and NH_4^+ concentrations followed the same pattern, together perhaps with an inhibition of P-adsorption onto FeOOH due to FeS formation, while oxygenation due to bio-irrigation restored the FeOOH adsorption capacity.

Kleeberg et al. (2000), discussing a restoration project of Lake Jabel (NE Germany), argued that the lake would not improve after a diversion of the nutrient input because less input of NO_3^- would lead to an increased P-release from the sediment. This P-release amounted to about 25% in two of the lake's basins. The continuing input of NO_3^- may also keep the H_2S concentration in the hypolimnion low by denitrification. Furthermore, the water retention time would increase from 1.3 to 2.3 years, probably causing an increased P-retention. The authors argued that P-removal from wastewater and in-lake measures are better possibilities.

Chapter 5

Methodological aspects

Though this be madness,
yet there is method in 't.
Hamlet, II ii (211)

5.1 Some analytical methods for N, P, S and Fe compounds discussed in the preceding chapters[1]

In this chapter some titrimetric and spectrophotometric methods for the determination of compounds discussed in the preceding chapters are presented. No methods are given for more sophisticated instruments like auto-analyser or C-H-N analyser, as the manuals for these procedures come with the equipment. It must, however, be strongly emphasized that not all these methods are suitable for sediment extracts or digests or interstitial water. The reduction of NO_3^-, e.g., with the Cd-reduction method is often strongly interfered with by organic matter.

5.1.1 DETERMINATION OF NH_3 WITH NESSLER'S REAGENT OR AS INDOPHENOLBLUE

Principle:

NH_3 is distilled from its solution at a slightly alkaline pH before the determination; it can then be estimated either by titration or by a spectrophotometric method using Nessler's reagent or an indophenolblue method, modified to avoid the use of phenol. The distillation is automatically included in the case of org-N determination, and usually in the case of NO_3^- (but see Section 5.1.2.2). The titration is more precise, but more time-consuming. With an automatic titrator with pH = 5 as the preset endpoint, the limit of detection is about $50 \mu g$ per sample.

In the Nessler determination HgI_2 is used, which forms a brown-yellowish colour with NH_3 in an alkaline medium. The colour depends on the alkalinity and the ratio Na/K, which

---

[1] Parts of this chapter are taken from Golterman *et al.*, 1978.

183

must therefore be standardized. As in a strongly alkaline medium $CaCO_3$ will precipitate, it is usually carried out after distillation, but under certain precautions Ca^{2+} can be chelated, e.g., with salicylate. (See Section 5.1.1.1.)

In the indophenolblue method a phenol and NH_3 are oxidized by $HClO$, forming a blue compound, indophenol. Na-nitroprusside and K-ferrocyanide are added as catalysts. Salicylate is used instead of phenol (Verdouw et al., 1978), as phenol is an unpleasant reagent to work with. The method is always used on lake water directly; there is, however, little experience with possible interferences.

The following modification is an optimization of Verdouw's method.

NH_3 (and NO_3^-) can be determined in lake water (even hard) without distillation with Nessler's reagent – see Section 5.1.1.1. There is so far little, though positive, experience, but the method should be tested carefully.

5.1.1.1 Determination of NH_3 with Nessler's reagent (with or without distillation)

Reagents:

Nessler's reagent:

N1) Dissolve 25 g of HgI_2 (red) and 20 g of KI in 500 ml of H_2O.

N2) Dissolve 100 g of $NaOH$ in 500 ml of H_2O.

N3a) Mix 1 vol N1 + 1 vol N2 and keep in a refrigerator. The mixed reagent is stable for at least 2 months.

N4) (Sodium salicylate solution) Dissolve 10 g of Na-salicylate in H_2O and make up to 100 ml in a measuring cylinder. The solution is stable in a refrigerator for at least 15 days.

Procedure:

NH_3 can be distilled from lake water or sediment (extract) with steam; the pH is rendered slightly alkaline with 1 or 2 ml borax. For the spectrophotometric determination of NH_3, mix 20 ml sample (either the distillate or the lake water), while shaking, with 1 or 2 ml of N4 in a 25 ml volumetric flask, followed, still swirling, by 1 ml mixed reagent N3a and make up to 25 ml. Measure the O.D. at 420 nm after 15–30 min. If the NH_3 has been distilled, no N4 is needed.

(If after addition of the Nessler's reagent a turbidity still occurs ($CaCO_3$), try an increased amount of N4.)

Reference: Golterman, 1991.

5.1.1.2 Determination of NH_3 with a modified indophenolblue method

Reagents:

N5) 40% solution of sodium salicylate in H_2O. This solution cannot be kept, and should be prepared daily. Use 'extra pure' as this may yield a lower blank than 'A.R.'.

N6a) Dissolve 1 g of $Na_2\{Fe(NO)(CN)_5\}.2H_2O$ (i.e., sodium nitroprusside) in 100 ml of H_2O. This solution can be kept in a clear bottle in a refrigerator for about 4 weeks; the normal colour 'orange–brown' turns greenish when it is deteriorated.

N6b) Dilute solution N6a 10 times precisely ($1.00\,ml \longrightarrow 10.0\,ml$) and add 100 mg of sodium dichloro-isocyanurate (= $Cl_2Na(CNO)_3.2H_2O$). This solution cannot be kept.

Procedure:

Mix 25–40 ml sample (lake water or distillate) in a 50 ml volumetric flask with 5 ml of N5. Add 1 ml of 10 M NaOH. Add 2 ml of the mixed reagent N6b.

Measure the O.D. at 650 nm after 60 min.

Notes:

- It is essential to run at least 2 standards with each series as, among other factors, temperature has some influence.
- Na dichloro-isocyanurate is expensive: do not prepare more than needed. It can be obtained from, e.g., Merck (10887) or Fluka (35915).
- NH_3 standards should be run at least once a month with the NH_3 titrations; with each series if the spectrophotometric methods for NH_3 are used.

Reference: Verdouw *et al.*, 1978.

5.1.2 DETERMINATION OF $\Sigma(NO_2^-$ AND $NO_3^-)$ AFTER REDUCTION WITH $TiCl_3$

Principle:

NO_3^- can be reduced to NH_3 either with Devarda's metal at $100\,°C$ or with $TiCl_3$, the latter method either at room temperature or at $100\,°C$; both reductions must be carried out at high pH. Reduction at room temperature has the advantage that less org-N will be hydrolysed. After reduction with $TiCl_3$ (in the cold), NH_3 can be determined with (Section 5.1.2.1) or without distillation (Section 5.1.2.2).

5.1.2.1 *With distillation*

Reagents:

N7) 10 M NaOH. Dissolve 40 g of NaOH in 100 ml H_2O.
N8) 15% $TiCl_3$ in strong HCl (commercially available).

Procedure:

Mix 50 ml water sample (or less, diluted to 50 ml) with 4 ml N7 in, e.g., a Kjeldahl flask. Add 1 or 2 ml N8, rinse quickly, attach **directly** to the distillator and trap the NH_3 in 2 ml 4% H_3BO_3. NH_3 in the distillate can be measured either by titration or spectrophotometrically.

Titration:

Add 0.02 M H^+ to the distillate till pH = 5 (pH meter or automatic titrator). Take care that there is still sufficient water present at the end of the distillation.

Run blanks using 50 ml H_2O and standards at least once a month.

Calculation:

1 mg N = 3.57 ml of 0.02 M H^+: let b = ml acid for blank and x = ml acid for sample, then $(x-b)/3.57$ = mg N per sample. Subtract already present NH_3 to obtain the $\Sigma(NO_2^-$ and $NO_3^-)$.

Notes:

- $TiCl_3$ *should be stocked in well-filled bottles* (because it oxidizes with the air). 1 ml is sufficient for 0.5 mg N; 2 ml is sufficient for 1 mg N.
- A 15% $TiCl_3$ solution in HCl is commercially available, but the concentration may be different for the different marks.

5.1.2.2 Determination of $\Sigma(NO_3^- + NO_2^-)$ *with Nessler's reagent, without distillation*

Principle:

Reduction of NO_3^- with $TiCl_3$ is followed by the Nessler method for the NH_3 produced. As after the $TiCl_3$ reduction the solution is already strongly alkaline Nessler's reagent is prepared with half the quantity of NaOH.

Reagents:

N9) Dissolve 50 g of NaOH in 500 ml H_2O.

N3b) Prepare as Nessler's reagent, but mix 1 vol N1 with 1 vol N9.

Procedure:

Mix 45 ml of the test solution (or a smaller quantity made up to 45 ml) with 3 ml N7 followed, still swirling, by 1 ml $TiCl_3$ (N8). Make up to 50 ml in a volumetric flask. After carefully mixing pour the black suspension into a 50 ml reagent tube and leave covered till the precipitate is sufficiently settled so that a 10 ml sample can be taken, or centrifuge in a closed bottle. Put the sample in a volumetric flask of 25 ml, add 1 ml of Nessler's reagent (N3b), and make up with H_2O. Measure the colour as above for NH_3.

Note:

- **For freshwater only**: Method not yet in common use; check your own samples to see whether it works correctly.

5.1.2.3 Determination of $\Sigma(NO_3^- + NO_2^-)$ *with the indophenolblue method*

The spectrophotometric procedure is a combination of methods 5.1.2.2 and 5.1.1.2.

Mix 45 ml sample or standard with 3 ml 10 M NaOH (N7) and add 1 ml $TiCl_3$ (under swirling). Mix and adjust to 50 ml and mix again. Allow to precipitate in 50 ml reagent tube. Use 10–15 ml of clear supernatant for NH_3 determination as in 5.1.1.2.

Note:

Use 0.6 ml NaOH instead of 1.0 ml 10 M NaOH as mentioned in 5.1.1.2.

5.1.3   Determination of NO_2^-

Principle:

In a strongly acid medium HNO_2 reacts with sulphanilamide to form a diazonium compound, which reacts quantitatively with N-(1-naphtyl)ethylenediamine.2HCl to form a strongly coloured azo compound. If any NO_2^- is left over at this stage it will destroy the reagent (N13) so that almost no colour will develop; the sample will appear to contain no NO_2^-. Possible excess of NO_2^- is therefore destroyed by adding ammonium sulphamate ($NH_2SO_3NH_4$) just before reagent (N13). This situation also allows the use of as near perfect reagent blank as may be obtained; ammonium sulphamate is added to the water sample as first (rather than as third) reagent. All NO_2^- is thus destroyed before colour development, and the selfcolour of the water sample is included in the blank. This is of importance in coloured waters, e.g., interstitial or humic rich waters.

Reagents:

N10) KNO_2 standard, 0.05 M (0.7 mg ml^{-1} of NO_2-N). Dissolve 1.064 g of KNO_2 (A.R., dried at 105 °C for 1 hr) in H_2O. Add 1 ml of 5 M NaOH and dilute to 250 ml. The solution can be stored, but must be checked oxidimetrically.

(To do this, mix 10.0 ml of N10 with 25.0 ml 0.1 M 1/5 $KMnO_4$ (standardized), acidify with 5 ml 2 M H$^+$. Wait 15 min and add a known excess of $(COOH)_2$. Then back-titrate with the $KMnO_4$ solution.)

N11) 6 M HCl.

N12) 0.2% Sulphanilamide. Dissolve 2 g of sulphanilamide in 1ℓ of H_2O.

N13) 0.1% N-(1-naphtyl)ethylenediamine.2HCl. Dissolve 0.1 g of N-(1-naphtyl)ethylenedi-amine.2HCl in 100 ml H_2O. The reagent can be kept in a refrigerator for at least 2 months.

N14) 5% Ammonium sulphamate. Dissolve 5 g of $NH_2SO_3NH_4$ in 100 ml H_2O. Store at room temperature in darkness.

Procedure:

Mix 40 ml sample (or an aliquot containing no more than 20 μg of NO_2-N) in a 50 ml volumetric flask. Add 2.5 ml N12 and 1 ml N11. After 3 min add 1 ml of solution N14 followed after 3 min by 1 ml N13. Bring up to volume.

Prepare a blank by adding 1 ml of reagent N14 and 1 ml N11 before N12 is added. Measure the O.D. near 530 nm of the reagent blank, standards (0–20 μg of NO_2-N) and samples after ≈ 15 min.

Storage of samples:

The NO_2^- concentration of samples can decrease rapidly if stored. When prolonged storage cannot be avoided, add the reagents N11 and N12 in the field. The sample can then be stored 24 hr at 4 °C. Filtration can be carried out after colour development.

5.1.4 DETERMINATION OF COD (ORG-C) AND ORG-N BY DIGESTION WITH $K_2Cr_2O_7$

Principle:

org-C compounds can be oxidized in a strongly acid medium according to reaction:

$$Cr_2O_7^{2-} + 14H^+ + 6e^- \longrightarrow 2Cr^{3+} + 7H_2O.$$

The oxidation is carried out at 130–135 °C in closed pots (Teflon, e.g., TuffTainer) in an oven. Cl$^-$ is oxidized as well and is, therefore, chelated with Hg^{2+}; Ag^+ is added as catalyst. The Cr^{3+} formed is green and can be measured by spectrophotometry.

Reagent:

CN1) (*Digestion mixture*) Add slowly 375 ml H_2SO_4 (s. g. = 1.98) to ±500 ml H_2O. Add while mixing 1 g $HgSO_4$ and 0.1 g Ag_2SO_4 and make up to 1ℓ. Add 49 g of $K_2Cr_2O_7$. **Never add H_2O to H_2SO_4!**

Procedure:

Bring sample in TuffTainer. Use ≈ 700–1000 mg of sediment sample (< 100 mg of plant material). Evaporate solutions till near dryness; if $CaCO_3$, H_2S or FeS are present, first add 1 or 2 ml 10 M H_2SO_4 (= 1:1) to remove CO_2 and/or H_2S. Add 25 ml or 35 ml digestion mixture CN1. (25 for small, 35 ml for large TuffTainers.) Close TuffTainer and heat for 3 hours at 135 °C. Open when at 100 °C and bring contents in 50 ml volumetric flask. Bring up to volume, mix well and if not clear, allow to settle completely. If clear: measure O.D. at 620 or 585 nm.

For org-N: Bring precisely 10 or 25 ml in a Kjeldahl flask, add small excess 10 M NaOH and determine the NH_3 after distillation.

Calculation:

1 meq. (e^- or as O_2) in 50 ml has an O.D. of 0.0922 at 620 nm; assuming a $CO_2/O_2 = 1$ we can also say: 3 mg C in 50 ml has an O.D. of 0.0922. Therefore, if b is the extinction of the blank and x is the O.D. of the sample, then

$$(x - b)/0.0922 = \text{mequivalents of } O_2$$

and

$$(x - b)/0.0922 \cdot 3 = \text{mg of org-C}.$$

Notes:

- If the concentration of org-C is low, measure at 585 nm (the absorption is about 20% higher), but the blank is relatively high! The decrease in the O.D. due to the consumption of the $Cr_2O_7^{2-}$ must be calculated.
- For brackish waters more $HgSO_4$ must be added: The Hg^{2+}/Cl^- molar-ratio should be about 10. It can be added in solid form to the sample in the TuffTainer.

Reference: Golterman *et al.*, 1978.

5.1.5 DETERMINATION OF UREA

Principle:

Urea must be enzymatically hydrolysed to NH_3 before determination; the NH_3 formed is measured by one of the above-mentioned procedures.

Reagent:

U1) Dissolve 2.5 g of Na_2-EDTA in 250 ml H_2O. Adjust the pH to 7. Add 50 mg of urease (e.g., Sigma, 70.000 μmolar units per g solid).

When the urease is dissolved, divide over 10 ml plastic tubes and deep-freeze the tubes not used directly.

Procedure:

Take 10–50 ml sample and incubate at room temperature, but not below 20 °C, with 1–5 ml of the reagent U1 during 30–120 min. The time depends on the quantity of urea present. Distill the NH_3 formed as described in Section 5.1.1.1. In the distillate the NH_3 originally present plus newly formed are determined together.

Run 2 controls with every series in the range of $50 \mu g$–1 mg of urea per sample.

5.1.6   DETERMINATION OF o-P AND TOT-P

Principle:

o-P is measured with the normal Murphy & Riley (1958) procedure, using molybdate mixed with sodium antimony tartrate followed by reduction with ascorbic acid. Tot-P in water, extracts and sediments can be determined after destruction by $K_2S_2O_8$ and H_2SO_4. The quantities needed depend on the material to be destructed.

Reagents:

P1) (*Molybdate-antimony solution*) Dissolve 4.8 g of $(NH_4)_6Mo_7O_{24}.4H_2O$ and 0.1 g of sodium antimony tartrate $(NaSbOC_4H_4O_6)$ in 400 ml of 2 M H_2SO_4 and make up to 500 ml with the same acid.

P2) (*Ascorbic acid*) Dissolve 2 g of ascorbic acid in 100 ml of H_2O. The solution is stable for 1 week if kept in a refrigerator.

5.1.6.1   Procedure for o-P

a) Put the sample (40 ml or less) in a 50 ml volumetric flask. Add 5 ml P1 followed by 2 ml P2, mix well and bring up to volume. O.D. is highest at 882 nm, but can be measured at lower wavelengths.

b) If the O.D. is below 0.050 but above 0.010, transfer the coloured sample quantitatively into a separatory funnel and extract with 10 ml *n*-hexanol. Discard the lower layer and transfer the hexanol into a 10 ml volumetric flask or precisely graduated cylinder. Adjust to 10 ml with a few drops of isopropanol which clears the solution. The O.D. is highest at 690 nm. The calibration curve should be checked more often than in a) above.

References: Murphy & Riley, 1962; Golterman *et al.*, 1978.

5.1.6.2   Procedure for Tot-P

The sample must be completely digested either with a wet method or by heating at 500 °C followed by an extraction with HCl. The wet method is easiest.

Wet method: Bring 100–200 mg of dry sediment or an equivalent of a sediment suspension in a 50 ml flask or TuffTainer. Add a few ml 2 M H_2SO_4, sufficient to neutralize $CaCO_3$ present. Add $K_2S_2O_8$; the quantity depends on the amount of org-C present. Preliminary trials are needed. Heat at 130 °C for 2–3 hr, cool and dilute to 100 ml. A few ml of this digest can be used with method 4.1a. Neutralization with (strong) NaOH is not needed and may cause losses as around the drops precipitation of $Fe(OOH) \approx P$ may occur. Method 5.1.6.1a tolerates a few ml of the acid digest.

For water the following quantities are sufficient: Add 2 ml 0.1 M H_2SO_4 to 40 ml lake water directly in a volumetric bottle of 50 ml. Add 1 g of $K_2S_2O_8$ and heat in an autoclave at 120 °C for 1–2 hr. Cool and add the o-P reagents.

Notes:

- $K_2S_2O_8$ produces H_2SO_4 upon decomposition; the digest is therefore more acid than the added H_2SO_4 alone.
- If a precipitate persists after bringing up to 100 ml this is probably $CaSO_4$. Try a further dilution or decant. H_2O_2 may replace the $K_2S_2O_8$ and will give fewer problems with $CaSO_4$ formation. Great care must then be taken that all H_2O_2 is destroyed as it interferes strongly with the o-P determination.

5.1.7   EXTRACTIONS OF FeOOH≈P AND CaCO₃≈P BY CHELATING COMPOUNDS

Principle:

FeOOH≈P is extracted from sediments with a buffered solution of Ca-EDTA + dithionite; the pH must be roughly adjusted to that of the sediment. $CaCO_3$≈P is then extracted from the residue with Na_2-EDTA. The method is developed for calcareous sediment; for more acid sediment some steps of the method should be modified.

Reagents:

Solution 'Ex1': 0.05 M Ca-EDTA. Dissolve 18.6 g of Na_2H_2-EDTA.2H_2O (Titriplex III) together with 7.35 g of $CaCl_2$.2H_2O in 1ℓ H_2O. Add Tris-buffer till the pH ≈ 9. After the addition of 1% Na-dithionite, just before the extraction, the pH must be 7–8, equal to the original sediment pH. Add to the samples without delay.

Solution 'Ex2': 0.1 M Na_2-EDTA. Dissolve 37.2 g of Na_2H_2-EDTA.2H_2O in 1ℓ of H_20. The pH is about 4.5.

Procedure:

A wet sample of sediment equal to ≈ 0.5 g of dw., containing no more than 8 mg of Fe as extractable FeOOH, is mixed with 50 ml of a buffered 0.05 M Ca-EDTA solution 'Ex1' and left for 1–2 hr under occasional shaking. After centrifuging, 1 ml (or less) is used for the Fe determination and no more than 4 ml for the P determination. The Fe is reduced with 1 ml 2% ascorbic acid (as used in the o-P determination); o-P is determined as in Section 5.1.6. The extraction is repeated till the last extract contains < 5% of the sum of the preceding ones. The pellet is then resuspended in a 0.1 M solution of Na_2-EDTA, solution 'B' (pH = 4.5), and centrifuged after 4–6 hr. If the Ca concentration of the sediment is high, several extractions may be necessary. Later extractions may last longer. Check always whether this is necessary. Use 2 ml for the o-P determination.

Notes:

- EDTA solutions have the disadvantage of interfering with the o-P and Fe determinations. For the o-P determination 2 ml 0.1 M of EDTA can be used without interference. For larger volumes, there are two options:
    - Heating (≈ 75 °C), which allows the normal procedure for up to 5 ml of 0.1 M Ca-EDTA.

– Using $25 \, g \, \ell^{-1}$ of $(NH_4)_6Mo_7O_{24}.4H_2O$ (instead of the usual $9.6 \, g \, \ell^{-1}$, Murphy & Riley, 1962) with 10 ml acetone added per 50 ml final volume and measuring the blue colour at 815 nm. With 7.5 ml EDTA 0.1 M, the calibration curve has a slope 1% less than the normal calibration curve. Golterman (1995) mentioned that with this stronger molybdate solution 10 ml EDTA can be present without influencing the O.D. This depends on the resolution of the spectrophotometer.

A different approach may be suggested, i.e., extraction of the yellow phosphate-molybdate complex with butanol or hexanol, followed by reduction to the blue colour in the alcoholic phase (Golterman & Würtz, 1961) which is not suitable for routine analysis. With sediments from the Camargue (France), several lakes in the Netherlands or in the National Park Doñana (Spain) this was not needed as sufficient P was always present to use small quantities of the two extractants.

- Dithionite may cause a greenish colour to develop when the molybdate solution (P1) is added. Usually this is no problem if the first drops of P1 are added slowly. When much dithionite is left over (e.g., in the blank) this can be destroyed by adding a few drops of acid and gently shaking.

- Dithionite solutions deteriorate upon standing. Add the dithionite in dry form, about 1% per sample. Even in the dry state, e.g., in pots which are often opened, dithionite deteriorates slowly. Store the stock therefore in small pots (used one by one) and check the activity regularly using a $Fe(OOH)$ suspension of $300 \, mg \, \ell^{-1}$ of Fe^{3+}.

- With the phenanthroline method the O.D. of Fe must be lower than 0.8; if higher, take a smaller sample. The Fe-phenanthroline colour develops slowly in the presence of EDTA, but not if 1 ml 2 M Na-acetate solution is added. Up to 2 ml 0.1 M EDTA can be used, if necessary, after dilution. As $Fe(OOH)$ is always present in much larger amounts than P, this dilution never poses a problem.

- EDTA (e.g., 10 ml) can be destroyed by boiling with ($1.5 \, g$ of $K_2S_2O_8 + H_2SO_4$) as for org-P determinations. Take care to use sufficient $K_2S_2O_8$ to ensure complete destruction; the pH of the extract to be destructed should be $\approx$ 3–8. The difference between the o-P and Tot-P in the extracts is supposed to be org-P.

If the molybdate with the higher concentration is used, it permits either larger volumes of the extractants for the colorimetric analysis (sediments with low $Fe(OOH) \approx P$ concentrations), or higher concentrations (sediments with high $Fe(OOH) \approx P$ concentrations). With 0.05 M of Ca-EDTA the extraction of Camargue sediments became more efficient, so that all $Fe(OOH) \approx P$ could be extracted in two extractions.

If no separation of $FeOOH \approx P$ and $CaCO_3 \approx P$ is needed, they can be extracted together with acid 0.05 M Na citrate plus dithionite. If much $CaCO_3$ is present the extractant/sediment ratio should be such that the pH does not increase over 4. The citric acid interferes with the o-P determination; with 4 ml the O.D. will be unaffected, but will be obtained more slowly than usual. The citric acid can be destroyed by boiling with ($K_2S_2O_8 + H_2SO_4$). 2 ml citric acid does not interfere with the Fe determination.

5.1.8 EXTRACTIONS OF ORG-P

After the extraction of inorganic phosphate two organic phosphate pools may be extracted:

A) by an acid extraction, and

B) by an alkaline extraction.

After removing the last EDTA extract for the extraction of $CaCO_3 \approx P$, the pellet is washed with H_2O; the Acid Soluble org-P (org-$P_{\rightarrow ac}$) can then be extracted with $0.25\,M\ H_2SO_4$ ($0.5\,M\ H^+$) during 30 min at room temperature. o-P can be measured without destruction using small volumes. The remaining pellet is then extracted with $1\,M\ NaOH$ at $90\,°C$ during 30 min. The extracted org-P ($P_{\rightarrow alk}$; mainly phytate plus humic-P) can be determined only after destruction with $K_2S_2O_8 + H_2SO_4$. (The phytate $\approx P$ can be measured after enzymatic hydrolysis, i. e., with phytase; humic compounds may interfere with this hydrolysis, but can be destroyed easily.)

5.1.9 DETERMINATION OF Fe^{2+} AND Fe^{3+}

Principle:

Both Fe^{2+} and Fe^{3+} can be measured with o-phenanthroline, Fe^{2+} without ascorbic acid and Fe^{3+} with ascorbic acid added.

Reagents:

Fe1) (*o-Phenanthroline solution*, 0.5%) Dissolve 0.5 g of o-phenanthroline in about 100 ml of $0.01\ M\ H_2SO_4$. This solution can be kept in the refrigerator for months.

Fe2) (*Ascorbic acid solution*, 2%) Dissolve 2 g of ascorbic acid in about 100 ml (as for the P-determination) (is equal to reagent P2).

Fe3) (Na-*acetate solution*, 2 M) Dissolve 16 g of Na-acetate in about 100 ml of H_2O.

5.1.9.1 Procedure for Fe^{2+}

Add 1 ml of reagent Fe1 (and 1 ml Fe3 for EDTA solutions) to 20 ml sample, the pH of which must be between 3 and 8. Fill up to 25 ml and read the O.D. at a wavelength between 490 and 510 nm after 30 min.

5.1.9.2 Procedure for Fe^{3+}

Add 1 ml of reagent Fe3, 1 ml of reagent Fe1, 1 ml of reagent Fe2 to 20 ml sample, the pH of which must be between 3 and 8. Fill up to 25 ml and read the O.D. at a wave length between 490 and 510 nm after 30 min.

Note:

- reagent Fe3 is needed in case of EDTA solutions. We normally take only 0.1 or 0.2 ml of the Ca-EDTA extracts. (Or 1 or 2 ml after 10-fold dilution.)

Interferences:

Humic acids may cause the colour to develop slowly (ionic Fe^{2+} takes about 30 min). The samples can, however, easily be left overnight. Correction for selfcolour of the water (sample) must be made, e.g., by measuring the O.D. of a sample blank without reagent Fe1.

5.1.9.3 Procedure for Tot-Fe

The sediment is digested as for Tot-P (see Section 5.1.6.2) and the digest is diluted to 50 or 100 ml. Take 0.25 or 1.0 ml as for Fe^{3+}, but add the reagent Fe3 before Fe1 and Fe2. Make a calibration curve covering the range of Fe = 2 to 100 μg per sample. Standards can be made with $(NH_4)_2SO_4.FeSO_4.6H_2O$ (Mohr's salt, which is a primary standard).

5.1.9.4 Testing Fe(OOH)≈P extractability

Procedure:

Dissolve a Fe^{3+} salt in water, making a solution of 300 mg of Fe^{3+}. Use, e.g., about 5.35 mmol $FeCl_3.6H_2O$ per litre H_2O. The solution is yellowish and nearly clear. Add under stirring 3 times 5.35 mmol of NaOH, e.g., as a 2 M solution. At the end the pH must be 8–8.5, which is not critical, but too high values must be avoided. The FeOOH will precipitate easily and rapidly. In the following days decant the water several times, e.g., each evening and morning during a week. The easiest is to put the suspension in a 1 or 2 litre measuring cylinder and decant the overlying water or use vacuum pump with pipette. The Cl^- concentration must be near 0. Suspend the FeOOH in H_2O; the brown suspension settles quickly.

Add 12 mg ℓ^{-1} of P (as KH_2PO_4) of which ≈ 11 mg will be adsorbed. For the extraction test use 5 ml of this suspension. The Fe(OOH) should go into solution when mixed with 25 ml of Ca-EDTA/dithionite extractant in 30–60 min. Fe and P can now be measured; the P should be recovered with 1 or 2% error.

5.1.10   DETERMINATION OF SO_4^{2-} AND H_2S + FeS

5.1.10.1 Determination of SO_4^{2-}

Principle:

SO_4^{2-} is precipitated with Ba^{2+}. The excess Ba^{2+} is precipitated with a known excess of CrO_4^{2-}. The first precipitate is made in a weakly acid medium with $BaCrO_4$, the second precipitate is made by increasing the pH. The remaining CrO_4^{2-} can be measured spectrophotometrically. The excess of Ba^{2+} over SO_4^{2-} need not be known; the excess of CrO_4^{2-} over Ba^{2+} must be constant; it is found in the blank.

Reagents:

S1) (*Mixed 0.005 M Ba/CrO₄ solution*) Dissolve 0.74 g of $K_2Cr_2O_7$ in about 400 ml of H_2O and 1.22 of $BaCl_2.2H_2O$ +3 ml 6 M HCl in 400 ml of water. Add the second solution to the first while swirling and make up to 1ℓ in a volumetric flask; the pH is then ≈ 2.

S2) (0.1 M *borate buffer*) Dissolve 38 g of $Na_2B_4O_7.10H_2O$ in H_2O and dilute to 1ℓ. Set pH = 9 with ≈ 8 ml of 6 M HCl.

S3) (*Sulphate stock solution*) Use a 10-fold dilution of a commercially available standardized solution of 0.1 M H_2SO_4.

Procedure:

Mix 10 ml of the mixed reagent S1 with 20 ml sample in a test tube (ø = 2–3 cm). If no precipitate occurs, rub with a glass rod against the inner surface of the tube to complete precipitation of $BaSO_4$. Leave for 30 min while swirling occasionally. Add, while swirling, 5 ml buffer solution, S2; the pH is then ≈ 8.5. Leave for 30 min. The precipitate may be left to settle or be centrifuged. Dilute 5 ml to 25 ml and read the O.D. at 365 nm. The colour is stable.

Make a calibration curve with 1–5 ml of 0.01 M SO_4^{2-}.

Notes:

The pH during the $BaSO_4$ precipitation is ≈ 3. As the $BaSO_4$ does not precipitate easily at this pH rubbing may be essential. In the second step the pH must be 8.5–9 to ensure that all Cr-anions are in the form of CrO_4^{2-}.

It is an advantage to have a slightly positive blank; therefore an excess of 3% of CrO_4^{2-} is used. If desired this may be made less, but a negative blank must be avoided. (Excess SO_4^{2-}.)

Golterman & De Graaf Bierbrauwer (1992), using Rhône water with a concentration of $SO_4^{2-} = 0.7$ mM, found a standard deviation of 0.6% ($N = 10$).

Reference: Golterman & De Graaf Bierbrauwer, 1992.

5.1.10.2   Determination of H_2S and FeS

Principle:

H_2S is flushed directly from the wet sediment. FeS is converted to H_2S by acidification. H_2S is trapped directly in a NaClO solution to which Zn-acetate is added. H_2S is oxidized to SO_4^{2-} according to the following reaction:

$$S^{2-} + 4ClO^- \longrightarrow SO_4^{2-} + 4Cl^-.$$

Either the excess NaClO is back-titrated with As_2O_3 and KIO_3 standardised solutions, or SO_4^{2-} is measured spectrophotometrically with method 5.1.10.1. The titration of NaClO is carried out by adding first KI and H^+, while after the production of the I_2 the pH is neutralized again by Na-acetate.

If much $CaCO_3$ is present in the sediment this can be removed with citric acid (at pH > 4.5 in order not to dissolve the ZnS or FeS) as the CO_2, formed from the $CaCO_3$ upon acidification, interferes with NaClO.

Sediment can be stored after addition of $ZnCl_2$, the quantity of which must be determined for each sediment.

Reagents:

S4) (≈ 0.013 M NaClO) Dilute commercial NaClO (bleaching or Javel water) 7 times. The solution can be kept in a refrigerator, but must be checked by titration with KIO_3.

S5) (≈ 0.5 M KI) Dissolve 8.3 g of KI in 100 ml of H_2O.

S6) $= 2$ M Na-acetate ($= Fe_3$).

S7) $= 0.5$ M and 4 M H_2SO_4.

S8) (≈ 0.07 M Zn-acetate) Dissolve 3.2 g of Zn-acetate in 250 ml of H_2O.

S9) $= 20\%$ citric acid.

S10) $= 1.5\%$ $ZnCl_2$ and $ZnCl_2$ dry.

S11) standardized 0.100 M As_2O_3.

S12) standardized 0.100 M KIO_3.

Procedure:

Sediment with or without added Zn-acetate is first flushed through with O_2-free N_2 and the gas is collected in a volumetric flask of 50 ml containing 20 or 40 ml NaClO (S4) plus 0.5 ml Zn-acetate (S8). Add 1 or 2 ml of S5, followed by 1 ml 0.5 M H_2SO_4 and 10 ml S6. The excess NaClO is then back-titrated with S11, using an automatic titrator or by simple titration with a Pt electrode. The correct potential (≈ 800 mV) of the endpoint must first be found with a standard titration of S12 with S11. The As_2O_3 solution is regularly standardized against KIO_3 (in presence of KI; the complete procedure for the KIO_3/As_2O_3 titration is found in every textbook on Analytical Chemistry, e.g., Kolthoff *et al.*, 1969).

Calculation:

The equivalent amount of NaClO is found in the blank and is equal to the sum of S^{2-} and (the measured amount) of As_2O_3.

Note:

- The oxidation of H_2S to SO_4^{2-} is 4 times more sensitive than the oxidation to S by I_2. SSO_4^{2-} can be estimated colorimetriccally as mentioned in Section 5.1.10.1, which is more sensitive, but less precise.

References: Golterman, 1996; Kolthoff *et al.*, 1969.

If the concentration of pyrite, FeS_2, must be measured it can be converted into H_2S by a strong reducing agent, e.g., metallic tin powder in 10 M HCl, or Cr^{2+} salts. (See, e.g., Luther *et al.*, 1992.) Cr^{2+} can be prepared by passing a 1 M $CrCl_3$ solution (in 1 M HCl) over a Jones reductor (Kolthoff *et al.*, 1969). Elemental S will be reduced as well. This can be estimated separately by a 24 hr extraction with acetone or ethanol. Neither FeS_2 nor S have been shown to play a role in the P-cycle; the role of FeS_2 in denitrification is uncertain.

5.2 Potential differences

5.2.1 REDOX POTENTIAL

When a rod of a metal (other than a noble metal) is placed in a solution of a salt of the same metal, a potential difference will appear between rod and solution. Metal ions tend to go into solution. As the metal ions are positive, the rod becomes electrically negative relative to the solution. This electric potential difference tends to hold back the positive metal ions, which are, moreover, already present in the solution. The rod will almost instantaneously reach equilibrium with the solution when as many metal ions are dissolving as are being deposited per unit of time.

According to Nernst, the potential difference E (in volts) depends on the tendency of the metal to dissolve and on the concentration of metal ions in solution, C, as given in Equation (5.1). When this concentration exceeds a certain value ('saturation value') the process is reversed.

$$E = \frac{RT}{nF} \ln \frac{K}{C},$$ (5.1)

where R = the molar gas constant, T = the absolute temperature, n = the number of electrons involved, F = the Faraday constant and K = a constant.

A metallic rod in a solution of its salt is called an *indicator electrode*. It is not possible to measure the potential of one electrode. Two rods must, therefore, be placed in two solutions with concentrations C_1 and C_2 and connected by a salt bridge. The potential difference between the two rods will be:

$$E = E_1 - E_2 = \frac{RT}{nF} \ln \frac{C_2}{C_1}. \tag{5.2}$$

There may also be a potential difference between the two solutions due to a difference in diffusion rate of anions or cations, a so-called diffusion potential or junction potential. By using an appropriate salt bridge between the two solutions the size of this potential may be made insignificant. The KCl salt bridge is used for this purpose within the calomel electrode.

The potential of an electrode is measured by comparing it with the potential of an arbitrary standard electrode. To obtain this standard potential the metallic rod is placed in a saturated solution of one of its salts. The most commonly used electrode consists of Hg as the metal in a slurry of Hg_2Cl_2 (calomel electrode) or Hg_2SO_4 in contact with 1 M or saturated KCl (or K_2SO_4).

This type of electrode is called the reference electrode. Equation (5.1) holds also for H_2-gas in contact with H^+. The H_2-gas is kept in equilibrium in the solution by bubbling H_2 at one atmosphere pressure over spongy Pt, from which the electrical connection is taken. The potential of this electrode depends only on the H^+ concentration of the solution (and the temperature). When this electrode is placed in a solution containing 1 M H^+, Equation (5.1) becomes:

$$E = \frac{RT}{nF} \ln K. \tag{5.3}$$

Measuring this potential relative to a calomel electrode gives:

$$E_0 = \frac{RT}{nF} \ln K - E_r, \tag{5.4}$$

where E_r is the potential of the calomel electrode. In a second solution, with $H^+ = C_H$, the potential becomes

$$E_c = \frac{RT}{nF} \ln \frac{K}{C_H} - E_r. \tag{5.5}$$

Subtracting Equation (5.5) from Equation (5.4) gives

$$E_c = E_o - \frac{RT}{nF} \ln C_H. \tag{5.6}$$

At 18 °C, $RT/nF = 0.0577$ (in which the conversion from ln to $\log_{10}$ is included. Then Equation (5.6) becomes:

$$-\log C_H = pH = \frac{E_c - E_0}{0.0577}. \tag{5.7}$$

Thus, when $E_c - E_0$ equals 57.7 mV the pH equals 1, when $E_c - E_0 = 2 \times 57.7$ mV, the pH = 2. This H-electrode can therefore be used as an indicator electrode for H^+ concentrations, with the same electrode in 1 M H^+ + 1 atm H_2 as reference electrode. The combination of the H_2-electrode and calomel electrode can be used for a direct measurement of the pH, after calibration in a buffer with a known pH.

It is more convenient to use a simpler H^+ sensitive electrode, the glass electrode. This is a small thin glass bulb which is filled with an electrolyte. The potential of this electrode depends on H^+ in the same way as the normal H_2-electrode. It must be calibrated against a buffer solution with a known pH.

The pH scale was originally defined by the pH of a 0.1 M KH-phthalate solution at 20 °C; this solution is useful for calibrating glass electrodes. The pH of this solution is only slightly temperature dependent. Nowadays the operational pH scale is defined by reference to five buffer solutions (McGlashan, 1971).

The actual numerical value of the potential can be defined in two ways. Either the normal H_2-electrode or the calomel electrode is arbitrarily set equal to zero. The first system is at present more commonly used.

5.2.2 OXIDATION-REDUCTION POTENTIALS

When a noble metal such as Pt is placed in a solution of, for example Fe^{3+} ions, the Pt becomes positively charged. The Fe^{3+} ions react with the electrons in the Pt-metal according to

$$Fe^{3+} + e^- \Longleftrightarrow Fe^{2+}. \tag{5.8}$$

When the Pt-rod is placed in a solution of Fe^{2+} ions, the rod will become negatively charged. The same processes occur when the Pt-rod is placed in solutions of salts which take up or give off electrons, e.g., Cr^{3+}, $Fe(CN)_6^{4-}$, MnO_4^-, $Cr_2O_7^{2-}$.

These potential differences are called oxidation–reduction potentials (= redox potentials). They depend on the nature and the concentration of the ions involved and are therefore fundamentally not different from the potentials described in Section 5.2.1.

When the Pt-rod is placed in a solution containing both Fe^{2+} and Fe^{3+} ions, the Pt-rod will become positive or negative depending on whether Fe^{3+} or Fe^{2+} ions are present in excess. At one ratio of Fe^{3+}/Fe^{2+} the Pt-rod remains neutral. The reaction velocities in reaction Equation (5.8) to the left or to the right are then equal so that the equilibrium constant

$$K = \frac{[Fe^{2+}]}{[Fe^{3+}]}. \tag{5.9}$$

The Pt-rod therefore remains neutral in a solution in which the ratio Fe^{3+}/Fe^{2+} equals K. The E.M.F. between two Pt-rods placed in two separate solutions (A and B) in which the concentrations of the oxidized ions are $[Ox_A]$ and $[Ox_B]$ respectively and the concentrations of the reduced ions $[Red_A]$ and $[Red_B]$, which solutions are in electric contact with each other, is:

$$E = \frac{RT}{nF} \ln \frac{[Ox_A]}{[Red_A]} - \frac{RT}{nF} \ln \frac{[Ox_B]}{[Red_B]}, \tag{5.10}$$

provided that the oxidation-reduction reaction is reversible. For reaction (5.8), Equation (5.10) becomes ($n = 1$)

$$E = \frac{RT}{nF} \ln \frac{[Fe_A^{3+}]}{[Fe_A^{2+}]} + \frac{RT}{nF} \ln \frac{[Fe_B^{2+}]}{[Fe_B^{3+}]}. \tag{5.11}$$

When solution B is in equilibrium Equation (5.11) becomes

$$E = \frac{RT}{F} \ln \frac{[Fe_A^{3+}]}{[Fe_A^{2+}]} + \frac{RT}{F} \ln K, \tag{5.12}$$

or in general, replacing RT/F by E_0

$$E = E_0 + \ln \frac{[Ox]}{[Red]}, \tag{5.13}$$

in which [Ox] means the concentration of the oxidized ions and [Red] that of the reduced ions. E_0 is the potential of the Pt electrode in a solution in which [Ox] = [Red], measured against, e.g., a calomel electrode. The lower the normal oxidation–reduction potential, the stronger the reducing capacity of the substance involved. It is customary to set the potential of the normal H_2-electrode equal to zero. A solution having a negative oxidation-reduction potential is called reducing (giving off electrons), a solution with a positive potential is oxidizing (taking up electrons).

5.2.3   pH DEPENDENT REDOX POTENTIALS

The redox potentials of some organic compounds are sensitive to the pH of the solution. This can be demonstrated with, e.g., the reaction

$$C_6H_4O_2 + 2H^+ + 2e^- \iff C_6H_6O_2 \tag{5.14}$$

quinone   hydroquinone.

If [quinone] = [hydroquinone], Equation (5.13) becomes

$$E = E_0 + \frac{RT}{F} \ln H^+ \qquad \text{or, at } 18\,°C, E = 0.704 + 0.0577 \log[H^+]. \tag{5.15}$$

By using quinhydrone – a slightly soluble compound formed by the combination of one molecule of quinone and one of hydroquinone – either the oxidation–reduction potential or the pH can be measured. Quinhydrone is, however, usually used only to calibrate Pt-electrodes, because for routine determinations of the pH the glass electrode is more convenient. When calibrating Pt electrodes quinhydrone is added to a buffer solution of known pH and the potential against the H_2-electrode is calculated with Equation (5.15):

$$E = 0.704 - 0.0577 pH. \tag{5.16}$$

If a saturated calomel electrode is used, this electrode will be negative with respect to the quinhydrone electrode and $E = 0.45 - 0.058\text{pH}$.

As hydroquinone is a weak acid ($K \approx 10^{-10}$), the dissociation becomes appreciable above pH = 8.5 and the hydroquinone electrode should not be used.

For purposes of comparison the potentials of pH sensitive systems are calculated to E_7, which is the potential the system would have at pH = 7. The calculation is carried out by assuming a rise of one pH unit to be equivalent to a decrease of 58 mV.

5.2.4 THE 'APPARENT' REDOX POTENTIAL IN MUD/WATER SYSTEMS

If one immerses a redox electrode in a natural water or mud system, a reading will be obtained. However, this reading is principally different from the thermodynamic as discussed above. The potential in natural water or sediment is not based on thermodynamically reversible chemical species and should, therefore, be called the 'apparent' redox potential. These 'apparent' potentials are caused by mixtures of inorganic (Fe^{2+} and HS^-) and organic compounds, most of which may not take part in oxidation–reduction reactions. Just as pH can be low without any buffering capacity, in redox systems one may find a very low redox potential, but no great reducing capacity. The 'apparent' potential usually falls between -0.1 V (O_2 free) and $+0.5$ V (O_2 saturated) relative to the calomel reference electrode. In theory the latter potential should have a value of 0.8 V. The reason for the discrepancy between 0.5 and 0.8 V is probably the fact that the Pt electrode becomes covered with a film of oxide and is no longer completely reversible.

The negative values are normally attained only in the sediment. Water or mud with a potential lower than 0.1 to 0.2 V is generally called reducing, although this is a relative description, as it also depends on the system to be reduced. Mortimer (1942) has given a list of approximate redox potential ranges within which certain reductions proceeded actively, as discussed in Section 2.3. (See Table 2.8.) However, in comparing different lakes one must be careful, as in one lake the redox may be measured in an Fe^{2+} dominated system and in another in an HS^- dominated system, while in both systems organic matter may play an undefined role. While the E_0 for the half reaction $\left[Fe^{3+} + e^- \Leftrightarrow Fe^{2+} \right]$ is 0.77 V, the E_0 for $\left[S + 2e^- \Leftrightarrow S^{2-} \right] = -0.48$ V and that for $[S + 2H^+ + 2e^- \Leftrightarrow HS^- + OH^-]$ is 0.14 V. (There is no reversible chemical reaction between H_2S and H_2SO_4.) The interpretation of an 'apparent' redox of, e.g., -0.1 V is, therefore, difficult if not impossible.

Pt-electrodes placed in mud containing S^{2-} may gradually be poisoned, and will not indicate the correct potential. This can be tested by placing the electrode in O_2 containing (tap) water; the potential must then be ≈ 0.5 V. A quinhydrone buffer is not suitable to check the possible S^{2-} effect as it is sufficiently strong to 'overrule' the effect of poisoning. Cleaning can be done by immersing in dilute $H_2SO_4/K_2Cr_2O_7$, or by dipping it in alcohol and flame it.

5.3 A note on diffusion calculations

5.3.1 INTRODUCTION

Diffusion of nutrients from the sediment layer into the water poses an important problem in the calculation of nutrient loadings of lakes.

The amount of a compound (J) diffusing through a given interface (q) is given by Fick's law:

$$J = -D_0 \frac{dc}{dx} \quad \text{or} \quad \frac{dc}{dt} = D_0 \frac{d^2c}{dx^2}, \tag{5.17a}$$

J in the dimensions m ℓ^{-2} s^{-1}, D_0 = the diffusion coefficient (ℓ^{-2} t^{-1}) and dc/dx = the concentration gradient in the x-direction, or

$$J = D_0 \cdot q \cdot \frac{dc}{dx} \cdot dt, \tag{5.17b}$$

which can be approximated for short periods and short pathways and per unit surface by:

$$J = -D_0 \cdot (c_1 - c_2) \cdot \frac{t}{l}, \tag{5.18}$$

where l = the length, c_1 and c_2 the concentrations of the compound in the two media and J = mole cm^{-2} s^{-1}. If J is expressed in mole cm^{-2} s^{-1} and c in mol cm^{-3}, x in cm, and t in sec, then D is in units of cm^{-2} s^{-1}.

5.3.2 THE DIFFUSION CONSTANT

There is much confusion over the molecular diffusion coefficient in limnological literature.

The basic equation for D_0 for a single ion is:

$$D_0 = \frac{RT}{F^2} \cdot \frac{\lambda}{|z|}, \tag{5.19}$$

in which R = the gas constant, F = the Faraday constant, T = the absolute temperature, λ = the equivalent conductance and $|z|$ = the ion valency. Both λ and D increase with temperature, D more than λ because of the influence of RT.

For a salt the D_0 is the result of the diffusion coefficients of the two ions:

$$D_{\text{salt}} = (z_+ + |z_-|) \frac{D_+ D_-}{z_+ \cdot D_+ + |z_-| \cdot D_-}. \tag{5.20}$$

For the value of D_0 the publication of Li & Gregory (1974) is often used, but these authors gave the D_0 for seawater at 25 °C, taken from the Landolt Börnstein (1969) tables. (The reference given by Li & Gregory to the Landolt Börnstein tables is only partially correct.) There is, however, a wide difference between, e.g., the values for K_2HPO_4 and Na_2HPO_4, the former being 1.1 10^{-5}, and the second 0.83 $\cdot$ 10^{-5}, cm^2 s^{-1}, a difference of $\approx$ 30%. As in freshwater Ca^{2+} is the dominant cation, a value for Ca^{2+} as compensating cation must be taken, which is not given in the Landolt Börnstein tables, and the value for seawater cannot be applied. There is also a difference between the values for NaH_2PO_4 and Na_2HPO_4, the former being 20% lower. There are only very few data on the influence of temperature on diffusion. The 'Handbook of Chemistry and Physics' (1984) gives a factor 1.6 for $15\,°C\,/2\,°C$ and 2.2 for $25\,°C\,/15\,°C$. An older edition only mentioned that D increases, non-linearly, 1–3% per $1\,°C$, which increase differs for the different ions. It is high for K^+, low for Ca^{2+}. Li & Gregory (loc. sup.) gave values of up to 4% per $1\,°C$, taken from the Landolt Börnstein tables, but in this work there are few data for nitrate and none for phosphate. The values for D_0 for KNO_3 at different temperatures come from several publications using different methods. The best available data set is for KCl, but these may not be extrapolated to KNO_3 and certainly not to $Ca(NO_3)_2$.

Lastly, there is a slight influence of the total salt concentration on D_0, which is not sufficient to be taken into account in limnological studies.

When ionic and temperature effects are taken together, there may be uncertainties of $>$ 50%, which is too much to permit comparisons between diffusion rates calculated with the concentration gradient and the release rates as measured in benthic chambers or cores.

5.3.3 PorositY AND tortuosity

Diffusion is affected by the porosity and the tortuosity of the sediments. Diffusion from one liquid layer into another is easy, but the particles in the sediment layer make the problem much more complicated. If with peepers or by slicing of cores the concentrations of the element X at increasing depth in the interstitial water are obtained, the diffusive flux can *theoretically* be calculated. Two phenomena make this more difficult than is often realized.

In the first place the surface of the liquid interface between sediment water and overlying water is limited. This can be approached by measuring the porosity Φ, i.e., the amount of space occupied by water and not by the solid. Most often it is calculated by comparing the total and dry weight of a given volume of sediment. E.g., Cermelj *et al.* (1997) calculated the porosity, Φ, as follows:

$$\Phi = \frac{(M_w/1.0)}{[(M_s/2.50) + (M_w/1.0)]}, \tag{5.21}$$

but often Φ is just equated to the ratio of dry over total weight. Using Equation (5.21) means that volumes are compared instead of weights. This approach seems more logical.

In the second place the pathway through which the compounds have to 'travel' is longer in a sediment than in a free liquid. This longer distance can be estimated by introducing the so-called tortuosity of the sediment. Tortuosity is defined as the average relative increment in distance that the compound must travel in the direction of the diffusion.

Tortuosity is related to porosity. Several equations have been proposed, with or without an adjustable parameter. One of the first cases is the so-called Archie's Law: $\tau^2 = \Phi^{1-n}$ in which n has to be adjusted to the sediment under study. The relation between D and D_{eff} is not standardized by different authors. In an excellent review on peepers, Urban *et al.* (1997)

TABLE 5.1: Diffusion correction factor from Equation (5.24) as function of porosity.

Φ	F	Φ	F	Φ	F
0.1	0.02	0.5	0.21	0.8	0.55
0.2	0.05	0.6	0.30	0.9	0.74
0.3	0.09	0.7	0.41	1.0	1.00
0.4	0.14				

used the equation:

$$D_{\text{eff}} = \Phi^n \cdot D, \tag{5.22}$$

where Φ = the porosity and n is a value between 1 and 3; the authors used 1, while values for D were taken from Li & Gregory (1974). One of the many solutions with an adjustable parameter, proposed by Manheim (1970) was used by Clavero *et al.* (1992) in a comparison between calculated flux and flux measured in benthic chambers. These authors calculated

$$D_{\text{sed}} = D_o \cdot \Phi \cdot F^{-1}, \tag{5.23}$$

F being one of the modified formation factors from the Archie relation: F was taken as $1.28 \cdot \Phi^{-2}$, but no experimental data for the mud in question were given to justify this choice. The agreement between calculated and observed flux was poor, showing the many uncertainties in this approach.

Boudreau (1996) listed 8 equations without an adjustable parameter and suggested that in Fick's law the molecular diffusion constant must be reduced by the square of the tortuosity, τ^2, to account for the longer pathways. Considering tortuosity estimations too time-consuming, and examining 8 theoretical tortuosity–porosity relations not containing an adjustable parameter, he suggested that the tortuosity can be derived from the porosity by the following relation:

$$\tau^2 \approx 1 - 2\ln(\Phi) = 1 - \ln(\Phi^2) = 1 + \ln\left(1/\Phi^2\right), \tag{5.24}$$

which is a numerical approximation satisfying the boundary conditions $\tau^2 > 1$ and $\lim(\tau) = 1$ for $\Phi \longrightarrow 1$.

Based on "both theory and dimensional reasoning", he suggests that Fick's law for sediments is written as:

$$J = -\frac{\Phi D_0}{\tau^2}\frac{dC}{dx} \tag{5.25}$$

or, with $\tau^2 \approx 1 - 2\ln(\Phi)$:

$$J = -\frac{\Phi}{1 - 2\ln(\Phi)} \cdot D_0 \cdot \frac{dC}{dx} \qquad \text{or} \quad J = -F \cdot D_0 \cdot \frac{dC}{dx}. \tag{5.26}$$

The values of $F = \Phi/(1 - 2\ln(\Phi))$ as function of Φ are as in Table 5.1. Whether tortuosity estimations are too time-consuming or not depends, of course, on the precision one wants to obtain.

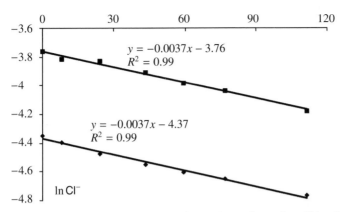

FIGURE 5.1: Chloride diffusion from a water layer into a clay sediment in a 60ℓ tank. Decrease in Cl^- concentration against time ($\sqrt[3]{hr^2}$).

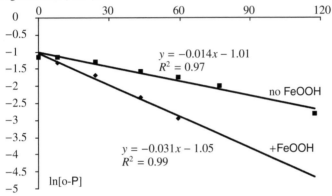

FIGURE 5.2: Phosphate diffusion from a water layer into a clay sediment in a 60ℓ tank, with or without FeOOH added. Decrease in o-P concentration against time ($\sqrt[3]{hr^2}$).

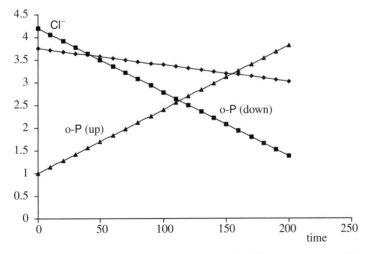

FIGURE 5.3: Model calculation of upward and downward P-diffusion using the diffusion coefficient obtained with Cl^-.

When measuring NH_4^+, NO_3^- or o-P flux in sediment cores, Cl^- can be added at the beginning of the experiment and the downward flux can be used to estimate the flux constant, even when the upward flux is needed. In the same way, during the 'incubation time' of 'peepers' a few cores may be taken and used for the estimation of the flux constant. Figure 5.1 and Figure 5.2 give the results of an experiment in a 60ℓ tank, in which Cl^- and o-P fluxes were compared. In these Figures, $\sqrt[3]{t^2}$ was chosen as unit of time, by which unit a linear relation between concentration and time is approached, in order to simplify calculations. The apparent difference between the two flux constants is caused by the adsorption of o-P onto the sediments, due to which the concentration of the o-P in the interstitial water remains low. Upward diffusion can be treated in the same way; an example is given in Figure 5.3 in which the P-release is calculated with the D_{eff} of Cl^-.

Panigatti & Maine (2003) found the same nearly linear relation between ln(o-P) and $\sqrt[3]{t^2}$ for the o-P adsorption onto sediment in water-sediment-plant (*Salvinia herzogii*) systems in sediment of a wetland of the Middle Paraná floodplain. If plants or N-compounds (NH_4^+ and NO_3^-) were added their influence could be demonstrated as the P-disappearance curves then fell below the curve for the controls.

5.3.4   CALCULATION OF DIFFUSIVE FLUX FROM 'PEEPERS'

The use of peepers (see Section 3.8.1.3) poses some practical problems which must be considered.

The thickness of the membrane is of great importance. Golterman (unpubl.) filled the middle compartment of some peepers with an agar–agar gel containing 1% fluoresceine, placed the peepers in reservoirs with mud and followed the diffusion of the fluoresceine. With Millipore membrane the diffusion was very limited, although detectable. With the thicker Gelman membrane, compartments above and below the labelled one began to fill rapidly and the water in the reservoir became strongly coloured. But with the correct membrane, peepers provide excellent profiles of interstitial components, see, e.g., Schindler *et al.* (1977).

Grigg *et al.* (1999) showed that in saline waters, when there is a wide density difference between cell fluid and pore water, convective motions may occur, disturbing concentrations in the pore water. This was shown by a simple experiment: by filling one of the middle chambers of a peeper with a fluoresceine solution made solid with 1% agar–agar, it was found that the fluoresceine arrived in higher concentrations in the cells just above the labelled one than in those below. Furthermore, it was shown experimentally that the time needed for equilibration (for Mg^{2+}: 4 d) was shorter than that estimated by a simple diffusion model; a model including convection approached this duration better. For o-P equilibration this duration will be longer, because then the dissolution rate from the particles must also be taken into account. The technique of labelling one chamber with fluoresceine could help to solve more problems in comparing 'peeper's' results with those of other devices.

Besides these practical aspects, two fundamental questions must be considered, viz., the estimation of the real concentration difference between interstitial and overlying water and the time scale.

In a few cases a real linear gradient can be found between overlying and interstitial water

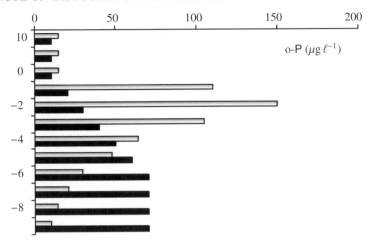

FIGURE 5.4: Two examples of interstitial water o-P concentrations in sediments from the Camargue, showing no clear slope.

in the few top cm's of the sediment. This is, however, not often the case and the question arises then to decide which concentration difference to consider. Figure 5.4 gives two examples where no linear gradients were present. Another, opposite, example of this problem is shown in the work of De Beer *et al.* (1991). By using a micro-electrode with a tip of $7\,\mu$m, these authors were able to measure NH_4^+ profiles in the overlying water and at intervals of 0.1 mm depth in the sediment. Obviously, there is no difference between the concentrations in the overlying water and at 0.1 mm interstitial water depth and thus these authors calculated a zero flux between these two layers.

In order to estimate the concentration gradient, Urban *et al.* (1997) calculated linear regression lines for those solutes of which the concentration changed sharply across the sediment/water interface (viz., NO_3^-, NO_3^-, $\Sigma\{PO_4\}$, Fe^{2+} and Mn^{2+}), while for other compounds a second- or third-order polynomial was fitted on the data points and the concentration gradient was set at the first derivative of the polynomial at $z = 0$. The fluxes were calculated per day, which introduces the problem of the time scale over which diffusive flux can be calculated. If the quantities transferred to the water layer are considerable, the concentration in the interstitial water decreases and must be replenished from the particle pool. This is, however, a time consuming process, so that the kinetics of the dissolution of the adsorbed compound must be known, which is usually not the case. Phosphate release from $Fe(OOH)\approx P$ is more rapid than from $CaCO_3\approx P$, but for neither of these two rates data are available, as they depend strongly on the characteristics of the sediment considered.

5.3.5   INFLUENCE OF EDDY DIFFUSION

Because of water movements caused by wind, bioturbation, swimming of fish, etc., the diffusion coefficient never equals the molecular diffusion coefficient, but must be increased by a factor which can easily be as much as 10. This makes it difficult to apply diffusion results as obtained by core or benthic chamber studies to the field situation. If work is carried out in enclosures in shallow lakes, a solution can be found by adding a conservative element or ion to the enclosure, e.g., Cl^-. This ion is known not to be involved in any chemical or biological reaction and reflects therefore the diffusion only. By comparison, the difference in decrease in NO_3^- concentration can be attributed to NO_3^- consuming processes. In a theoretical study, Golterman (1995) found a good description of the decrease in concentration of a conservative anion in the overlying water, $C_{(t)}$, with time by applying the following equation[2] to the numerically generated data:

$$C_{(t)} = C_0 \cdot \left\{ \frac{B}{A+B} \cdot e^{-\alpha t^b} + \frac{A}{A+B} \right\}, \tag{5.27}$$

in which: A = height of the water column (dm); B = height of the mud column (dm); C_0 = initial concentration in water; and a and b are constants. Comparing this curve with, e.g., that of o-P, the adsorption of this compound onto the sediment adsorption system could be quantified. (See Figure 5.2 and Figure 5.3.) P-diffusion seems faster than Cl^- diffusion, but in reality is not. The difference between the two curves is due to the P-adsorption onto the sediment, which keeps the interstitial concentration low in contrast to that of Cl^-. The Cl^- diffusion can also be used to calculate diffusion factors for P-release, as porosity and tortuosity have the same value in up- and downward fluxes.

For deeper lakes this is obviously not possible, although it can be done in benthic chambers. Benoit & Hemond (1996) used data on heat budget to calculate the heat exchange between sediment and overlying water. In this way an *in situ* diffusion constant can be obtained, as the effect of eddy movements is the same as that on chemical compounds. For this approach very precise measurements of the temperature are needed, as the difference between these two layers is always small. The authors used an instrument with a precision of 0.2 °C, which can easily be improved nowadays. As other factors than heat transport by diffusion are included in the observed temperatures, corrections are needed and demand a good insight into these processes. A relatively simple computer programme can be used. The authors obtained a precision ranging from 3–30%. The article must be carefully studied before use, as in this short summary sufficient details cannot be given.

In limnology the layer of per- or bioturbation is mostly assumed to be no more than 10 cm, but no studies are available. Mulsow & Boudreau (1998) presented two models describing interaction between bioturbation and porosity in marine environments correctly, but these have not yet been applied to freshwater environments, although they are probably applicable to deeper lakes. Boudreau (1998) found a limited surface zone of marine sediments caused by deposit-feeding organisms with a world-wide mean of 9.8 cm and a standard deviation of 4.5 cm. A simplified model taking into account the feedback between food abundance, food availability and the intensity of bioturbation leads to an estimate of 9.7 cm. The proposed equation is:

$$G = G_0 \left(1 - \frac{x}{L} \right)^2, \qquad \text{with } L = 4 \sqrt{\frac{3D \cdot G_0}{8k}}, \tag{5.28}$$

---

[2]Equation proposed by M. F. L. Golterman.

with G = organic matter concentration; G_0 = concentration of compound X at distance 0; L = mixed depth; k = lability of organic matter. $D \cdot G_0$ cannot be measured, but was taken as the average value of $D \cdot G$: $D \cdot G_0 = 3D_B$, where D_B = the mixing or biodiffusion coefficient ($\ell^2 \, t^{-1}$).

For further details the original paper should be consulted.

Further studies on vertical diffusivity are: Wüest, Piepke & Van Senden (2000) and Ravens, Kocsis & Wüest (2000).

References

Roman numbers following a reference refer to the relevant chapters in this book. 'P' means Preface.

Abdollahi H. & D. B. Nedwell (1979): 'Seasonal temperature as a factor influencing bacterial sulfate reduction in a saltmarsh sediment', *Microb. Ecol.* **5**, 73–79.　　　　　　**II**

Alaoui-Mhamdi, M. & Lofti Aleya (1995): 'Assessment of the eutrophication of Al Massira reservoir (Morocco) by means of a survey of the biogeochemical balance of phosphate', *Hydrobiol.* **297**, 75–82.　　　　　　**III**

Ambühl, H. (1969): 'Die neueste Entwicklung des Vierwaldstättersees (Lake of Lucerne)', *Verh. int. Ver. Theor. Angew. Limnol.* **17**, 219–230.　　　　　　**II**

Andersen, F. Ø. & H. S. Jensen (1991): 'The influence of chironomids on decomposition of organic matter and nutrient exchange in a lake sediment', *Proc. int. Soc. Limn.* **24**, 3051–3055.　　　　　　**IV**

Andersen, F. Ø. & H. S. Jensen (1992): 'Regeneration of inorganic phosphorus and nitrogen from decomposition of seston in a freshwater sediment', *Hydrobiol.* **228**, 71–81.　　**III**

Andersen, J. M. (1982): 'Effect of nitrate concentration in lake water on phosphate release from the sediment', *Wat. Res.* **16**, 1119–1126.　　　　　　**IV**

Andersen, J. M. (1985): 'Significance of denitrification on the strategy for preserving lakes and coastal areas against eutrophication', In: *Denitrification in the nitrogen cycle*, Golterman, H. L. (Ed.), Nato conference series I: Ecology, Plenum Press, New York and London, 294 pp.　　　　　　**IV**

Anderson, L. D. & M. L. Delaney (2000): 'Sequential extraction and analysis of phosphorus in marine sediments: Streamlining of the SEDEX procedure', *Limnol. Oceanogr.* **45**, 509–515.　　　　　　**III**

Baas Becking, L. G. M. (1934): *Geobiologie*, Van Stockum & Zoon, den Haag, 263 pp.　　**P**

Baldwin, D. S. (1996a): 'Effects of exposure to air and subsequent drying on the phosphate sorption characteristics of sediments from a eutrophic reservoir', *Limnol. Oceanogr.* **41**, 1725–1732.　　　　　　**III**

Baldwin, D. S. (1996b): 'The phosphorus composition of a diverse series of Australian sediments', *Hydrobiol.* **335**, 63–73.　　　　　　**III**

Baldwin, D. S., A. M. Mitchell & G. N. Rees (2000): 'The effects *of in situ* drying on sediment-phosphate interactions in sediments from an old wetland', *Hydrobiol.* **431**, 3–12.   **III**

Barbanti, A., M. C. Bergamini, F. Frascari, S. Miserocchi & G. Rosso (1994): 'Critical aspects of sedimentary phosphorus chemical fractionation', *J. Envir. Qual.* **23**, 1093–1102.   **III**

Barbanti, A., V. U. Ceccherelli, F. Frascari, G. Reggiani & G. Rosso (1992): 'Nutrient regeneration processes in bottom sediments in a Po Delta lagoon (Italy) and the role of bioturbation in determining the fluxes at the sediment-water interface', *Hydrobiol.* **228**, 1–21.   **III**

Bayly, I. A. E. & W. D. Williams (1972): 'The major ions of some lakes and other waters in Queensland, Australia', *Aust. J. Mar. Freshwat. Res.* **23**, 121–131.   **II**

Benoit, G. & H. F. Hemond (1996): 'Vertical eddy diffusion calculated by the flux gradient method. Significance of sediment – water heat exchange', *Limnol. Oceanogr.* **41**, 157–168.   **II, V**

Berelson, W. M., D. Heggie, A. Longmore, T. Kilgore, G. Nicholson & G. Skyring (1998): 'Benthic nutrient recycling in Port Phillip Bay, Australia', *Estuarine, Coastal and Shelf Science* **46**, 917–934.   **IV**

Berner, E. K. & R. A. Berner (1987): *The Global Water Cycle: Geochemistry and environment*, Prentice Hall, Englewood Cliffs, NY., 397 pp.   **II**

Bertuzzi, A., J. Faganeli, C. Welker & A. Brambati (1997): 'Benthic fluxes of dissolved inorganic carbon, nutrients and oxygen in the Gulf of Trieste (Northern Adriatic)', *Wat. Air Soil Pollut.* **99**, 305–314.   **III**

Best, E. P. H., J. H. A. Dassen, J. J. Boon & G. Wiegers (1990): 'Studies on decomposition of *Ceratophyllum demersum* litter under laboratory and field conditions: Losses of dry mass and nutrients, qualitative changes in organic compounds and consequences for ambient water and sediments', *Hydrobiol.* **194**, 91–114.   **II**

Binnerup, S. J., K. Jensen, N. P. Revsbech, M. H. Jensen & J. Sørensen (1992): 'Denitrification, dissimilatory reduction of nitrate to ammonia, and nitrification in a bioturbated estuarine sediment as measured with ^{15}N and microsensor techniques', *Appl. Envir. Microbiol.* **58**, 303–313.   **IV**

Blackburn, T. H. & K. Henriksen (1983): 'Nitrogen cycling in different types of sediments from Danish waters', *Limnol. Oceanogr.* **28**, 477-493.   **IV**

Blicher-Mathiesen, G. & C. C. Hoffmann (1999): 'Denitrification as a sink for dissolved nitrous oxide in a freshwater riparian fen', *J. Envir. Qual.* **28**, 257–262.   **IV**

Bloesch, J. (1994a): 'Editorial: Sediment resuspension in lakes', *Hydrobiol.* **284**, 1–3.   **III**

Bloesch, J. (1994b): 'A review of methods used to measure sediment resuspension', *Hydrobiol.* **284**, 13–18.   **III**

Bloesch, J. & N. M. Burns (1980): 'A critical review of sediment trap technique', *Schweiz. Z. Hydrol.* **42**, 15–55.   **III**

Boers, P. C. M., J. W. Th. Bongers, A. G. Wisselo & Th. E. Cappenberg (1984): 'Loosdrecht Lakes restoration projekt: Sediment phosphorus distribution and release from the sediments', *Verh. int. Ver. Limnol.* **22**, 842–847.   **III**

Boers, P. C. M. & O. van Hese (1988): 'Phosphorus release from the peaty sediments of the

Loosdrecht Lakes (The Netherlands)', *Wat. Res.* **22**, 355–363. **III**

Boers, P. C. M. & F. de Bles (1991): 'Ion concentrations in interstitial water as indicators for phosphorus release processes and reactions', *Wat. Res.* **25**, 591–598. **III**

Boers, P., J. van der Does, M. Quaak, J. van der Vlugt & P. Walker (1992): 'Fixation of phosphorus in lake sediments using iron(III)chloride: experiences, expectations', *Hydrobiol.* **233**, 211–212. **III**

Bonanni, P., R. Caprioli, E. Ghiara, C. Mignuzzi, C. Orlandi, G. Paganin, & A. Monti (1992): 'Sediment and interstitial water chemistry of the Orbetello lagoon (Grosseto, Italy); nutrient diffusion across the water-sediment interface', *Hydrobiol.* **235/236**, 553–568. **III**

Bonetto, C., F. Minzoni & H. L. Golterman (1988): 'The Nitrogen cycle in shallow water sediment systems of rice fields. Part II. Fractionation and bioavailability of organic nitrogen compounds', *Hydrobiol.* **159**, 203–210. **IV**

Bonetto, C., L. de Cabo, N. Gabellone, A. Vinocur, J. Donadelli & F. Unrein (1994): 'Nutrient dynamics in the deltaic floodplain of the lower Paraná River', *Arch. Hydrobiol.* **131**, 277–295. **III**

Bonin, P. P. Omnes & A. Chalamet (1998): 'Simultaneous occurrence of denitrification and nitrate ammonification in sediments of the French Mediterranean Coast', *Hydrobiol.* **389**, 169–182. **IV**

Boon, P. I., A. Mitchell & K. Lee (1995): 'Methanogenesis in the sediments of an Australian freshwater wetland: Comparison with aerobic decay, and factors controlling methanogenesis', *FEMS Microbiol. Ecol.* **18**, 175–190. **II**

Boon, P. I., A. Mitchell & K. Lee (1997): 'Effects of wetting and drying on methane emissions from ephemeral floodplain wetlands in south-eastern Australia', *Hydrobiol.* **357**, 73–87. **IV**

Boström, B., M. Jansson & C. Forsberg (1982): 'Phosphorus release from lake sediments', *Arch. Hydrobiol. Beih. Ergebn. Limnol.* **18**, 5–59. **III**

Boström, B., I. Ahlgren & R. Bell (1985): 'Internal nutrient loading in a eutrophic lake reflected in seasonal variations of some sediment parameters', *Verh. int. Ver. Limnol.* **22**, 3335–3339. **III**

Boström, B., G. Persson & B. Broberg (1988): 'Bioavailability of different phosphorus forms in freshwater systems', *Hydrobiol.* **170**, 133–155. **III**

Boudreau, B. P. (1996): 'The diffusive tortuosity of fine grained unlithified sediments', *Geochim. Cosmochim. Acta* **60**, 3139–3142. **V**

Boudreau, B. P. (1998): 'Mean mixed depth of sediments:The wherefore and why', *Limnol. Oceanogr.* **43**, 524–526. **V**

Boudreau, B. P., D. E. Canfield & A.Mucci (1992): 'Early diagenesis in a marine sapropel, Mangrove Lake, Bermuda', *Limnol. Oceanogr.* **37**, 1738–1753. **V**

Bowes, M. J. & W. A. House (2001): 'Phosphorus and dissolved silicon dynamics ion the River Swale catchment, UK: a mass-balance approach', *Hydrol. Process.* **15**, 261–280. **II**

Brooks, A. S. & D. N. Edgington (1994): 'Biogeochemical control of phosphorus cycling and primary production in Lake Michigan', *Limnol. Oceanogr.* **39**, 961–968. **III**

Brunberg, A.-K., & B. Boström (1992): 'Coupling between benthic biomass of Microcystis and phosphorus release from the sediments of a highly eutrophic lake', *Hydrobiol.* **235/236**, 375–385.   **III**

Bruning, C. & S. P. Klapwijk (1984): 'Application of derivative spectroscopy in bioassays estimating algal available phosphate in lake sediments', *Verh. int. Ver. Limnol.* **22**, 172–178.   **III**

Buresh, R. J. & W. H. Patrick (1981): 'Nitrate reduction to ammonia and organic nitrogen in an estuarine sediment', *Soil Biol. Biochem.* **13**, 279–283.   **IV**

Burke, A. J., P. A. Waller & W. F. Pickering (1989): 'The evaluation of inorganic phosphate species in salt water lake sediments', *Chem. Speciation and Bioavailability* **1**, 47–57. **III**

Burns N. M. & C. Ross (1972a, b): 'Project Hypo, Papers (8) & (9)', In: *Project Hypo*, Burns N. M. & C. Ross (Eds.), Canada Centre for Inland Waters, Burlington, Ont., pp. 85-119 & 120-126.   **III**

Burns N. M. & C. Ross (1972c): 'Oxygen-nutrient relationships within the central basin of Lake Erie', In: *Nutrients in natural waters*, Allen, H. E. & J. R. Kramer (Eds.), Wiley, New York, pp. 193–250.   **III**

Carignan, R. & J. Kalff (1979): 'Quantification of the sediment phosphorus available to aquatic macrophytes', *J. Fish. Res. Board Can.* **36**, 1002–1005.   **III**

Carignan, R. & P. Vaithiyanathan (1999): 'Phosphorus availability in the Paraná floodplain lakes (Argentina): Influences of pH and phosphate buffering by fluvial sediments', *Limnol. Oceanogr.* **44**, 1540–1548.   **III**

Carmouze, J.-P. (1969): 'La salure globale et les salures spécifiques des eaux du Lac Tchad', *Cah. O.R.S.T.O.M., Sér. Hydrobiol.* **3**, 3–14.   **II**

Carpenter, S. R., and M. S. Adams (1979): 'Effects of nutrients and temperature on decomposition of Myriophyllum spicatum in a hard-water eutrophic lake', *Limnol. Oceanogr.* **24**, 520–529.   **IV**

Carr, G. M. & P. A. Chambers (1998): 'Macrophyte growth and sediment phosphorus and nitrogen in a Canadian prairie river', *Freshw. Biol.* **39**, 525–536.   **III**

Casey, H. (1969): 'The chemical composition of some southern English chalk streams and its relation to discharge', In: *The Year Book*, , Association of the River Authorities, London, pp. 100–113.   **II**

Casey, H. & P. V. R. Newton (1972): 'The chemical composition and flow of the South Winterbourne in Dorset', *Freshwat. Biol.* **2**, 229–234.   **II**

Casey, H. & P. V. R. Newton (1973): 'The chemical composition and flow of the R. Frome and its main tributaries', *Freshwat. Biol.* **3**, 317–333.   **II**

Caskey, W. H. & J. M. Tiedje (1980): 'The reduction of nitrate to ammonium by a *Clostridium* sp. isolated from soil', *J. Gen. Microbiol.* **119**, 217–223.   **IV**

Cermelj, B., A. Bertuzzi & J. Faganelli (1997): 'Modelling of pore water nutrient distribution and benthic fluxes in shallow coastal waters (Gulf of Trieste, Northern Adriatic)', *Wat. Air Soil Pollut.* **99**, 435–444.   **III**

Chalamet, A. (1985): 'Effects of environmental factors on denitrification', In: *Denitrification in the Nitrogen cycle*, Golterman, H. L. (Ed.), Nato Conference Series I: Ecology **Vol. 9**, Plenum Press, New York, London, pp. 7–29.   **IV**

Chang, S. C. & M. L. Jackson (1957): 'Fractionating of soil phosphorus', *Soil Sci.* **84**, 133–144.   **III**

Christensen, P. B., L. P. Nielsen, J. Sørensen & N. P. Revsbech (1990): 'Denitrification in nitrate-rich streams: Diurnal and seasonal variation related to benthic oxygen metabolism', *Limnol. Oceanogr.* **35**, 640–652.   **IV**

Clavero, V., J. A. Fernández & F. Xavier Niell (1992): ' Bioturbation by *Nereis* sp. and its effects on the phosphate flux across the sediment-water interface in the Palmones River estuary', *Hydrobiol.* **235/236**, 387–392.   **III**

Clavero, V., M. J. García-Sanchez, F. Xavier & J. A. Fernandez (1997): 'Influence of sulfate enrichment on the carbon dioxide and phosphate fluxes across the sediment-water interface', *Hydrobiol.* **345**, 59–65.   **III**

Clavero, V., J. J. Izquierdo, J. A. Fernandez & F. X. Niell (1999): 'Influence of bacterial density on the exchange of phosphate between sediment and overlying water', *Hydrobiol.* **392**, 55–63.   **III**

Clavero, V., J. J. Izquierdo, J. A. Fernandez & F. X. Niell (2000): 'Seasonal fluxes of phosphate and ammonia across the sediment-water interface in a shallow small estuary (Palmones River, southern Spain)', *Mar. Ecol. Progess Series* **198**, 51–60.   **III**

Clymo, R. S. & H.L.Golterman (1985): 'Precision and accuracy of the determination of the ionic product of hydroxy-apatite', *Hydrobiol.* **126**, 31–34.   **III**

Cornwell, J. C., W. M. Kemp & T. M. Kana (1999): 'Denitrification in coastal ecosystems: methods, environmental controls, and ecosystem level controls, a review', *Aquatic Ecology* **33**, 41–54.   **IV**

Cranwell, P. (1975): 'Environmental organic chemistry of rivers and lakes, both water and sediment', In: *Envir. Chem.*, G. Englinton (Ed.), **Vol. 1**, Chemical Society, London, pp. 22–54.   **I**

Cranwell, P. (1976a): 'Organic geochemistry of lake sediments', In: *Environmental Biogeochemistry*, J. O. Nriagu (Ed.), **Vol. 1**, Ann Arbor Science Publishers. Ann Arbor, MI, pp. 75–88.   **I**

Cranwell, P. (1976b): 'Decomposition of aquatic biota and sediment formation: organic compounds in detritus resulting from microbial attack on the alga *Ceratium hirundinella*', *Freshw. Biol.* **6**, 41–48.   **I**

Cranwell, P. (1977): 'Organic compounds as indicators of allochthonous and autochthonous input to lake sediments', In: *Interactions between sediments and fresh water*, Golterman, H. L. (Ed.), Junk/Pudoc, The Hague, pp. 133–141.   **I**

Damiani, V. (1972): 'Studio di un ambiente fluvio-lacustre sulla base di una analisi granulometrica dei sediment: fiume Toce e bacino della isole Borromee (Lago Maggiore)', *Mem. Ist. Ital. Idrobiol.* **29**, 37–95.   **I**

Damiani, V., Ferrario, A., Gavelli, G., & R. L. Thomas (1977): 'Trace metal composition and fractionation of Mn, Fe, S, P, Ba and Si in the Bay of Quinte freshwater ferromanganese concretions, Lake Ontario', In: Golterman (Ed.), 1977, pp. 83–93.   **I**

Danen-Louwerse, H., L. Lijklema & M. Coenraats (1993): 'Iron content of sediment and phosphate adsorption process', *Hydrobiol.* **253**, 311–317.   **III**

Dapples, E. C. (1959): *Basic Geology*, John Wiley & Sons, New York, 609 pp.   **I**

Davidsson, T. E., R. Stepanauskas & L. Leonardson (1997): 'Vertical patterns of nitrogen transformations during infiltration in two wetland soils', *Appl. Envir. Microbiol.* **63**, 3648–3656.   **IV**

Davis, R. B., D. L. Thurlow & F. E. Brewster (1975): 'Effects of burrowing tubificid worms on the exchange of phosphorus between lake sediment and overlying water', *Verh. int. Ver. Limnol* **19**, 382–394.   **III**

Davison W. (1985): 'Conceptual models for transport at the redox boundary', In: *Chemical processes in lakes*, W. Stumm (Ed.), Wiley-Interscience Publ., New York, pp. 31–55.   **II**

Davison W. & S. I. Heaney (1978): 'Ferrous iron-sulfide interactions in anoxic hypolimnetic waters', *Limnol. Oceanogr.* **23**, 1194–1200.   **II**

Davison, W., C. Woof & E. Rigg (1982): 'The dynamics of iron and manganese in a seasonally anoxic lake; direct measurement of fluxes using sediment traps', *Limnol. Oceanogr.* **27**, 987–1003.   **II**

De Beer, D., Sweerts, J. P. R. A. & J. C. van den Heuvel (1991): 'Microelectrode measurement of ammonium profiles in freshwater sediments', *FEMS-Microbiol. Ecol.* **86**, 1-6.   **V, IV**

De Bie, M. J. M., M. Starink, H. T. S. Boschker, J. J. Peene & H. J. Laanbroek (2002): 'Nitrification in the Schelde estuary: methodological aspects and factors influencing its activity', *FEMS Microbiol. Ecol.* **42**, 99–107.   **IV**

De Graaf Bierbrauwer-Würtz, I. M. & H. L. Golterman (1989): 'Fosfaatfracties in de bodem van een aantal Nederlandse meren (in Dutch, with English summary)', H_2O **22**, 411–414.   **III**

De Groot, C. J. (1990): 'Some remarks on the presence of organic phosphates in sediments', *Hydrobiol.* **207**, 303–309.   **III**

De Groot, C. J. (1991): 'The influence of FeS on the inorganic phosphate system in sediments', *Verh. int. Ver. Limnol.* **23**, 3029–3035.   **III**

De Groot, C. J. & H. L. Golterman (1990): 'Sequential fractionation of sediment phosphate', *Hydrobiol* **192**, 143-149.   **III**

De Groot, C. J. & H. L. Golterman (1993): 'On the presence of organic phosphate in some Camargue sediments: evidence for the importance of phytate', *Hydrobiol.* **252**, 117–126. **III**

De Groot, C. J. & A. C. Fabre (1993): 'The impact of desiccation of a freshwater marsh (Garcines Nord, Camargue, France) on the sediment-water-vegetation interactions. Part 3: The fractional composition and the phosphate adsorption characteristics of the sediment', *Hydrobiol.* **252**, 105–116.   **III**

De Groot, C. J. & C. Van Wijck (1993): 'The impact of desiccation of a freshwater marsh (Garcines Nord, Camargue, France) on the sediment-water-vegetation interactions. Part 1: The sediment chemistry', *Hydrobiol.* **252**, 83–94.   **III**

De Haan, R. I. Jones & K. Salonen (1990): 'Abiotic transformations of iron and phosphate in humic lake water revealed by double-isotope labeling and gel filtration', *Limnol. Oceanogr.* **35**, 491–497.   **III**

De Haan, H., T. de Boer & J. Voerman (1991): 'Size distribution of dissolved organic carbon and aluminium in alkaline and humic lakes', *Hydrobiol. Bull.* **24**, 145–151.   **III**

De Jonge, V. N. & L. A. Villerius (1989): 'Possible role of carbonate dissolution in estuarine phosphate dynamics', *Limnol. Oceanogr.* **34**, 332–340.   **III**

De Kanel, J. & J. W. Morse (1978): 'The chemistry of orthophosphate uptake from seawater onto calcite and aragonite', *Geochim. et Cosmochim. Acta* **42**, 1335–1340.   **III**

De Vicente, I., L. Serrano, V. Amores, V. Clavero & L. Cruz-Pizarro (2002): 'Sediment phosphate fractionation and interstitial water phosphate concentration in two coastal lagoons (Albufera de Adra, SE Spain)', *Hydrobiol.* , in press.   **III**

Díaz-Espelo, A., L. Serrano & J. Toja (1999): 'Changes in sediment phosphate composition of seasonal ponds during filling', *Hydrobiol.* **392**, 21–28.   **III**

Dittrich, M. & R. Koschel (2002): 'Interactions between calcite precipitation (natural and artificial) and phosphorus cycle in the hardwater lake', *Hydrobiol.* **469**, 49–57.   **III**

Dominik, J., A. Mangini & G. Müller (1981): 'Determination of recent deposition rates in Lake Constance with radioisotopic methods', *Sedimentology* **28**, 653–677.   **I**

Dorich, R. A., D. W. Nelson & L. E. Sommers (1984): 'Availability of phosphorus to algae from eroded soil fractions', *Agriculture, Ecosystems and Environment* **11**, 253–264.   **III**

Dorich, R. A., D. W. Nelson & L. E. Sommers (1985): ' Estimating algal available phosphorus in suspended sediments by chemical extraction', *J. of Envir. Qual.* **14**, 400–405.   **III**

Eckert, W., A. Nishri & R. Parparov (1997): 'Factors regulating the flux of phosphate at the sediment-water interface of a sub-tropical calcareous lake: a simulation study with intact sediment cores', *Wat. Air Soil Pollut.* **99**, 401–409.   **III**

Edmondson, W. T. (1986): 'The sedimentary record of the eutrophication of Lake Washington', *Proc. Nat. Acad. Sci. USA* **71**, 5093–5095.   **III**

Eglington, G. (1975): *Environmental Chemistry*, Chem. Soc., London, 199 pp.   **I**

Einsele, W. (1936): 'Ueber die Beziehungen des Eisenkreislaufs zum Phosphatekreislauf im eutrophen See', *Arch. Hydriobiol.* **29**, 664–686.   **III**

Einsele, W. (1937): 'Physikalisch-chemische Betrachtung einiger Probleme des limnischen Mangan- und Eisenkreislaufs', *Verh. int. Ver. Limnol.* **5**, 69–84.   **III**

Einsele, W. (1938): 'Ueber chemische und kolloidchemische Vorgange in Eisen-Phosphate Systemen unter limnochemischen und limnogeologischen Gesichtspunkten', *Arch. Hydrobiol.* **33**, 361–387.   **III**

Einsele, W. & H. Vetter (1938): 'Untersuchungen über die Entwicklung der physikalischen und chemischen Verhältnisse im Jahreszyklus in einem mässig eutrophen See (Schleinsee bei Langenargen)', *Int. Rev. Hydrobiol.* **36**, 285–324.   **III**

Ekholm, P. (1994): 'Bioavailability of phosphorus in agricultural loaded rivers in southern Finland', *Hydrobiol.* **287**, 179–194.   **III**

Ekholm, P. & K. Krogerus (2003): 'Determining algal-available phosphorus of differing origin: routine phosphorus analyses versus algal assays', *Hydrobiol.* **492**, 29–42.   **III**

Ekholm, P., O. Malve & T. Kirkkala (1997): 'Internal and external load as regulators of nutrient concentrations in the agricultural loaded Lake Pyhäjärvi (soutwest Finland)', *Hydrobiol.* **345**, 3–14.   **III**

El Habr, H. & H. L. Golterman (1987): 'Input of nutrients and suspended matter into the

Golfe du Lion and the Camargue by the river Rhône', *Sciences de l'Eau* **6**, 393–402. **III**

El Habr, H. & H. L. Golterman (1990): 'In vitro and in situ studies on nitrate disappearance in water-sediment systems of the Camargue (southern France)', *Hydrobiol.* **192**, 223–232. **IV**

Englund, J. O., Jørgensen, P., Roaldset, E. & P. Aargaard (1977): 'Composition of water and sediments in Lake Mjøsa, South Norway, in relation to weathering process', In: Golterman (Ed.), 1977, pp. 125–132. **I**

Entz, B. (1959): 'Chemische Characterisierung der Gewässer in der Umgebung des Balatonsees und chemische Verhältnisse des Balatonwassers', *Annal. Biol. Tihany* **26**, 131–201. **II**

Fabre, A. (1988): 'Experimental studies on some factors influencing phosphorus solubilization in connection with the drawdown of a reservoir', *Hydrobiol.* **159**, 153–158. **III**

Fabre, A. (1992): 'Inorganic phosphate in exposed sediments of the River Garonne', *Hydrobiol.* **228**, 37–42. **III**

Fabre, A., A. Dauta, A. Qotbi & V. Baldy (1996): 'Relation between algal available phosphate in the sediments of the River Garonne and chemically determined phosphate fractions', *Hydrobiol.* **335**, 43–48. **III**

Faust, S. D. & O. M. Aly (1981): *Chemistry of natural waters*, Ann Arbor science, Ann Arbor, Michigan, 400 pp. **I**

Feibicke, M. (1997): 'Impact of nitrate addition on phosphorus availability in sediment and water column and on plankton biomass', *Wat. Air Soil Pollut.* **99**, 445–456. **IV**

Fleischer, M. & W. O. Robinson (1963): 'Some problems of the geochemistry of fluorine', In: *Studies in analytical chemistry (Special publications of the Royal Society of Canada, No 6.)*, D. M. Shaw (Ed.), University of Toronto Press, Toronto, 139 pp. **II**

Ford, P. W., N. J. Grigg, P. R. Teasdale, & I. T. Webster (1998): 'Pore water sampling with peepers', In: *Proceedings of Workshop on Sampling nutrients in aquatic ecosystems: Collecting valid and representative samples*, K. N. Markwort and W. A. Maher (Eds.), CRC for Freshwat. Ecol., Canberra, pp. 138–153. **III**

Forés, E., M. Menéndez & F. A. Comin (1988): 'Rice straw decomposition in rice-field soil', *Plant and Soil* **109**, 145–146. **IV**

French, R. H. (1983): 'Lake modeling: State of the art', *CRC Critical Reviews in Envir. Control* **13**, 311. **III**

Froelich, P. N. (1988): 'Kinetic control of dissolved phosphate in natural rivers and estuaries: A primer on the phosphate buffer mechanism', *Limnol. Oceanogr.* **33**, 649–668. **III**

Fukuhara. H. & M. Sakamoto (1987): 'Enhancement of inorganic nitrogen and phosphate release from lake sediment by tubificid worms and chironomid larvae', *Oikos* **48**, 321–320. **IV**

Furrer, O. J. & R. Gächter (1972): 'Der Beitrag der Landwirtschaft zur Eutrophierung der Gewässer in der Schweiz. II', *Schweiz. Z. Hydrol.* **30**, 71–92. **II**

Gächter, R. & O. J. Furrer (1972): 'Der Beitrag der Landwirtschaft zur Eutrophierung der Gewässer in der Schweiz. I', *Schweiz. Z. Hydrol.* **30**, 41–70. **II**

Gächter, R. & J. S. Meyer (1993): 'The role of microorganisms in mobilization and fixation of phosphorus in sediments', *Hydrobiol.* **253**, 103–121. **III**

Gächter, R. & B. Müller (2003): 'Why the phosphorus retention of lakes does not necessarily depend on the oxygen supply to their sediment surface', *Limnol. Oceanogr.* **48**, 929–933.   **III**

Garcia, J.-L. (1975): 'Evaluation de la dénitrification dans les rizières par la méthode de réduction de N_2O', *Soil Biol. Biochem.* **7**, 251–256.   **IV**

Garcia, J.-L. (1977): 'Evaluation de la dénitrification par la mesure de l'activité oxyde nitreux reductase. Etude complémentaire', *Cah. Orstom, sér. Biol.* **XII**, 89–95.   **IV**

Garcia, J.-L. (1978): 'Etude de la dénitrification dans les sols des rizières du Sénégal', *Cah. Orstom, sér. Biol.* **XIII**, 117–127.   **IV**

García-Gil, J. & H. L. Golterman (1993): 'Influence of FeS on the denitrification rate in sediments from Camargue (France)', *FEMS Microbiol. Ecol.* **13**, 85–92.   **IV**

García-Ruiz, R., S. N. Pattinson & B.A. Whitton (1998a): 'Kinetic parameters of denitrification in a river continuum', *Appl. Envir. Microbiol.* **64**, 2533–2538.   **IV**

García-Ruiz, R. S., S. N. Pattinson, & B. A. Whitton (1998b): 'Denitrification and nitrous oxide production in sediments of the Wiske, a lowland eutrophic river', *Sci. Total Envir.* **210/211**, 307–320.   **IV**

García-Ruiz, R. S., S. N. Pattinson, & B. A. Whitton (1998c): 'Denitrification in sediments of the freshwater tidal Yorkshire Ouse', *Sci. Total Envir.* **210/211**, 321–327.   **IV**

García-Ruiz, R. S., S. N. Pattinson, & B. A. Whitton (1998d): 'Denitrification in river sediments: relationships between process rate and properties of water and sediment', *Freshwat. Biol.* **39**, 467–476.   **IV**

Garnier, J., B. Leporcq, N. Sanchez & X. Philippon (1999): 'Biogeochemical mass-balances (C, N, P, Si) in three large reservoirs of the Seine Basin (France)', *Biogeochem.* **47**, 119–146.   **III, IV**

Gilbert, F., G. Stora & P. Bonin (1998): 'Influence of bioturbation on denitrification Activity in Mediterranean coastal sediments: an in situ experimental approach', *Mar. Ecol. Prgr. Ser.* **163**, 99–107.   **IV**

Goedkoop, W. & R. K. Johnson (1996): 'Pelagic-benthic coupling: Profundal benthic community response to spring diatom deposition in mesotrophic Lake Erken', *Limnol. Ocenogr.* **41**, 636–647.   **III**

Goedkoop, W. & E. Törnblom (1996): 'Seasonal fluctuations in benthic bacterial production and abundance in Lake Erken', *Arch. Hydrobiol. Spec. Issues Adv. Limnol.* **48**, 197–205.   **III**

Goedkoop, W. & K. Pettersson (2000): 'Seasonal changes in sediment phosphorus forms in relation to sedimentation and benthic bacterial biomass in Lake Erken', *Hydrobiol.* **431**, 41–50.   **III**

Goldberg, S. & G. Sposito (1984): 'A chemical model of phosphate adsorption by soils. 1. Reference oxide minerals', *Soil Sci. Soc. Am. J.* **48**, 772-778.   **III**

Goldschmidt, V. M. (1937): 'The principle of distribution of chemical elements in minerals and rocks', *J. Chem. Soc. Lond.* **1937**, 655–673.   **II**

Golterman, H. L. (1967): 'Influence of the mud on the chemistry of water in relation to productivity', In: *Chemical environment in the aquatic habitat; Proceedings of an IBP-*

Symposium held in Amsterdam and Nieuwersluis, 10-16 October 1966, H. L. Golterman and R. S. Clymo (Eds.), Amsterdam, Noord-Holland, pp. 297-313.   **III**

Golterman, H. L. (1973a): 'Deposition of river silts in the Rhine and Meuse delta', *Freshwat. Biol.* **3**, 267–281.   **II**

Golterman, H. L. (1973b): 'Natural phosphate sources in relation to phosphate budgets: a contribution to the understanding of eutrophication', *Wat. Res.* **7**, 3–17.   **II**

Golterman, H. L. (1973c): 'Vertical movement of phosphate in freshwater', In: *Environmental Phosphorus Handbook*, E. D. Griffith and A. Beeton (Eds.), John Wiley, New York, N.Y., pp. 509-538.   **III**

Golterman, H. L. (1975a): 'Chemistry', In: *River ecology*, B.A. Whitton (ed.), Blackwell, Oxford, etc., 1975. Studies in Ecology **2**, pp. 39-80.   **I**

Golterman, H. L. (1975b): *Physiological Limnology. An approach to the Physiology of Lake Ecosystems*, Elsevier, Amsterdam, Oxford, New York, 489 pp.   **I, IV**

Golterman, H. L. (1976a): 'Zonation of mineralization in stratifying lakes', In: *The role of terrestrial and aquatic organisms in decomposition processes; 17th Symposium of the British Ecological Society, 15-18 April 1975*, J.M. Anderson and A. Macfadyen (Eds.), Blackwell, Oxford, etc., pp. 3–22.   **II**

Golterman, H. L. (1976): 'Zonation of mineralization in stratifying lakes', In: *The role of terrestrial and aquatic organisms in decomposition processes; 17th Symposium of the British Ecological Society, 15-18 April 1975*, J. M. Anderson & A. Macfadyen (Eds.), Blackwell, Oxford, etc., pp. 3–22.   **II**

Golterman, H. L. (1976b): 'The sediments and their importance, relative to the inflows, as a source of nutrients for the growth of algae', In: *Proceedings of a Symposium "The effects of storage on water quality", Reading University, March 1975, organized by the Water Research Centre*, , Medmenham Laboratory, Medmenham, pp. 47–59.   **III**

Golterman, H. L. (1977): 'Sediments as a source of phosphate for algal growth', In: Golterman, H. L. (Ed.), 1977, pp. 286–293.   **III**

Golterman, H. L. (Ed.) (1977): *Interactions between sediments and freshwater. Proceedings of an International Symposium held at Amsterdam, 6-10 September 1976*, Junk/Pudoc, The Hague, 473 pp.   **III**

Golterman, H. L. (1980): 'Phosphate models, a gap to bridge', *Hydrobiol.* **72**, 61–71.   **III**

Golterman, H. L. (1982a): 'Loading concentration models in shallow lakes', *Hydrobiol.* **91**, 169–174.   **III**

Golterman, H. L. (1982b): 'Differential extraction of sediment phosphates with NTA solutions', *Hydrobiol.* **92**, 683–687.   **III**

Golterman, H. L. (1984): 'Sediments, modifying and equilibrating factors in the chemistry of freshwaters', *Verh. int. Ver. Limnol.* **22**, 23–59.   **I, III**

Golterman, H. L. (Ed.) (1985): *Denitrification in the N-cycle*, Nato Conference series; Series I: Ecology, Plenum Press, New York & London, 294 pp.   **IV**

Golterman, H. L. (1988): 'The calcium- and iron bound phosphate phase diagram', *Hydrobiol.* **159**, 149–151.   **III**

Golterman, H. L. (1991a): 'Reflections on post O.E.C.D. eutrophication models', *Hydrobiol.*

218, 167–176.   **III**

Golterman, H. L. (1991b): 'The influence of FeS on the denitrification rate', *Verh. int. Ver. Limnol.* **23**, 3025–3028.   **IV**

Golterman, H. L. (1991c): 'Direct Nesslerization of ammonia and nitrate in fresh-water', *Annls. Limnol.* **27**, 99–101.   **V**

Golterman, H. L. (1994): 'Nouvelles connaisssances des formes du phosphate: conséquences sur le cycle du phosphate dans les sédiments des eaux peu profondes', *Annls. Limnol.* **30**, 221–232.   **III**

Golterman, H. L. (1995a): 'The role of the ironhydroxide-phosphate-sulphide system in the phosphate exchange between sediments and overlying water', *Hydrobiol.* **297**, 43–54. **III**

Golterman, H. L. (1995b): 'Theoretical aspects of adsorption of ortho-phosphate onto iron-hydroxide', *Hydrobiol.* **315**, 59–68.   **III**

Golterman, H. L. (1995c): 'The labyrinth of nutrient cycles and buffers in wetlands: results based on research in the Camargue (Southern France)', *Hydrobiol.* **315**, 39–58.   **III, IV**

Golterman, H. L. (1995d): 'Remarks on numerical and analytical methods to calculate diffusion in water/sediment systems', *Hydrobiol.* **315**, 69–88.   **IV**

Golterman, H. L. (1996a): 'Fractionation of sediment phosphate with chelating compounds: a simplification, and comparison with other methods', *Hydrobiol.* **335**, 87–95.   **III**

Golterman, H. L. (1996b): 'A volumetric or colorimetric determination of FeS and H_2S in sediments by oxidation to sulphate', *Hydrobiol.* **335**, 83–86.   **V**

Golterman, H. L. (1998): 'The distribution of phosphate over iron-bound and calcium-bound phosphate in stratified sediments', *Hydrobiol.* **364**, 75–81.   **III**

Golterman, H. L. (1999): 'Quantification of P-flux through shallow, agricultural and natural waters as found in wetlands of the Camargue (S-France)', *Hydrobiol.* **392**, 29–39.   **III**

Golterman, H. L. (2000): 'Denitrification and a numerical modelling approach for shallow waters', *Hydrobiol.* **431**, 93–104.   **IV**

Golterman, H. L. (2001): 'Phosphate release from anoxic sediments or 'What did Mortimer really write?'', *Hydrobiol.* **450**, 99–106.   **III**

Golterman, H. L., C. C. Bakels & J. Jakobs-Mögelin (1969): 'Availability of mud phosphates for the growth of algae', *Verh. int. Ver. Limnol.* **17**, 467–479.   **III**

Golterman, H. L., A. B. Viner, & G. F. Lee (1977): 'Preface', In: *Interactions between Sediments and Freshwater*, Golterman, H. L. (Ed.), Junk, the Hague, pp. 1–9.   **III**

Golterman, H. L., R. S. Clymo & M. A. M. Ohnstad (1978): *Methods for physical and chemical analysis of fresh waters*, = IBP Handbook **8**, 3rd edition, Blackwell Scientific Publ., Oxford, 214 pp.   **V**

Golterman H. L. & Kouwe, F. A. (1980): 'Chemical budgets and nutrient pathways', In: *Functioning of freshwater ecosystems (Chapter 4)*, E. D. Le Cren & R. H. Lowe-McConnell (Eds.), IBP 22, Cambridge University Press, London, 588 pp.   **II**

Golterman, H. L, P. Sly & R. L. Thomas (1983): *Study of the relationship between water quality and sediment transport*, UNESCO, Technical Papers in Hydrology **26** (Unesco, Paris), 232 pp.   **I**

Golterman, H. L. & M. L. Meyer (1985): 'The geochemistry of two hard water rivers, the Rhine and the Rhone Part 1–4', *Hydrobiol.* **126**, 3-10; 11–19; 21–24; 25–29.   **III**

Golterman, H. L. & A. Booman (1988): 'Sequential extraction of iron-phosphate and calcium-phosphate from sediments by chelating agents', *Verh. int. Ver. Limnol.* **23**, 904–909.   **III**

Golterman, H. L., F. Minzoni & C. Bonetto (1988): 'The Nitrogen cycle in shallow water sediment systems of rice fields. Part III: The influence of N-application on the yield of rice', *Hydrobiol.* **159**, 211–217.   **IV**

Golterman, H. L & N. T. De Oude (1991): 'Eutrophication of Lakes, Rivers and Coastal Seas', In: *The Handbook of Environmental Chemistry Vol. 5A*, Hutzinger, O. (Ed.), Springer Verlag, Berlin, Heidelberg, page 79–124.   **P**

Golterman, H. L. & I. M. De Graaf Bierbrauwer (1992): 'Colorimetric determination of sulphate in freshwater with a chromate reagent', *Hydrobiol.* **228**, 111–115.   **V**

Golterman, H. L. & C. J. De Groot (1994): 'Nouvelles connaissances des formes du phosphate: conséquences sur le cycle du phosphate dans les sédiments des eaux douces peu profondes', *Annls Limnol.* **30**, 221–232.   **III**

Golterman, H. L., P. Bruijn, J. G. M. Schouffoer, E. Dumoulin (1998): 'Urea fertilization and the N-cycle of rice-fields in the Camargue (S. France)', *Hydrobiol.* **384**, 7–20.   **IV**

Golterman H. L., Joëlle Paing, Laura Serrano, Elena Gomez (1998): 'Presence of and phosphate release from polyphosphates or phytate phosphate in lake sediments', *Hydrobiol.* **364**, 99–104.   **III**

Gonsiorczyk, T., P. Casper & R. Koschel (1997): 'Variations of phosphorus release from sediments in stratified lakes', *Wat. Air Soil Pollut.* **99**, 427–434.   **III**

Gorham, E. (1958): 'The influence and importance of daily weather conditions in the supply of chloride, sulphate and other ions to fresh waters from atmospheric precipitation', *Philos. Trans. R. Soc. Lond., Ser. B. Biol. Sci.* **241**, 509–538.   **II**

Gran, V. & H. Pitkänen (1999): 'Denitrification in estuarine sediments in the eastern Gulf of Finland, Baltic Sea', *Hydrobiol.* **393**, 107–115.   **IV**

Granéli, W. (1998): 'Internal phosphorus loading in Lake Ringsjön', *Hydrobiol.* **404**, 19–26.   **III**

Grenz, C., Th. Moutin, B. Picot & H. Massé (1991): 'Comparaison de deux méthodes de mesure de flux de nutriments à l'interface eau-sédiment: methode des peepers et méthode des chambres benthiques', *C. R. Acad. Sci Paris serie III* **313**, 239–244.   **III**

Grigg, N. J., I. T. Webster & P. W. Ford (1999): 'Pore-water convection induced by peeper emplacement in saline sediment', *Limnol. Oceanogr.* **44**, 425–430.   **III**

Grobbelaar J. U. (1983): 'Availability to algae of N and P adsorbed on suspended solids in turbid waters of the Amazon River', *Arch. Hydrobiol.* **96**, 302–316.   **III**

Grobler D. C. & E. Davies (1979): 'The availability of sediment phosphate to algae', *Water SA* **5**, 114–123.   **III**

Grobler D. C. & E. Davies (1981): 'Sediments as a source of phosphate; A study of 38 impoundments', *Water SA* **7**, 54–60.   **III**

Grossart, H.-P. & M. Simon (1993): 'Limnetic macroscopic organic aggregates (lake snow): occurrence, characteristics, and microbial dynamics in Lake Constance', *Limnol. Oceanogr.*

38, 532–546.   **II**

Grossart, H.-P. & M. Simon (1998): 'Significance of limnetic aggregates (lake snow) for the sinking flux of particulate organic matter', *Aquat. Microbiol. Ecol.* **15**, 115–125.   **II**

Güde, H., M. Seidel, P. Teiber & M. Weyhmüller (2001): 'P-release from littoral sediments in Lake Constance', *Proc. int. Soc. Limn.* **27**, 2624–2627.   **III**

Gunnars, A. & S. Blomquist (1997): 'Phosphate exchange across the sediment-water interface when shifting from anoxic to oxic conditions – a comparison of freshwater and brackish-marine systems', *Biogeochem.* **37**, 203–266.   **III**

Gunnars, A., S. Blomqvist, P. Johansson & C. Armstrong (2002): 'Formation of Fe(III) oxy-hydroxide colloids in freshwater and brackish seawater, with incorporation of phosphate and calcium', *Geochim. et Cosmochim. Acta* **66**, 745–758.   **III**

Gächter, R. & B. Müller (2003): 'Why the phosphorus retention of lakes does not necessarily depend on the oxygen supply to their sediment surface', *Limnol. Oceanogr.* **48**, 929–933.   **III**

Hadas, O. & R. Pinkas (1992): 'Sulphate-reduction process in sediments of Lake Kinneret', *Israel. Hydrobiol.* **235/236**, 295–301.   **II**

Handbook of Chemistry and Physics (1984): 64th ed., CRC Press, Boca Raton, Florida, pp. F46.   **V**

Hanna, M. (1989): 'Biologically available phosphorus: Estimation and prediction using an anion-exchange resin', *Can. J. Fish. Aquat. Sci.* **46**, 638–643.   **III**

Hansen, K., S. Mouridsen & E. Kristensen (1998): 'The impact of *Chironomus plumosus* larvae on organic matter decay and nutrient (N, P) exchange in a shallow eutrophic lake sediment following a phytoplankton sedimentation', *Hydrobiol.* **364**, 65–74.   **III**

Heaney, S. I., W. J. P. Smyly and J. F. Talling (1986): 'Interactions of Physical, Chemical and Biological Processes in Depth and Time within a Productive English Lake during Summer Stratification', *Int. Rev. ges. Hydrobiol.* **71**, 441–494.   **III**

Hegemann, D. A., A. H. Johnson & J. D. Keenan (1983): 'Determination of Algal-available Phosphorus on Soil and Sediment: A Review and Analysis', *J. Envir. Qual.* **12**, 12–16. **III**

Heggie, D.T., G.W. Skyring, J. Orchardo, A.R. Longmore, G.J. Nicholson & W.M. Berelson (1999): 'Denitrification and denitrifying efficiencies in sediments of Port Phillip Bay: direct determinations of biogenic N₂ and N-metabolite fluxes with implications for water quality', *Mar. Freshwat. Res.* **50**, 589–596.   **IV**

Herut, B., T. Zohary, R. D. Robarts & N. Kress (1999): 'Adsorption of dissolved phosphate onto loess particles in surface and deep Eastern Mediterranean water', *Mar. Chem.* **64**, 253–265.   **III**

Hesslein, R. H. (1976): 'An in situ sampler for close interval pore water studies', *Limnol. Oceanogr.* **21**, 912–914.   **III**

Hieltjes, A. H. M. & L. Lijklema (1980): 'Fractionation of inorganic phosphates in calcareous sediments', *J. Envir. Qual.* **9**, 405–407.   **III**

Hill, B. H. (1979): 'Uptake and release of nutrients by aquatic macrophytes', *Aquat. Bot.* **7**, 87–93.   **IV**

Hines, M. E., S. L. Knollmeyer & J. B. Tugel (1989): 'Sulfate reduction and other sedimentary biogeochemistry in a northern New England salt marsh', *Limnol. Oceanogr.* **34**, 578–590.   **II**

Hopkinson, C. S., A. E. Giblin, J. Tucker & R. H. Garritt (1999): 'Benthic metabolism and nutrient cycling along an estuarine salinity gradient', *Estuaries* **22**, 825–843.   **IV**

Hosomi, M., M. Okada & R, Sudo (1981): 'Release of phosphorus from sediments', *Verh. int. Ver. Limnol.* **21**, 628–633.   **III**

House, W. A. (1990): 'The prediction of phosphate coprecipitation with calcite in freshwaters', *Wat. Res.* **24**, 1017–1023.   **III**

House, W. A. & L. Donaldson (1986): 'Adsorption of phosphate on calcite', *J. Colloid Interface Sci.* **112**, 309–324.   **III**

House, W. A., H. Casey, L. Donaldson, & S. Smith (1986): 'Factors affecting the coprecipitation of phosphorus with calcite in hardwaters', *Wat. Res.* **20**, 917–922.   **III**

House, W. A., F. H. Denison & P. D. Armitage (1995): 'Comparison of the uptake of inorganic phosphorus to a suspended and stream-bed sediment', *Wat. Res.* **29**, 767–779.   **III**

House, W. A. & F. H. Denison (1997): 'Nutrient dynamics in a lowland stream impacted by sewage effluent: Great Ouse, England', *Sci. Total Envir.* **205**, 25–49.   **III**

House, W. A., T. D. Jickells, A. C. Edwards, K. E. Praska & F. H. Denison (1998): 'Reactions of phosphorus with sediments in fresh and marine waters', *Soil Use and Management* **14**, 139–146.   **III**

House, W. A. & M. S. Warwick (1998): 'A mass-balance approach to quantifying the importance of in-stream processes during nutrient transport in a large river catchment', *Science of the Total Envir.* **210/211**, 139–152.   **III**

House, W. A. & M. S. Warwick (1999): 'Interactions of phosphorus with sediments in the River Swale, Yorkshire, UK', *Hydrol. Process.* **13**, 1103–1115.   **III, IV**

Huettl, R. C., R. C. Wendt & R. B. Corey (1979): 'Prediction of algal-available phosphorus in runoff sediments', *J. Envir. Qual.* **8**, 130–132.   **III**

Hupfer, M., R. Gächter & H. Rüegger (1995): 'Polyphosphate in lake sediments: 31P NMR spectroscopy as a tool for its identification', *Limnol. Oceanogr.* **40**, 610–617.   **III**

Hupfer, M., R. Gächter & R. Giovanoli (1995): 'Transformation of phosphorus species in settling seston and during early sediment diagenesis', *Aquatic Sciences* **57**, 305–324.   **III**

Huttula, T. (1994): 'Suspended sediment transport in Lake Säkylän Pyhäjärvi', *Aqua Fennica* **24**, 171–185.   **I**

Jackson, M. L. (1958): *Soil chemical analysis*, Prentice Hall, Englewood Cliffs, 498 pp. **III**

Jacoby, J. M., D. D. Lynch, E. B. Welch & M. A. Perkins (1982): 'Internal phosphorus loading in a shallow eutrophic lake', *Wat. Res.* **16**, 911-919.   **III**

James, W. F. & J. W. Barko (1997): 'Net and gross sedimentation in relation to the phosphorus budget of Eau Galle Reservoir, Wisconsin', *Hydrobiol.* **345**, 15–20.   **III**

Jáuregui, J. & J. A. García Sanchez (1993): 'Fractionation of sedimentary phosphorus: A comparison of four methods', *Verh. int. Ver. Limnol.* **25**, 1150–1152.   **III**

Jaúrequi, J. & J. Toja (1993): 'Dinámica del fósforo en lagunas temporales del P.N. de Doñana', In: *Actas VI Congreso Español de Limnología*, L. Cruz-Pizarro y otros (Eds.),

Granada (Spain), pp. 99–106. **IV**

Jensen, H. S. & F. Ø. Andersen (1992): 'Importance of temperature, nitrate, and pH for phosphate release from aerobic sediments of four shallow, eutrophic lakes', *Limnol. Oceanogr.* **37**, 577–589. **III**

Jensen, H. S., P. Kristensen, E. Jeppesen & A. Skytthe (1992): 'Iron: phosphorus ratio in surface sediment as an indicator of phosphate release from aerobic sediments in shallow lakes', *Hydrobiol.* **235/236**, 731–743. **III**

Jensen, H. S. & B. Thamdrup (1993): 'Iron-bound phosphorus in marine sediments as measured by bicarbonate-dithionite extraction', *Hydrobiol.* **253**, 47–59. **III**

Jensen, H. S., P. B. Mortensen, F. Ø. Andersen, E. Rasmussen & A. Jensen (1995): 'Phosphorus cycling in a coastal marine sediment, Aarhus Bay, Denmark', *Limnol. Oceanogr.* **40**, 908–917. **III**

Jensen, H. S., K. J. McGlathery, R. Marino & R. W. Howarth (1998): 'Forms and availability of sediment phosphorus in carbonate sand of Bermuda seagrass beds', *Limnol. Ocenogr.* **43**, 799–810. **III**

Jensen, M. H., E. Lomstein & J. Sørensen (1990): 'Benthid NH and NO flux following sedimentation of a spring phytoplankton bloom in Aarhus Bight, Denmark', *Mar. Ecol. prog. Series* **61**, 87–96. **IV**

Jones, B. F., C. J. Bowser (1978): 'The mineralogy and related chemistry of lake sediments', In: *Lakes; chemistry, geology, physics*, Lerman, A. (Ed.), Springer Verlag, New-York/Heidelberg, pp. 179–235. **I**

Jones, J. G. (1979): 'Microbial nitrate reduction in freshwater sediments', *J. gen. Microbiol.* **115**, 27–35. **IV**

Jones, J. G. (1982): 'Activities of aerobic and anaerobic bacteria in lake sediments and their effect on the water column', In: *Sediment Microbiology*, D. B. Nedwell & C. M. Brown (Eds.), Academic Press, London, pp. 107–145. **II**

Jones, J. G. (1985): 'Microbes and microbial processes in lakes', *Phil. Trans. R. Soc. Lond. A* **315**, 3–17. **II**

Jones, J. G. & B. M. Simon (1980): 'Decomposition processes in the profundal region of Blelham Tarn and the Lund tubes', *J. Ecol.* **68**, 493–512. **IV**

Jones, J. G. & B. M. Simon (1981): 'Differences in microbial decomposition in profundal and littoral lake sediments, with particular reference to the nitrogen cycle', *J. Gen. Microbiol.* **123**, 297–312. **IV**

Jones, J. G., B. M. Simon & R. W. Horsley (1982): 'Microbiological sources of ammonia in freshwater sediments', *J. gen. Microbiol.* **128**, 2823–2831. **IV**

Jones, J. G., S. Gardener & B. M. Simon (1983): 'Bacterial reduction of Ferric iron in a stratifid eutrophic lake', *J. of Gen. Microbiol.* **129**, 131–139. **II**

Jones, J. G., S. Gardener & B. M. Simon (1984): 'Reduction of ferric iron by heterotrophic bacteria in lake sediments', *J. of Gen. Microbiol.* **130**, 45–51. **II**

Jones, J. G. & B. M. Simon (1985): 'Interaction of acetogens and methanogens in anaerobic freshwater sediments', *Appl. Envir. Microbiol.* **49**, 944–948. **II**

Jørgensen, K. S. (1989): 'Annual pattern of denitrification and nitrate ammonification in estuarine sediment', *Appl. Envir. Microbiol.* **55**, 1841–1847.   **IV**

Jørgensen, K. S. & J. Sørensen (1985): 'Seasonal cycles of O_2, nitrate and sulphate reduction in estuarine sediments', *Mar. Ecol.–Progr. Ser.* **24**, 65–74.   **IV**

Jørgensen, K. S. & J. Sørensen (1988): 'Two annual maxima of nitrate reduction and denitrification in estuarine sediment, Norsminde Fjord, Denmark', *Mar. Ecol.–Progr. Ser.* **48**, 147–155.   **IV**

Kamiyama, K. (1978): 'Studies on the release of ammonium nitrogen from the bottom sediments in freshwater regions. IV. A model for ammonium nitrogen movement in the surface layer of sediments', *Jap. J. Limnol.* **39**, 181-188.   **IV**

Kamiyama. K., S. Okuda, & A. Kawai (1978): 'Studies on the release of ammonium nitrogen from the bottom sediments in freshwater regions. III. Ammonium nitrogen in the sediments of different water regions', *Jap. J. Limnol.* **39**, 176-180.   **IV**

Kana T. M., M. B. Sullivan, J. C. Cornwell & K. M. Groszkowski (1998): 'Denitrification in estuarine sediments determined by membrane inlet mass spectrometry', *Limnol. Oceanogr.* **43**, 334–339.   **IV**

Kaplan, W., and I. Valiela (1979): 'Denitrification in a salt marsh ecosystem', *Limnol. Oceanogr.* **24**, 726–734.   **IV**

Kelly-Gerreyn, B. A., M. Trimmer & D. J. Hydes (2001): 'A diagenetic model discriminating denitrification and dissimilatory nitrate reduction to ammonium in a temperate estuarine sediment', *Mar. Ecol.–Progr. Ser.* **220**, 33–46.   **IV**

Kelso, B. H. L., R. V. Smith & R. J. Laughlin (1999): 'Effects of carbon substrates on nitrite accumulation in freshwater sediments', *Appl. Envir. Microbiol.* **65**, 61–66.   **IV**

Kemp, A. L. W. & N. S. Harper (1976): 'Sedimentation rates and a revised sediment budget for Lake Erie', *J. Great Lakes Res.* **3**, 221–233.   **I**

Kemp, A. L. W., G. A. Macinnis. & N. S. Harper (1977): 'Sedimentation rates and a sediment budget for Lake Ontario', *J. Great Lakes Res.* **2**, 324–340.   **I**

Kemp, A. L. W., C. I. Dell & N. S. Harper (1978): 'Sedimentation rates and a sediment budget for Lake Superior', *J. Great Lakes Res.* **4**, 276–287.   **I**

Kemp, W. M., P. Sampou, J. Caffrey, M. Mayer, K. Henriksen & W. R. Boyton (1990): 'Ammonium recycling versus denitrification in Chesapeake Bay sediments', *Limnol. Oceanogr.* **35**, 1545–1563.   **IV**

Keskitalo, J. & P. Eloranta (Eds.) (1999): *Limnology of humic waters*, Backhuys Publishers, Leiden, The Netherlands, 292 pp.   **I**

King, G. M. (1988): 'Patterns of sulfate reduction and the sulfur cycle in a South Carolina salt marsh', *Limnol. Oceanogr.* **33**, 376–390.   **II**

Klapwijk, S. P. & C. Bruning (1984): 'Available Phosphorus in the Sediments of Eight Lakes in the Netherlands', In: *Proceedings of the Third International Symposium on Interactions Between Sediments and Water*, P. Sly (Ed.), Springer Verlag, pp. 391–398.   **III**

Kleeberg, A. (1997): 'Interactions between benthic phosphorus release and sulfur cycling in lake Scharmützelsee (Germany)', *Wat. Air Soil Pollut.* **99**, 391–399.   **III**

Kleeberg, A. (2002): 'Phosphorus sedimentation in seasonal anoxic Lake Scharmützel, NE Germany', *Hydrobiol.* **472**, 53–65.   **III**

Kleeberg, A., D. Jendritzki & B. Nixdorf (1999): 'Surficial sediment composition as a record of environmental changes in the catchment of shallow Lake Petersdorf, Brandenburg, Germany', *Hydrobiol.* **408/409**, 183–192.   **IV**

Kleeberg, A. & H. Schubert (2000): 'Vertical gradients in particle distribution and its elemental composition under oxic and anoxic conditions in a eutrophic lake, Scharmützelsee, NE Germany', *Arch. Hydrobiol.* **148**, 187–207.   **IV**

Kleeberg, A., B. Nixdorf & J. Mathes (2000): 'Lake Jabel restoration project: Phosphorus status and possibilities and limitations of diversion of its nutrient-rich main inflow', *Lakes & Reservoirs: Research and Management* **5**, 23–33.   **III**

Knowles, R. (1979): 'Denitrification, acetylene reduction and methane metabolism in lake sediment exposed to acetylene', *Appl. Envir. Microbiol.* **38**, 486–493.   **IV**

Knuuttila, S., O.-P. Pietiläinen & L. Kauppi (1994): ' Nutrient balances and phytoplankton dynamics in two agriculturally loaded shallow lakes', *Hydrobiol.* **275/276**, 359–369. **III**

Kölle, W., P. Werner, O. Strebel & J. Böttcher (1983): 'Denitrification in einem reduzierende Grundwasserleiter', *Vom Wasser* **61**, 125–147.   **IV**

Kolthoff, I. M., E. B. Sandell, E. J. Meehan & S. Bruckenstein (1969): *Quantitative chemical analysis*, The Macmillan Cy, London, 1199 pp.   **V**

Kristensen, P., M. Søndergaard & E. Jeppesen (1992): 'Resuspension in a shallow eutrophic lake', *Hydrobiol.* **228**, 101–109.   **III**

Kristjansson, J. K., P. Schönheit & R. K. Thauer (1982): 'Different K_s values for hydrogen of methanogenic bacteria and sulphate reducing bacteria: an explanation for the apparent inhibition of methanogenesis by sulphate', *Arch. Microbiol.* **131**, 278–282.   **II**

Kulaev, I. S. (1979): *The biochemistry of inorganic polyphosphates*, Wiley.   **III**

Kulaev, I. S. & V. M. Vagabov (1983): 'Polyphosphate metabolism in micro-organisms', *Advances in Microbial Physiology* **24**, 83–171.   **III**

Kuznetsov, S. I. (1970): *The microflora of lakes and its geochemical activity*, Oppenheimer C. H. (Ed.), University of Texas Press, Austin & London, 503 pp.   **III, IV**

Lal, V. B. & S. Banerji (1977): 'A multivariate prediction equation for rate of sedimentation in North Indian reservoirs', In: Golterman, H. L. (Ed.), 1977, pp. 183–188.   **I**

Landolt Börnstein (1969): *Zahlenwerte und Functionen aus Physik, Chemie, Astronomie, Geophysik, Technik*, Elektrische Eigenschaften II **5** (1960), 258–263; Transport Phenomene I (1969), 612–631, Springer Verlag, .   **V**

Laursen, A. E. & S. P. Seitzinger (2002): 'Measurement of denitrification in rivers: an integrated, whole reach approach', *Hydrobiol.* **484**, 67–81.   **IV**

Lee, G. F., W. C. Sonzogni, & R. D. Spear (1977): 'Significance of oxic vs anoxic conditions for lake Mendota sediment', In: *Interactions between Sediments and Freshwater*, Golterman H. L. (Ed.), Junk, the Hague, pp. 294–306.   **III**

Li, Y. H., & S. Gregory (1974): 'Diffusion of ions in sea water and in deep-sea sediments', *Geochim. Cosmochim Acta* **38**, 703–714.   **V**

Löfgren, S. & S.-O. Ryding (1985): 'Apatite solubility and microbial activities as regulators of internal loading in shallow, eutrophic lakes', *Verh. intern. Ver. Limnol.* **22**, 3329–3334.   **III**

Logan, T. (1977): 'Forms and sediment associations of nutrients (C, N and P), pesticides and metals; nutrients–N', In: *The fluvial transport of sediment-associated nutrients and contaminants*, Shear, H. & A.E.P.Watson (Eds.), Internat. Joint Commission, Windsor, Ontario, 309 pp.   **IV**

Lorenzen, J., L.H. Larsen,, T. Kjaer & N.-P. Revsbech (1998): 'Biosensor determination of the microscale distribution of nitrate, nitrate assimilation, nitrification, and denitrification in a diatom-inhabited freshwater sediment', *Appl. Envir. Microbiol.* **64**, 3264–3269.   **IV**

Luczak, C., M.-A. Janquin & A. Kupka (1997): 'Simple standard procedure for the routine determination of organic matter in marine sediment', *Hydrobiol.* **345**, 87 -94.   **IV**

Luther, G. W., J. E. Kostka, T. M. Church, B. Sulzberger & W. Stumm (1992): 'Seasonal iron cycling in the salt-marsh sedimentary environment: The importance of ligand complexes with Fe(II) and Fe(III) in the dissolution of Fe(III) minerals and pyrite, respectively', *Mar. Chem.* **40**, 81–103.   **V**

Lijklema, L. (1977): 'The role of iron in the exchange of phosphate between water and sediment', In: *Interactions between Sediments and Freshwater*, Golterman, H. L. (Ed.), Junk,the Hague, pp. 313–317.   **III**

Lijklema, L. (1980): 'Interaction of ortho-phosphate with iron(III) and aluminium hydroxides', *Envir. Sci. Technol.* **14**, 537–541.   **III**

Lijklema, L.& A.H.M. Hieltjes (1982): 'A dynamic phosphate budget model for a eutrophic lake', *Hydrobiol.* **91**, 227–233.   **III**

Masaaki Hosomi, Mitsumasa Okada & Ryuichi Sudo (1982): 'Release of phosphorus from lake sediments', *Envir. Int.* **7**, 93.   **III**

Mackereth, F.J.H (1966): 'Some chemical observations on post-glacial lake sediments', *Philos. Trans. R. Soc. Lond., Ser. B.* **250**, 165–213.   **IV**

Mäkelä, K. & L. Tuominen (2003): 'Pore water nutrient profiles and dynamics in soft bottoms of the northern Baltic Sea', *Hydrobiol.* **492**, 43–53.   **IV**

Manheim, F. T. (1970): 'The diffusion of ions in unconsolidated sediment', *Earth Planet. Sci. Lett.* **9**, 307–309.   **V**

Mann, K. H. (1972): 'Macrophyte production and detritus food chains in coastal waters', In: *Detritus and its role in aquatic ecosystems*, U. Melchiori-Santolini & J. W. Hoptom (Eds.), Memorie dell' Istituto Italiano di Idrobiologia **29**, Suppl. 13–16, pp. 353–384. **II**

Maran, S., G. Ciceri & W. Martinotti (1995): 'Mathematical models for estimating fluxes at the sediment-water interface in benthic chamber experiments', *Hydrobiol.* **297**, 67–74. **III**

Masahiro Suzumura & Akiyoshi Kamatani (1993): 'Isolation and determination of inositol hexaphosphate in sediments from Tokyo Bay', *Geochim. et Cosmochim Acta* **57**, 2197–2202.   **III**

Masahiro Suzumura & Akiyoshi Kamatani (1995a): 'Mineralization of inositol hexaphosphate in aerobic and anaerobic marine sediments: Implications for the phosphorus cycle', *Geochim. et Cosmochim. Acta* **59**, 1021–1026.   **III**

Masahiro Suzumura & Akiyoshi Kamatani (1995b): 'Origin and distribution of inositol hexaphosphate in estuarine and coastal sediments', *Limnol. Oceanogr.* **40**, 1254–1261.   **III**

Masumi Yamamuro & Isao Koike (1998): 'Concentrations of nitrogen in sandy sediments of a eutrophic lagoon', *Hydrobiol.* **386**, 37–44.   **IV**

Matisoff, G., J. Fisher Berton & P.L.McGall (1981): 'Kinetics of nutrients and metal release from decomposing lake sediments', *Geochim. Cosmochim. Acta* **45**, 2333-2347.   **III**

Mayer, T. D. & W. M., Jarrell (1999): 'Phosphorus sorption during iron(II) oxidation in the presence of silica', *Wat. Res.* **34**, 3949–3956.   **III**

Mayer, T., C. Ptacek & L. Zanini (1999): 'Sediments as source of nutrients to hypereutrophic marshes of Point Pelee, Ont. Canada', *Wat. Res.* **33**, 1460–1470.   **III**

McAuliffe, T. F., R. J. Lukatelich, A. J. McComb & S. Qiu (1998): 'Nitrate applications to control phosphorus release from sediments of a shallow eutrophic estuary: an experimental evaluation', *Mar. Freshwat. Res.* **49**, 463–473.   **IV**

McComb, A. J., S. Qiu, R. J. Lukatelich & T. F. McAuliffe (1998): 'Spatial and temporal heterogeneity of sediment phosphorus in the Peel-Harvey estuarine system', *Estuarine, Coastal and Shelf Science* **47**, 561–577.   **III**

McGlashan, M. L. (1971): *Physico-Chemical Quantities and Units*, Royal Institute of Chemistry London, 116 pp.   **V**

Menéndez, M., E. Forés & F. A. Comin (1989): 'Ruppia cirrhosa – Decomposition in a coastal temperate lagoon as affected by macroinvertebrates', *Arch. Hydrobiol.* **117**, 39–48.   **IV**

Mengis, M., R. Gächter, B. Wehrli & S. Bernasconi (1997): 'Nitrogen elimination in two deep eutrophic lakes', *Limnol. Oceanogr.* **42**, 1530–1543.   **IV**

Mesnage, V., & B. Picot (1995): 'The distribution of phosphate in sediments and its relation with eutrophication of a mediterranean coastal lagoon', *Hydrobiol.* **297**, 29–41.   **III**

Messer, J. & P. Brezonik (1983): 'Comparison of denitrification rate estimation techniques in a large shallow lake', *Wat. Res.* **17**, 631–640.   **IV**

Meybeck, M. (1976): 'Total mineral dissolved transport by world major rivers', *Hydrolog. Sci. Bull. Sci. Hydrolog.* **XXI, 2/6**, 265–284.   **I**

Meybeck, M. (1977): 'Dissolved and suspended matter carried by rivers: composition, time and space variations, and world balance', In: Golterman, H. L. (Ed.), 1977, pp. 25–32. **I**

Middelburg, J. J., K. Soetart, P. M. J. Herman (1996a): 'Evaluation of the nitrogen isotope-pairing method for measuring benthic denitrification: A simulation analysis', *Limnol. Oceanogr.* **41**, 1839–1844.   **IV**

Middelburg, J. J., K. Soetart & P. M. J. Herman (1996b): 'Reply to the comment by Nielsen *et al.*', *Limnol. Oceanogr.* **41**, 1846–1847.   **IV**

Miltenburg J. C. & H. L. Golterman (1998): 'The energy of the adsorption of o-phosphate onto ferric hydroxide', *Hydrobiol.* **364**, 93–97.   **III**

Minzoni, F., C. Bonetto & H. L. Golterman (1988): 'The Nitrogen cycle in shallow water sediment systems of rice fields. Part I: The denitrification process', *Hydrobiol.* **159**, 189–202.   **IV**

Mitchell, A. M. & D. S. Baldwin (1998): 'The effect of desiccation/oxidation on the potential for bacterial mediated P release from sediments', *Limnol. Oceangr.* **43**, 21–27. **IV**

Mitchell, A. M. & D. S. Baldwin (1999): 'The effects of sediment desiccation on the potential for nitrification, denitrification and methanogenesis in an Australian reservoir', *Hydrobiol.*

392, 3–11.   **II, IV**

Mortimer, C. H. (1941): 'The exchange of dissolved substances between mud and water in lakes', *J. Ecol.* **29**, 280–329.   **III**

Mortimer, C. H. (1942): 'The exchange of dissolved substances between mud and water in lakes', *J. Ecol.* **30**, 147–201.   **III**

Mortimer, C. H. (1971): 'Chemical exchanges between sediments and water in the Great Lakes – speculations on probable regulatory mechanisms', *Limnol. Oceanogr.* **16**, 387–404.   **II, III**

Mountfort, D. O. & R. A. Asher (1981): 'Role of sulfate reduction versus methanogenesis in terminal carbon flow in polluted intertidal sediment of Waimea inlet, Nelson, New Zealand', *Appl. Envir. Microbiol.* **42**, 252–258.   **II**

Moutin, T., B. Picot, M. C. Ximenes & J. Bontoux (1993): 'Seasonal variations of P compounds and their concentrations in two coastal lagoons (Herault, France)', *Hydrobiol.* **252**, 45–59.   **III**

Moutin, T., P. Raimbault, H. L. Golterman & B. Coste (1998): 'The input of nutrients by the Rhone river into the Mediterranean Sea: recent observations and comparison with earlier data', *Hydrobiol.* **373/374**, 237–246.   **I, III**

Müller, G. (1966): 'Die Sedimentbildung im Bodensee', *Naturwissenschaften* **53**, 237–247. **I**

Müller, G. (1977): 'Schadstoff–Untersuchungen an datierten Sedimentkernen aus dem Bodensee', *Z. Natürforsch.* **32**, 920–925.   **IV**

Müller, G. & U. Förstner (1968a): 'Sedimenttransport im Mündungsgebiet des Alpenrheins', *Geol. Rundsch.* **58**, 229–259.   **I**

Müller, G. & U. Förstner (1968b): 'General relationship between suspended sediment concentration and water discharge in the Alpenrhein and some other rivers', *Nature* **217**, 244–245.   **I**

Mulsow, S., Boudreau, B. P. & Smith, J. N. (1998): 'Bioturbation and porosity gradients', *Limnol. Oceanogr.* **43**, 1–9.   **V**

Murphy, J. & J. P. Riley (1962): 'A modified single-solution method for the determination of phosphate in natural waters', *Analytica chim. Acta* **27**, 31.   **V**

Newell, B. S., B. Morgan & J. Cundy (1967): 'The determination of urea in seawater', *J. Mar. Res.* **25**, 201–202.   **IV**

Nicholson, G. J. & A. R. Longmore (1999): 'Causes of observed temporal variability of nutrient fluxes from a southern Australian marine embayment', *Mar. Freshwat. Res.* **50**, 581–588.   **IV**

Nielsen, L. P. (1992): 'Denitrification in sediment determined from nitrogen isotope pairing', *FEMS Microbiol. Ecol.* **86**, 357–362.   **IV**

Nielsen, L. P., N. Risgaard-Petersen, S. Rysgaard & T. H. Blackburn (1996): 'Reply to the note by Middelburg *et al.*', *Limnol Oceanogr.* **41**, 1845–1846.   **IV**

Nissenbaum, A. (1979): 'Phosphorus in marine and non-marine humic substances', *Geochim. and Cosmochim Acta* **43**, 1973–1978.   **III**

Nürnberg, G. K (1987): 'A comparison of internal phosphorus loads in lakes with anoxic hypolimnia: Laboratory incubation versus in situ hypolimnetic phosphorus accumulation', *Limnol. Oceanogr.* **32**, 1160–1164.   **III**

Nürnberg, G. K (1988): 'Prediction of phosphorus release rates from total and reductant-soluble phosphorus in anoxic sediments', *Can. J. Fish. Aquat. Sci.* **45**, 453–462.   **III**

Nürnberg, G. K. & R. H. Peeters (1984): 'The importance of internal phosphorus load to the eutrophication of lakes with anoxic hypolimnia', *Proc. int. Soc. Limnol.* **22**, 190–194. **III**

Nürnberg, G. K., M. Shaw, P. J. Dillon & D. J. MaQuenn (1986): 'Internal phosphorus load in an oligotrophic Precambian Shield Lake with an anoxic Hypolimnion', *Can. J. Fish. Aquat. Sci.* **43**, 574–580.   **III**

O.E.C.D. (1982): *Eutrophication of waters: Monitoring, Assessment and Control. Report of the O.E.C.D. Cooperative programme on Eutrophication*, Prepared by Vollenweider, R.A. & J. Kerekes, O.E.C.D., Paris, 164 pp.   **IV**

Olila, O. G. & K. R. Reddy (1997): 'Influence of redox potential on phosphate-uptake by sediments in two sub-tropical eutrophic lakes', *Hydrobiol.* **345**, 45–57.   **III**

Orive, I., M. Elliot & V. N. De Jonge (2002): 'Proceedings of the 31st Symposium of the Estuarine and Coastal Sciences Association, held in Bilbao, Spain, 3–7 July 2000', *Hydrobiol.* **475/476**.   **P**

Olsen S. (1964): 'Phosphate equilibrium between reduced sediments and water, laboratory experiments with radioactive phosphorus', *Verh. int. Ver. Limnol.* **13**, 915–922.   **III**

Olsen, S. R., C. V. Cole, F. S. Watanabe & L. A. Dean (1954): 'Estimation of available phosphorus in soils by extraction with sodium bicarbonate', *U.S. Dept. Agric. Circular* **939**.   **III**

Olsen, S. R. & L. E. Sommers (1982): 'Phosphorus', In: *Methods of soil analysis, part 2*, A. L. Page *et al.* (Eds.), American Society of Agronomy, pp. 403–430.   **III**

Oremland, U. S. & S. Polcin (1982): 'Methanogenesis and sulphate reduction: Competitive and noncompetitive substrates in estuarine sediments', *Appl. Envir. Microbiol.* **44**, 1270–1276.   **II**

Paludan, C. & H. S. Jensen (1995): 'Sequential extraction of phosphorus in freshwater wetland and lake sediment: Significance of humic acids', *Wetlands* **15**, 365–373.   **IV**

Paludan, C. & J. T. Morris (1999): 'Distribution and speciation of phosphorus along a salinity gradient in intertidal marsh sediments', *Biogeochemistry* **45**, 197–221.   **III**

Panigatti, M. C. & M. A. Maine (2003): 'Influence of nitrogen species (NH_4^+ and NO_3^-) on the dynamics of P in water-sediment-*Salvinia herzogii* systems', *Hydrobiol.* **492**, 151–157. **V**

Pardo, P., J. F. López-Sánchez, G. Rauret, V. Ruban, H. Muntaur & Ph. Quevauviller (1999): 'Study of the stability of extractable phosphate content in a candidate reference material using a modified Williams extraction procedure', *The Analyst* **124**, 407–411.   **III**

Parfitt, R.L. (1989): 'Phosphate reactions with natural allophane, ferrihydrite and goethite', *J. Soil Sci.* **40**, 359–369.   **III**

Parfitt, R. L., J. D. Russel & V. C. Farmer (1976): 'Confirmation of the surface structures of goethite and phosphated goethite by infrared spectroscopy', *J. Chem. Soc. Farad. Trans.* **72**, 1082–1087.   **III**

Pattinson, S.N., R. Garcia-Ruiz, & B. A. Whitton (1998): 'Spatial and seasonal variation in denitrification in the Swale-Ouse system, a river continuum', *Sci. Total Envir.* **210-211**, 289–305.   **IV**

Pauer, J. J. & M. T. Auer (2000): 'Nitrification in the water column and sediment of a hypereutrophic lake and adjoining river system', *Wat. Res.* **34**, 1247–1254.   **IV**

Penn, M. R. & M. T. Auer (1997): 'Seasonal variability in phosphorus speciation and deposition in a calcareous, eutrophic lake', *Marine Geology* **139**, 47–59.   **III**

Pettersson, K. & V. Istvanovics (1988): 'Sediment phosphorus in Lake Balaton–forms and mobility', *Arch. Hydrobiol. Beih. Ergebn. Limnol.* **30**, 25–41.   **III**

Pettersson, K., B. Boström & O.-S. Jacobsen (1988): 'Phosphorus in sediments–speciation and analysis', *Hydrobiol.* **170**, 91–101.   **III**

Pizarro, M. J., J. Hammerly, M. A. Maine & N. Suñe (1992): 'Phosphate adsorption on bottom sediments of the Rio de la Plata', *Hydrobiol.* **228**, 43–54.   **III**

Poon, C. P. C. (1977): 'Nutrient exchange kinetics in sediment-water interface', *Prog. Wat. Tech.* **XX**, 881-895.   **IV**

Postma, D. (1982): 'Formation of siderite and vivianite in brackish and freshwater swamp sediments', *Am. J. Sci.* **282**, 1151–1183.   **II**

Postma, D. (1983): 'Pyrite and siderite oxidation in swamp sediments', *J. Soil Sci.* **34**, 163–182.   **II**

Postma, H. (1967): 'Sediment transport and sedimentation in the estuarine environment', *Estuaries. Publ. Amer. Assoc. Adv. Sci. (Lauff, G. H. (Ed.))* **83**, 158–179.   **I**

Povoledo, D. & H. L. Golterman (1975): *Humic substances; Their structure and function in the biosphere. Proceedings of an international meeting held at Nieuwersluis, the Netherlands, May 29–31, 1972*, Centre for Agricultural Publishing and Documentation, Wageningen, 368 pp.   **I, IV**

Prairie, Y., C. de Montigny & P. Giorgio (2002): 'Phosphorus release from anoxic sediments: Mortimers paradigma revisited', *Proc. int. Soc. Limnol.* **27**, 4013–4021.   **III**

Premazzi, G. & G. Zanon (1984): 'Availability of sediment P in Lake Lugano', *Verh. int. Ver. Limnol.* **22**, 1113–1118.   **III**

Psenner, R., R. Pucsko & M. Sager (1985): 'Die fractionierung organischer und anorganischer Phosphorverbindungen von Sedimenten', *Arch. Hydrobiol./Suppl.* **70**, 111–155.   **III**

Psenner, R. & R. Puckso (1988): 'Phosphorus fractionation: advantages and limits of the method for the study of sediment P origins and interactions', *Arch. Hydrobiol. Beih. Ergeb. Limnol.* **30**, 43–59.   **III**

Qiu Song & A. McComb (1994): 'Effects of oxygen concentration on phosphorus release from reflooded air-dried wetland sediments', *Aust. J. Mar. Freshw. Res.* **45**, 1319–1328. **III**

Qiu Song & A. McComb (1995): 'Planktonic and microbial contributions to phosphorus release from fresh and air-dried sediments', *Mar. Freshw. Res.* **46**, 1039–1045.   **III**

Qiu Song & A. McComb (1996): 'Drying-induced stimulation of ammonium release and nitrification in reflooded lake sediment', *Mar. Freshw. Res.* **47**, 531–536.   **IV**

Qiu Song & A. McComb (2000): 'Properties of sediment phosphorus in seven wetlands of the Swan Coastal Plain, S-W. Australia', *Wetlands* **20**, 267–279.   **III**

Qiu, Song & A. J. McComb (2002): 'Interrelations between iron extractability and phosphate sorption in reflooded air-dried sediments', *Hydrobiol.* **472**, 39–44. **III**

Ravens, T. V., O. Kocsis & A. Wüest (2000): 'Small-scale turbulence and vertical mixing in Lake Baikal', *Limnol. Oceanogr.* **45**, 159–173. **V**

Reynolds, C. S. & P. S. Davies (2001): 'Sources and bioavailability of phosphorus fractions in freshwaters: a British perspective', *Biol. Rev.* **76**, 27–64. **P**

Reddy, K. R., R. E. Jessup & P. S. C. Rao (1988): 'Nitrogen dynamics in a eutrophic lake sediment', *Hydrobiol.* **159**, 177–188. **IV**

Redshaw, C. J., C. F. Mason, C. R. Hayes & R.D. Roberts (1990): 'Factors influencing phosphate exchange across the sediment/water interface of eutrophic reservoirs', *Hydrobiol* **192**, 233–245. **III**

Rekohainen,S., P. Ekholm, B. Ulen & A. Gustafson (1997): 'Phosphorus losses from agriculture to surface waters in the Nordic countries', In: *Phosphorus Loss from Soil to Water*, H. Tunney *et al.* (Eds.), , . **II**

Revsberg, N. P. & J. Sørensen (Eds.) (1990): *Denitrification in soil and sediment*, FEMS symposium series **56**, Plenum Press, New York, 349 pp. **IV**

Richardson, C. J. (1985): 'Mechanisms controlling phosphorus retention capacity in freshwater wetlands', *Science* **228**, 1424–1427. **III**

Richardson, C. J. & P. E. Marshall (1986): 'Processes controlling movement, storage and export of phosphorus in a fen peatland', *Ecological Monographs* **56**, 279–302. **III**

Ridame, C. & C. Guieu (2002): 'Saharan input of phosphate to the oligotrophicwater of the open western Mediterranean Sea', *Limnol. Oceanogr.* **47**, 856–869. **III**

Riemsdijk, W. H. van, F. A. Weststrate & J. Beek (1977): 'Phosphates in soils treated with sewage water. III. Kinetic studies on the reaction of phosphate with Aluminium compounds', *J. Envir. Qual.* **6**, 26–29. **III**

Ringbom, A. (1963): *Complexation in Analytical Chemistry*, Interscience, New York. **III**

Ripl, W. (1976): 'Biochemical oxidation of polluted lake sediment. A new restoration method', *Ambio* **5**, 132–135. **IV**

Ripl, W., L. Leonardsson, G. Lindmark, G. Andersen & G. Cronberg (1979): 'Optimering av reningsverk/recipientsystem', *Vatten* **35**, 96–103. **IV**

Risgaard-Petersen, N. (2003): 'Coupled nitrification-denitrification in autotrophic and heterotrophic estuarine sediments: On the influence of benthic microalgae', *Limnol. Oceanogr.* **48**, 93–105. **IV**

Risgaard-Petersen, N., T. Dalsgaard, S. Rysgaard, P. B. Christensen, J. Borum, K. MaGlathery & L. P. Nielsen (1998): 'Nitrogen balance of a temperate eelgrass *Zosteria marina* bed', *Mar. Ecol. Progress Series* **174**, 281–291. **IV**

Robbe, D. (1975): *Influence des matières en suspension sur la qualité des eaux de surface (In French)*, Report of Laboratoire des Ponts et Chaussées, Paris, 123 pp. **I**

Roden, E. E. & J. W. Edmonds (1997): 'Phosphate mobilisation in anaerobic sediments: microbial Fe(III) oxide reduction versus iron-sulphide formation', *Arch. Hydrobiol.* **139**, 347–378. **III**

Rodhe, W. (1949): 'The ionic composition of lake waters', *Verh. int. Ver. Theor. Angew. Limnol.* **10**, 377–386.   **II**

Romero-Gonzalez, M. E., E. Zambrano, J. Mesa & H. Ledo de Medina (2001): 'Fractional composition of phosphorus in sediments from a tropical river (Catatumbo river, Venezuela)', *Hydrobiol.* **450**, 47–55.   **III**

Rozan, T. F., M. Taillefert, R. E. Trouwborst, B. T. Glazer, S. Ma, J. Herszage, L. M. Valdes, K. S. Price & G. W. Luther (2002): 'Iron-sulfur-phosphorus cycling in the sediments of a shallow coastal bay: Implications for sediment nutrient release and benthic macroalgal blooms', *Limnol. Oceanogr.* **47**, 1346–1354.   **II**

Russell, E. W. (1973): *Soil conditions and plant growth*, 10th ed., Longman, London & New York, 840 pp.   **IV**

Rust, C. M., C. M. Aelion & J. R. V. Flora (2000): 'Control of pH during denitrification in subsurface microcosms using encapsulated buffer', *Wat. Res.* **34**, 1447–1454.   **IV**

Ruttenberg, K. C. (1992): 'Development of a sequential extraction method for different forms of phosphorus in marine sediments', *Limnol. Oceanogr.* **37**, 1460–1482.   **III**

Rysgaard, S., N. Risgaard-Petersen, N. P. Sloth, K. Jensen & L. P. Nielsen (1994): ' Oxygen regulation of nitrification in sediments', *Limnol. Oceanogr.* **3**, 1643–1652.   **IV**

Rysgaard, S., P. Thastum, T. Dalsgaard, P. B. Christensen & N. P. Sloth (1999): 'Effects of salinity on NH adsorption capacity, Nitrification and Denitrification in Danish Estuarine sediments', *Estuaries* **22**, 21–30.   **IV**

Saunders, D. L. & J. Kalff (2001): 'Nitrogen retention in wetlands, lakes and rivers', *Hydrobiol.* **443**, 205–212.   **IV**

Schindler, D. W., R. Hesslein & G. Kipphut (1977): 'Interactions between sediments and overlying waters in an experimentally eutrophied Precambrian shield lake', In: *Interactions between sediments and freshwater. Proceedings of a symposium Sept. 6–10, 1975 held in Amsterdam*, Golterman, H. L. (Ed.), Junk Publishers, the Hague, .   **IV**

Schroeder, F., D. Klages, G. Blöcker, H. Vajen-Finnern & H.-D. Knauth (1992): 'The application of a laboratory apparatus for the study of nutrient fluxes between sediment and water', *Hydrobiol.* **235/236**, 545–552.   **IV**

Schuiling, R.D (1977): 'Source and composition of lake sediments', In: Golterman, H. L. (Ed.), 1977, pp. !!??!!.   **I**

Seitzinger, S. P (1988): 'Denitrification in freshwater and coastal marine ecosystems: Ecological and geochemical significance', *Limnol. Oceanogr.* **33**, 702–724.   **IV**

Seitzinger, S. P., L. P. Nielsen, J. Caffrey & P. Christensen (1993): 'Denitrification measurements in aquatic sediments: A comparison of three methods', *Biogeochem.* **23**, 147–167.   **IV**

Selig, U. & G. Schlungbaum (2002): 'Longitudinal patterns of phosphorus and phosphorus binding in sediment of a lowland lake-river system', *Hydrobiol.* **472**, 67–76.   **III**

Senior, E., E. B. Lindström, J. M. Banat, & D. B. Nedwell (1982): 'Sulfate reduction and methanogenesis in the sediment of a saltmarsh on the east coast of the United Kingdom', *Appl. Envir. Mictrobiol.* **43**, 987–996.   **II**

Serrano, L. (1992): 'Leaching from vegetation of soluble polyphenolic compounds, and their abundance in temporary ponds in the Doñana National Park (SW Spain)', *Hydrobiol.* **229**, 43–50. **IV**

Serrano, L., M. Reina, E. de Verdi, J. Toja & H. L. Golterman (2000): 'Determination of the sediment phosphate composition by the EDTA method of fractionation', *Limnetica* **19**, 199–204. **III**

Serrano, L., I. Caldaza-Bujak & J. Toja (2003): 'Variability of the sediment phosphate composition of a temporary pond (Doñana National Park, SW Spain)', *Hydrobiol.* **492**, 159–169. **III**

Sinke, A. J., C. A. A. Cornelese, P. Keizer, O. F. R. van Tongeren, & T. E. Cappenberg (1990): 'Mineralization, pore water chemistry and phosphorus release from peaty sediments in the eutrophic Loosdrecht lakes, The Netherlands', *Freshwat. Biol.* **23**, 587–599. **III**

Sioli, H. (1967): 'Studies in Amazonian waters. Atas do Simpósio sôbre a Biota Amazônia, Rio de Janeiro', *Limnologia* **3**, 9–50. **II**

Sioli, H. (1968): 'Hydrochemistry and geology in the Brazilian Amazon region', *Amazonia Kiel* **1**, 267–277. **II**

Slomp, C. P., W. van Raaphorst, J. F. P. Malschaert, A. Kok & A. J. J. Sandee (1993): 'The effect of deposition of organic matter on phosphorus dynamics in experimental marine systems', *Hydrobiol.* **253**, 83–98. **III**

Slomp, C. P., J. F. P. Malschaert & W. Van Raaphorst (1998): 'The role of adsorption in sediment-water exchange of phosphate in North Sea continental margin sediments', *Limnol. Oceanogr.* **43**, 832–846. **III**

Sly, P. (1983): 'Sedimentology and Geochemistry of Recent Sediments off the Mouth of the Niagara River, Lake Ontario', *J. Great Lakes Res.* **9**, 134–159. **I**

Smith, E. A., C. I. Mayfield & P. T. S. Wong (1978): 'Naturally occurring apatite as a source of orthophosphate for growth of bacteria and algae', *Microb. Ecol.* **4**, 105–107. **III**

Søndergaard, M (1990): 'Pore water dynamics in the sediment of a shallow and hypertrophic lake', *Hydrobiol* **192**, 247–258. **III**

Søndergaard, M., P. Kristensen & E. Jeppesen (1992): 'Phosphorus release from resuspended sediment in the shallow and windexposed Lake Arresø, Denmark', *Hydrobiol.* **228**, 91–99. **III**

Søndergaard, M., J. Windolf & E. Jeppesen (1996): 'Phosphorus fractions and profiles in the sediment of shallow Danish lakes as related to phosphorus load, sediment composition and lake chemistry', *Wat. Res.* **30**, 992–1002. **III**

Sonzogni, W. C., S. C. Chapra, D. E. Armstrong & T. J. Logan (1982): 'Bioavailability of Phosphorus Inputs to Lakes', *J. Envir. Qual.* **11**, 555–563. **III**

Sørensen, J. (1978a): 'Capacity for denitrification and reduction of nitrate to ammonia in a coastal marine sediment', *Appl. Envir. Microbiol.* **35**, 301–305. **IV**

Sørensen, J. (1978b): 'Denitrification rates in a marine sediment as measured by the acetylene inhibition technique', *Appl. Envir. Microbiol.* **36**, 139–143. **IV**

Stabel, H.-H. (1986): 'Impact of sedimentation on the phosphorus content of the euphotic zone of Lake Constance', *Verh. Intern. Verein. Limnol.* **22**, 964–969. **III**

Stabel, H.-H. (1986): 'Calcite precipitation in Lake Constance: Chemical equilibrium, sedimentation, and nucleation by algae', *Limnol. Oceanogr.* **31**, 1081–1093.   **III**

Stabel, H.-H. & U. Münster (1977): 'On the structure of soluble organic substances in sediments of Lake Plusse', In: Golterman, H. L. (Ed.), 1977, pp. 156–160.   **I**

Steinhart, G. S., G. E. Likens & P. M. Groffman (2001): 'Denitrification in stream sediments
in five northeastern (USA) streams', *Proc. int. Soc. Limnol.* **27**, 1331–1336.   **IV**

Stepanauskas, R., E. T. Davidsson & L. Leonardson (1996): 'Nitrogen transformations in
wetland soil cores measured by ^{15}N isotope pairing and dilution at four infiltration rates',
Appl. Envir. Microbiol. **62**, 2345–2351.   **IV**

Stephen, D., B. Moss & G. Phillips (1997): 'Do rooted macrophytes increase sediment phosphorus release?', *Hydrobiol.* **342/343**, 27–34.   **III**

Stief, P. & D. Neumann (1998): 'Nitrite formation in sediment cores from nitrate-enriched
running waters', *Arch. Hydrobiol.* **142**, 153–169.   **IV**

Stumm, W., R. Kummert & L. Sigg (1980): 'A ligand exchange model for the adsorption of
inorganic and organic ligands at hydrous oxide interfaces', *Croatica Chemica Acta* **53**,
291–312.   **III**

Stumm, W. & J. J. Morgan (1981): *Aquatic chemistry*, 2nd ed., Wiley-Interscience, New York,
780 pp.   **III**

Sumi, T. & I. Koike (1990): 'Estimation of ammonification and ammonium assimilation in
surficial coastal and estuarine sediments', *Limnol. Oceanogr.* **35**, 270–286.   **IV**

Sunborg, A. (1973): 'Significance of fluvial processes and sedimentation', In: *Fluvial processes and sedimentation. Proc. Hydrol. Symp. Univ. Alberta, Edmonton. Can. Dept. Envir., Inland Water Direct., Ottawa*, pp. 1-10, , .   **I**

Sundareshwar, P. V. & J. T. Morris (1999): 'Phosphorus sorption characteristics of intertidal marsh sediments along an estuarine salinity gradient', *Limnol. Oceanogr.* **44**, 1693–
1701.   **III**

Sutherland, R. A. (1998): 'Loss-on-ignition estimates of organic matter and relationships to
organic carbon in fluvial bed sediments', *Hydrobiol.* **389**, 153–167.   **IV**

Suzumura, M. & A. Kamatani (1995): 'Mineralization of inositol hexaphosphate in aerobic
and anaerobic marine sediments: Implications for the phosphorus cycle', *Geochim. Cosmochim. Acta* **59**, 1021–1026.   **III**

Svensson, J. M. (1997): 'Influence of *Chironomus plumosus* larvae on ammonium flux and
denitrification (measured by the acetylene blockage- and the isotope pairing-technique)
in eutrophic lake sediment', *Hydrobiol.* **346**, 157–168.   **IV**

Svensson, J. M., G. M. Carrer & M. Bocci (2000): 'Nitrogen cycling in sediments of the
Lagoon of Venice, Italy', *Mar. Ecol. Prog. Ser.* **199**, 1–11.   **IV**

Svensson, J. M., A. Enrich-Prast & L. Leonardson (2001): 'Nitrification and denitrification
in a eutrophic lake sediment bioturbated by oligochaetes', *Aquat Microb. Ecol.* **23**, 177–
186.   **IV**

Syers, J. K. & D. Curtin (1989): 'Inorganic reactions controlling phosphate cycling', In: *Phos-
phorus cycles in terrestrial and aquatic ecosystems. Proceedings of a SCOPE workshop,
May 1988. Czerniejewo, Poland*, Thiessen, H. (Ed.), Saskatchewan Institute of Pdedology,
Saskatoon, Camada, .   **III**

Takahashi, M. & Y. Saijo (1982): 'Nitrogen metabolism in Lake Kizaki, Japan. II', *Arch. Hydrobiol.* **92**, 359–376. **IV**

Talling, J. F. & I. B. Talling (1965): 'The chemical composition of African lake waters', *Int. Rev. Ges. Hydrobiol.* **50**, 421–463. **II**

Terwindt, J. H. J (1977): 'Deposition, transportation and erosion of mud', In: Golterman, H. L. (Ed.), 1977, pp. 19–24. **I**

Thomas, E. A (1965): 'Phosphat-Elimination in der Belebtschlammanlage von Männedorf und Phosphat-Fixation in See- und Klärschlamm', *Vierteljahrsschr. Naturforsch. Ges. Zürich* **110**, 419–434. **III**

Tiedje, J. M., S. Simkins & P. G. Groffman (1989): ' Perspectives on measurements of denitrification in the field including recommended protocols for acetylene methods', *Plant and Soil* **115**, 261–284. **IV**

Tillmans, J. & O. Heublein (1912): 'Uber die Kohlensäure der natürlichen Wässer', *Gesundheits-Ingenieur* **35**, 669–677. **II**

Tirén, T. & K. Petterson (1985): 'The influence of nitrate on the phosphorus flux to and from oxygen depleted lake sediments', *Hydrobiol.* **120**, 207–223. **IV**

Tuominen, L., T. Kairesalo, H. Hartikainen & P. Tallberg (1996): 'Nutrient fluxes and microbial activity in sediment enriched with seston', *Hydrobiol.* **335**, 19–31. **III**

Tuominen, L., K. Mäkelä, K. K. Lehtonen, H. Haahti, S. Hietanen & J. Kuparinen (1999): 'Nutrient fluxes, Porewater profiles and denitrification in sediment influenced by algal sedimentation and bioturbation by *Monoporeia affinis*', *Estuarine, Coastal and Shelf Science* **49**, 83–97. **IV**

Törnblom, E. & E. Rydin (1998): 'Bacterial and phosphorus dynamics in profundal Lake Erken sediments following the deposition of diatoms: a laboratory study', *Hydrobiol.* **364**, 55–63. **III**

Turner, R. R., E. A. Laws & R. C. Harris (1983): 'Nutrient retention and transformation in relation to hydraulic flushing in a small impoundment', *Freshwat. Biol.* **13**, 113-127. **IV**

Turner, B. L. & P. M. Haygarth (2001): 'Phosphorus immobilisation in rewetted soils', *Nature* **411**, 258. **III**

Ullman, W. Q. & R. C. Aller (1989): 'Nutrient release rates from the sediments of Saginaw Bay, Lake Huron', *Hydrobiol.* **171**, 127–140. **IV**

Ulrich, K.-U. (1997): 'Effects of land use in the drainage area on phosphorus binding and mobility in the sediments of four drinking-water reservoirs', *Hydrobiol.* **345**, 21–38. **III**

Urban, N. R., P. L. Brezonik, L. A. Baker & L. A. Sherman (1994): 'Sulfate reduction and diffusion in sediments of Little Rock Lake, Wisconsin', *Limnol. Oceanogr.* **39**, 797–815. **II**

Urban, N. R., C. Dinkel & B. Wehrli (1997): 'Solute transfer across the sediment surface of a eutrophic lake: Pore water profiles from dialysis samplers', *Aquat. Sci.* **59**, 1–25. **V**

Uusitalo, R. & Markku Yli-Halla (1999): 'Estimating errors associated with extracting phosphorus using Iron Oxide and resin methods', *J. Envir. Qual.* **28**, 1891- 1897. **III**

Valiela, I., & J. M. Teal (1979): 'The N-budget of a salt marsh ecosystem', *Nature* **280**, 652–657. **IV**

Val Klump, J. & C. S. Martens (1989): 'The seasonality of nutrient regeneration in an organic-rich coastal sediment: Kinetic modeling of changing pore-water nutrient and sulfate distribution', *Limnol. Oceanogr.* **34**, 559–577. **IV**

Van De Kraats, J. A. (Ed.) (1999): *Fifth scientific and technical review: Farming without harming*, Rijkswaterstaat, Lelystad, 231 pp. **II**

Van Gemerden, H. (1967): *On the bacterial sulfur cycle of inland waters*, J. H. Plasmans, the Hague (Thesis, Univ. Leiden), 110 pp. **II**

Van Wazer, J. R. (Ed.) (1961): *Phosphorus and its compounds*, **Vol 2**. Technology, biological functions and application, Wiley-Interscience, New York, 1091 pp. **I**

Verdouw, H., C. J. A. Van Echteld & E. M. J. Dekkers (1978): 'Ammonia determination based on indophenol formation with sodium salicylate', *Wat. Res.* **12**, 399–402. **V**

Verdouw, H. & E. M. J. Dekkers (1980): 'Iron and manganese in Lake Vechten (The Netherlands); dynamics and role in the cycle of reducing power', *Arch. Hydrobiol.* **89**, 509–532. **II**

Vicente, E. & M. R. Miracle (1992): 'The coastal lagoon Albufera de Valencia: An ecosystem under stress', *Limnetica* **8**, 87–100. **II**

Vidal, M., J. A. Morgui, M. Latasa, J. Romero & J. Camp (1992): 'Factors controlling spatial variability in ammonium release within an estuarine bay (Alfacs Bay, Ebro Delta, N. W. Mediterranean)', *Hydrobiol.* **235/236**, 519–525. **IV**

Vidal, M., J. A. Morgui, M. Latasa, J. Romero & J. Camp (1997): 'Factors controlling seasonal variability of benthic ammonium release and oxygen upyake in Alfacs Bay (Ebro, NW Mesditerranean)', *Hydrobiol.* **350**, 169–178. **IV**

Villar, C. A., L. de Cabo, P. Vaithiyanathan & C. Bonetto (1998): 'River-floodplain interactions: nutrient concentrations in the Lower Paraná River', *Arch. Hydrobiol.* **142**, 433–450. **IV**

Villar, C., L. de Cabo, P. Vaithiyanathan & C. Bonetto (1999): 'Pore water N and P concentration in a floodplain marsh of the Lower Parana River', *Hydrobiol.* **392**, 65–71. **III**

Vollenweider, R. A. (1975): 'Input-output models, with special reference to the phosphorus loading concept in limnology', *Schweiz. Z. Hydrol.* **37**, 53–84. **III**

Waara, T., M. Jansson & K. Petterson (1993): 'Phosphorus composition and release in sediment bacteria of the genus *Pseudomonas* during aerobic and anaerobic conditions', *Hydrobiol.* **253**, 131–140. **III**

Wagner, G. (1972): 'Stratifikation der Sedimenten und Sedimentationsrate im Bodensee', *Verh. int. Ver. Theor. Angew. Limnol.* **18**, 475–481. **III**

Wagner, G. (1976): 'Simulationsmodelle der Seeneutrophierung, dargestellt am Beispiel des Bodensee-Obersees', *Arch. Hydrobiol.* **78**, 1–41. **III**

Waller, P.A. & W. F. Pickering (1992): 'Determination of 'labile' phosphate in lake sediments using anion exchange resins: a critical evaluation', *Chem. Speciation and Bioavailability* **4**, 59–68. **III**

Walling, D. E. (1977): 'Natural sheet and channel erosion of unconsolidated source material', In: *The fluvial transport of sediment-associated nutrients and contaminants*, Shear, H. & A. E. P. Watson (Eds.), Intern. Joint Commiss. Great Lakes Regional Office, Windsor,

Ont., pp. 11–33.   I

Wanatabe, Y. & S. Tsunogai (1984): 'Adsorption-desorption control of phosphate in anoxic sediment of a coastal sea, Funka Bay, Japan', *Mar. Chem.* **15**, 71–83.   III

Watts, C. J. (2000a): 'The effect of organic matter on sedimentary phosphorus release in an Australian reservoir', *Hydrobiol.* **431**, 13–25.   III

Watts, C. J. (2000b): 'Seasonal phosphorus release from exposed, re-inundated littoral sediments of two Australian reservoirs', *Hydrobiol.* **431**, 27–39.   III

Weaver, C. E. & L. D. Pollard (1975): *The chemistry of clay minerals*, Elsevier Scientific Publishing Cy., Amsterdam, Oxford, New York, 213 pp.   I

Weber W. J. & W. Stumm (1963): 'Mechanism of hydrogen ion buffering in natural waters', *J. Am. Wat. Works Assoc.* **55**, 1553–1578.   II

Wetzel, R. G (1972): 'The role of carbon in hard water marl lakes', In: *Nutrients and eutrophication: The limiting Nutrient Controversy*, G. E. Likens (Ed.), Spec. Symp., Am. Soc. Limnol. Oceanogr. **1**, 84–97.   II

Wetzel, R. G. & A. Otzuki (1974): 'Allochthonous organic carbon of a marl lake', *Arch. Hydrobiol.* **73**, 31–56.   II

Wieder, R. K. & G. E. Lang (1982): 'A critique of the analytical methods used in examining decomposition data obtained from litter bags', *Ecology* **63**, 1636–1642.   IV

Wieland, E., P. Lienemann, S. Bollhalder, A. Lück & P. H. Santschi (2001): 'Composition and transport of settling particles: relative importance of vertical and lateral pathways', *Aquat. Sci.* **63**, 123–149.   I

Williams, J. D. H., J. M. Jaquet & R. L. Thomas (1976): 'Forms of phosphorus in the surficial sediments of Lake Erie', *J. Fish. Res. Bd. Can.* **33**, 430–439.   III

Williams, J. D. H., H. Shear & R. L. Thomas (1980): 'Availability to *Scenedesmus quadricauda* of different forms of phosphorus in sedimentary materials from the Great Lakes', *Limnol. Oceanogr.* **25**, 1–11.   III

Williams W. D. & Hang Fong Wan (1972): 'Some distinctive features of Australian inland waters', *Wat. Res.* **6**, 829–836.   II

Winfrey, M. R. & D. M. Ward (1983): 'Substrates for sulfate reduction and methane production in intertidal sediments', *Appl. Envir. Microbiol.* **45**, 193–199.   II

Wodka, M. C., S. W. Effer & C. T. Driscoll (1985): 'Phosphorus deposition from the epilimnion of Onongada Lake', *Limnol. Oceanogr.* **30**, 833–843.   III

Wolf, A. M., D. E. Bake, H. B. Pionke & H. Kunishi (1985): 'Soils tests for estimating labile soluble and algae available phosphorus in agricultural soils', *J. Envir. Qual.* **14**, 341–348.   III

Wright, J. C. & I. K. Mills (1967): 'Productivity studies on the Madison river, Yellowstone National Park', *Limnol. Oceanogr.* **12**, 568–5767.   II

Wüest, A., G. Piepke & D. C. Van Senden (2000): 'Turbulent kinetic energy balance as a tool for estimating vertical diffusivity in wind-forced stratified waters', *Limnol. Oceanogr.* **45**, 1388–1400.   V

Young, T. C. & J. V. DePinto (1982): 'Algal-availability of particulate phosphorus from diffuse and point sources in the lower Great Lakes Basin', *Hydrobiol.* **91**, 111–119.   **III**

Zeikus, J. G. & M. R. Winfrey (1976): 'Temperature limitation of methanogenesis in aquatic sediments', *Appl. Envir. Microbiol.* **31**, 99–107.   **II**

Index